Geopolitics, Culture, and the Scientific Imaginary in Latin America

Geopolitics, Culture, and the Scientific Imaginary in
LATIN AMERICA

Edited by María del Pilar Blanco and Joanna Page

University of Florida Press

Gainesville

Publication of the paperback edition made possible by a Sustaining the Humanities through the American Rescue Plan grant from the National Endowment for the Humanities.

First cloth printing, 2020
First paperback printing, 2023

28 27 26 25 24 23   6 5 4 3 2 1

Library of Congress Cataloging-in-Publication Data
Names: Blanco, María del Pilar, editor. | Page, Joanna, 1974– editor.
Title: Geopolitics, culture, and the scientific imaginary in Latin America
   / edited by María del Pilar Blanco and Joanna Page.
Description: Gainesville : University of Florida Press, 2020. | Includes
   bibliographical references and index.
Identifiers: LCCN 2019034407 (print) | LCCN 2019034408 (ebook) | ISBN
   9781683401483 (hardback) | ISBN 9781683401766 (pdf) | ISBN 9781683403876 (pbk.)
Subjects: LCSH: Science—Latin America—History. | Science—Social
   aspects—Latin America—History. | Science—Political aspects—Latin
   America—History. | Science and the humanities—Latin America—History.
Classification: LCC Q127.L38 G46 2020 (print) | LCC Q127.L38 (ebook) |
   DDC 338.98/06—dc23
LC record available at https://lccn.loc.gov/2019034407
LC ebook record available at https://lccn.loc.gov/2019034408

University of Florida Press
2046 NE Waldo Road
Suite 2100
Gainesville, FL 32609
http://upress.ufl.edu

UF PRESS

UNIVERSITY
OF FLORIDA

Contents

Figures

Acknowledgments

The idea for this book emerged from exchanges with academics, writers, and artists from around the world who met in a series of encounters that took place in the United Kingdom and Latin America between 2014 and 2016 and were funded by an international research network grant from the Arts and Humanities Research Council (AHRC) in the United Kingdom. The editors gratefully acknowledge the assistance of the AHRC in making these conversations possible and would also like to thank their home institutions, the University of Oxford and the University of Cambridge, for their help in hosting some of the gatherings. Warm thanks also go to the network participants and audiences for their enthusiasm and for posing important questions about reading science within and across disciplines and to the contributors to this volume for sharing their research; we are incredibly fortunate to have encountered such wonderful collegiality and openness while putting the book together.

Catriona McAllister, Geoff Maguire, and Lucy Bollington played important roles in coordinating the network and preparing this manuscript. We would also like to express our deep gratitude to the Cuban artist Yoan Capote for his generosity in allowing us to use his work *Open Mind* as our cover image. Finally, we are grateful for the support and guidance of Stephanye Hunter at the University of Florida Press.

Note on Translations

All translations, unless otherwise noted, are those of the authors of the respective chapters.

Introduction

Reimagining Science in Latin America

MARÍA DEL PILAR BLANCO AND JOANNA PAGE

The transnational transfers of ideas, technologies, materials, and people that have shaped the history of science in Latin America are marked, as in any region, by asymmetries of power. These are often replicated or even magnified in the narratives we have forged about that history. The journeys to Latin America of some of Europe's most famous naturalists (Humboldt and Darwin, for example) are often depicted as the heroic overcoming by European science of savage local terrains and ways of life. Those epic explorers are recast, in other narratives, as the forerunners of (neo)colonial exploitation in the history of the ransacking of Latin America's mineral riches to pay for European imperial ventures, repeated in the often-illegal plundering of the region's dinosaur fossils to swell museum collections in Europe and North America. In such accounts, Latin America becomes the arena for European adventures, the testing ground for new scientific theories, or the passive victim of colonial profiteering, but rarely a place of innovation. It is certainly the case that over the centuries the flow of natural resources, data, and expertise from Latin America to more developed regions has generally been to the benefit of those regions and has not reduced an imbalance of power that dates back to the colonial period.

Despite serious financial, institutional, and political disadvantages, however, many Latin American scientists have succeeded in pioneering significant advances. Our histories need to be nuanced enough to register these while exploring the powerful geopolitical forces at play that would impose limits on Latin

American science or seek to define its objectives. In this volume we investigate the impact of such forces while charting how scientific knowledge in Latin America has been creatively and critically forged at the intersections between indigenous and foreign cultures and put to use in projects of a political or cultural nature, often with emancipatory ends. We also highlight the rich epistemological insights that have emerged from Latin America's sustained engagement in and reflection on scientific practices. Our aim is not only to rethink the place of Latin America in the history of science and the scientific imagination but also to understand in greater depth the geopolitics of knowledge as it has been produced and deployed in the region. Further, we also bring to the fore the significance of Latin American contributions over time in (re)thinking science within interdisciplinary and intercultural contexts, a theme that is rapidly gaining importance in today's world.

The chapter authors detail local engagements with technology and the natural world in Latin America across time and reveal the social, political, and economic conditions that have led to the relative obscurity of such research in a world history of science. Comparative thinking is an important feature in this volume, as it helps situate the issue of Latin American scientific innovation within the global currents of science and understand the particular inequalities they produce and reproduce. The asymmetries that govern the global production of scientific knowledge have certainly affected the kind of science that is possible "at the periphery," to use the term adopted by many Latin American historians of science. While examining a number of cases from the colonial times to the present, we propose a critical understanding of how such asymmetries have operated. To give an example, the history of science in Latin America has been bound up, since colonization, with that of Spain, sharing its peripheral status in the global history of science. As Miruna Achim states, until recently, science in Spain and its empire was regarded as "slow, backwards, anachronistic, and derivative" when compared to that of the rest of Europe.[1] This representation is now beginning to be challenged with greater attention to the "dynamic and multiple" exchanges that characterized the production and dissemination of scientific knowledge in the colonial era and to the particular forms taken by colonial science, thoroughly embedded as it was in "the Spanish political, commercial, economic, cultural, medical, and religious enterprise from the era of the Conquest onward."[2] A number of chapters in this volume contribute to this new thrust in scholarship on colonial Spanish and Latin American science.

The dynamic and diverse nature of these exchanges continues, of course, into the nineteenth century and beyond. Irina Podgorny, writing on the field of comparative animal anatomy in the mid-nineteenth century, draws attention to the interweaving of "transversal" and "vertical" communication among a number of different political, scientific, and commercial agents.[3] Like that of many historians of Latin American science, her work questions the central dichotomy between local and global that often still governs our understanding of the production of knowledge in and of Latin America.[4]

Another pivotal feature of this book is the authors' engagement with evolving perceptions of the hegemony of science in the modern world and with shifting conceptions of science's autonomy. The mid-twentieth century saw the development, in a number of regions, of important theorizations of the social dimensions of science and the growth and consolidation of history of science as a discipline. Thomas Kuhn's *Structure of Scientific Revolutions* (1962) offered a new account of scientific discovery that inspired further studies on the extent to which scientific practice is affected by the political, social, and cultural contexts in which scientists are located.[5] This productive dismantling of science's self-sufficiency has helped pave the way for further theorizations of science such as those by Pierre Bourdieu, François Lyotard, and Bruno Latour that have insisted on the constructed and situated nature of scientific knowledge. Such approaches have sometimes been misunderstood as entirely dismissing the referentiality of science, which is not the case; to argue for the social construction of science in the way that these theorists and others do is not to deny the existence of the material world beyond the laboratory but to emphasize the ways in which knowledge of that world is inescapably mediated by scientific beliefs and practices, which are ultimately social and cultural in nature.

The extent to which scientific practices are shaped by extrascientific factors is a theme that becomes central in theorizations and histories of science produced in Latin America in the second half of the twentieth century. These point to the coloniality of power and knowledge that continues to shape the production of science in and about Latin America and offer a set of evolving reflections on the region's complex position within the realm of global science.

The Geopolitics of Science in Latin America

The geopolitics of scientific knowledge in Latin America comes sharply into focus in writings published by Latin American scientists and historians of science during the twentieth century and beyond. Reflecting in 1983 about the social context of scientific research in Venezuela in a text that would become highly influential in the field of science and technology studies in Latin America, Hebe Vessuri outlines three main areas in which the impact of socioeconomic conditions in Latin America on the development of science can be seen. First, the development of genuinely new concepts is hampered by the economic risks involved, meaning that scientific communities at the periphery tend to be more "conservative," dedicating themselves to working through concepts that originate elsewhere.[6] Second, and for related reasons, the kind of science developed tends to be "applied" rather than "pure," with a greater orientation toward social purposes, which again means responding to agendas that are first developed in other contexts.[7] And third, Vessuri asserts, the relative weakness of the social sciences in Venezuela, for example, seriously limits the potential to develop scientific activities that would respond to the region's specific needs, within a clear political program that looks beyond immediate circumstances.[8]

The Argentine scientist Oscar Varsavsky explains in a seminal essay published in 1969 that the internationalization of science hides the asymmetry that has governed its global community, which has unconditionally accepted the leadership of the Northern Hemisphere in all matters. It is in Europe, Russia, and the United States that topics are chosen for research and the most promising methods identified, and it is there also that the contributions of scientists are evaluated.[9] Pablo Kreimer argues that the current-day integration of scientific researchers from peripheral countries in international science should be regarded as an "integración subordinada" (subordinate or dependent integration), to the extent that they are often forced to adopt lines of inquiry developed elsewhere and to subject their work to the evaluation of interlocutors located beyond the nation's borders.[10] Given the enormous differential in funding for the sciences, Kreimer characterizes the international division of labor into transnational "megarredes" (mega-networks) in which highly specialized technical work is farmed out to the periphery as a system that is effectively one of subcontracting.[11]

To these huge disparities are added the disruptions to scientific activity caused by national politics or economic problems, whose impact on science is particularly acute given that most science in Latin America has been funded by the state. The widespread experience of political repression and exile has interrupted the formation of national and regional scientific communities and made of Latin American science a nomadic and diasporic phenomenon. In this regard, an emblematic case would be the infamous Noche de los Bastones Largos (Night of the Long Batons) in 1966 that inaugurated a period of repres-

María del Pilar Blanco and Joanna Page

sion against academics under Juan Carlos Onganía's regime in Argentina. In the aftermath of that event, 8,600 university professors resigned from their positions at the Universidad de Buenos Aires alone, with no fewer than 77 percent resigning from the School of Exact Sciences,[12] and a number of these emigrated to other Latin American countries, the United States, Canada, and Europe.[13] This example reflects a broader history of emigration in Latin American science in the twentieth and twenty-first centuries. The Argentine biochemist César Milstein, who received the Nobel Prize in Medicine in 1984 for his work on antibodies, is only one of many outstanding Latin American scientists who have pursued their careers overseas. Milstein spent most of his career in Cambridge, England, leaving Buenos Aires definitively in 1962 as a result of the political persecution of liberal intellectuals and scientists.[14] In more recent decades it has not been political oppression so much as economic necessity that has prompted the large-scale emigration of scientists from Latin America to better-funded institutes in Europe and the United States.

Notwithstanding the very real difficulties of working in the natural sciences at the periphery in the twentieth and twenty-first centuries, it is undoubtedly the case that Latin America has been the site of some important advances that are sometimes cited in global histories of science and technology and more frequently in Latin American ones. Figures often mentioned include the Cuban epidemiologist Carlos Juan Finlay (1833–1915), who conducted pioneering research on the causes of yellow fever, and the Venezuelan Jacinto Convit (1913–2014), who was credited with transforming the landscape of epidemiology when he invented a vaccine to combat leprosy. Other Latin Americans broke ground in the fields of medicine and biotechnology during the twentieth century. In the 1920s and 1930s, the Brazilian Manuel Dias de Abreu (1894–1962) developed what is now known as abreugraphy, an X-ray system for the quick diagnosis of tuberculosis. Some years later, the Argentine Bernardo Houssay (1887–1971) won the Nobel Prize in Medicine for his work on the interrelation between sugar metabolism and the pituitary gland, findings that revolutionized diabetes research. In physics, the Peruvian Santiago Antúnez de Mayolo (1887–1967) wrote *Hipótesis sobre la constitución de la materia* (Hypothesis about the constitution of matter, 1924), in which he argues for the existence of a neutral element in the atom, a theory that was confirmed years later with the discovery of the neutron. In the field of technology, at the age of only seventeen, the Mexican Guillermo González Camarena (1917–1965) invented one of the earliest color transmission systems for television.

A Relational Understanding of Scientific Knowledge

Accounts of individual discoveries such as these often reveal little about the exchanges and collaborations that underpin all advances in science and technology. Nevertheless, they constitute an attempt to recover a history of Latin American science that has been continually overlooked or dismissed. Significantly, European and US erasures of Latin American contributions to science are often repeated by Latin American scholars. As a number of historians have recently observed, Latin American accounts of the history of science from the 1950s and 1960s were keen to stress the dependency of the region and therefore contributed to a vision of its backwardness.[15] The persistence even within Latin America of views that scientific knowledge and superior technologies are imported into Latin America from the global North "do not acknowledge that innovation, invention, and discovery take many forms, occur in multiple contexts, and travel in many directions, nor do they acknowledge that diverse communities use scientific ideas and technologies in different ways."[16] In the absence of robust, long-term public policies regarding science and technology, Diego Hurtado argues, the evolution of knowledge has been governed by "un *ethos* consecuente de la 'asimilación,' la 'copia' o el 'trasplante'" (an ethos consistent with "assimilation," "imitation," or "transplantation").[17]

Perspectives from "decolonial" thinkers may throw light on the exclusion of Latin America from the discourses of European modernity, strongly invested as these discourses remain in science and rationalism as the motors of progress. The philosopher Enrique Dussel has argued that European "modernity" is founded on a relation of alterity with the non-European other; it creates the "periphery" that affirms its own position at the "center."[18] He deconstructs the "fallacy of developmentalism" arising from Hegel's understanding of world history, according to which "the path of European development must be followed unilaterally by every other culture."[19] Aníbal Quijano likewise argues that Europeans have succeeded in creating a temporality in which they represent the culmination of history, the most advanced, modern, and civilized of the human species, and the "exclusive bearers, creators, and protagonists of that modernity," while the rest of the world belongs to the evolutionary past.[20] This framework for understanding modernity, powerful as it is, is based on historical error. As Quijano points out, echoing Dussel, if modernity is associated with the advanced, the rational, the scientific, and the secular, then many of these characteristics were in abundance in earlier civilizations such as those of China,

India, and Greece as well as Maya and Aztec cultures.[21] The Eurocentric version of modernity ignores the fact that the inequality of relations between center and periphery is not a historical accident but *constitutive* of the center and of modernity, as the colonies provided much of the wealth and means of expansion that underpinned that modernity. Quijano underscores that it was only "the free labour of American Indians, blacks and mestizos, with their advanced technology in mining and agriculture, and with their products such as gold, silver, potatoes, tomatoes, and tobacco," that made modernization in Europe possible.[22]

Decolonial thinking allows for a rewiring of persistent perceptions about the places and times in which knowledge originates and where and how it is disseminated. A number of contributions in this volume depart in this way from center-periphery thinking, emphasizing that Latin American reality, from the colonial period onward, requires a relational understanding of how scientific knowledge is produced that in turn can travel and be applied elsewhere. This is the case with Heidi Scott's chapter on the colonial development of geological knowledge in the Andean region. This knowledge emerges from years of local mining experience that connected white miners to their indigenous counterparts and the metropolis. Drawing on archives that detail many facets of mining in this area, Scott shows how an understanding of nature is conditioned by an extensive web of everyday social and experimental practices, demonstrating the extent to which "production of knowledge was both localized and highly relational" (119, this volume).

The role of indigenous knowledge has been consistently discounted in accounts of science in Latin America, from the colonial period to the present. This occludes, for example, the contact between indigenous and European cultures in the centuries of Spanish and Portuguese colonization that resulted in myriad exchanges of information about the American region's geology and botany and the use of autochthonous resources for medical treatment. The historian Juan José Saldaña argues that the Spanish conquistadors situated their main viceroyalties in Mexico and Peru precisely because the societies they found there had well-established political, social, and cultural systems.[23] The contributions of native American cultures to the fields of astronomy and mathematics have been a particular point of interest for science historians and anthropologists in past decades.[24] In this volume, Julio Prieto discusses the artist Jorge Eduardo Eielson's recovery of part of this legacy, namely the Incan *khipu* method of numeration. Historians including Eduardo Estrella have established the importance of innovations in the medical sciences produced by pre-Columbian cultures like the Aztecs and the Incas, who were skillful in surgical and treatment practices.[25] The power imbalances underlying colonialism, however, more often than not resulted in the usurpation rather than equal commerce of this knowledge. The violence of this exchange also entailed a lack of regard for the spiritual importance the indigenous populations attached to their complex understanding of the natural world. Illustrating the inequality of these relations, Jorge Cañizares-Esguerra has recently highlighted the case of the royal physician Francisco Hernández, whose celebrated plant descriptions, developed on a scientific expedition to Mexico (1570–1577), borrowed extensively from the work of Nahua intellectuals.[26] In her chapter in this book, Yarí Pérez Marín clearly demonstrates the extent to which European practices drew on indigenous knowledge in the early colonial period. In her reading of Agustín Farfán's *Tractado breue de Anothomia y Chirugia* (1579), she traces the increased recognition in a Spanish medical text of indigenous treatment practices, revealing an active exchange—and actual commerce—underpinning medical knowledge and materials in colonial times.

During the postcolonial era, as different republics consolidated notions of nation and *patria*, the indigenous subject—and the racial diversity of the region, for that matter—represented a problem for intellectuals and politicians who sought to whitewash the image of Latin American nations in the modern world. In the first decades of the twentieth century, Mexico and Peru, among other Latin American countries, began to address the indigenous question in the realm of politics as well as in the development of modern archaeology. In the case of Peru, as Manuel Burga explains, a new constitution established in 1920 restored indigenous rights that had been stripped away in 1821.[27] This shift in domestic policy in different nations was reflected in modernized approaches to the recovery and preservation of indigenous artifacts and by extension the development of museology. An illustrious example of this drive is found in the work of Manuel Gamio (1883–1960), who, instead of relating Mexico's antiquity to a Greco-Roman or Egyptian past as his nineteenth-century predecessors did, sought connections between an indigenous past and the populations that made up a large part of the population in Mexico in his time.[28] In the first decades of the twentieth century, political movements across the region worked to recognize the contributions of indigenous populations to national cultures that had for so long insisted on segregationist agendas.

One facet of this impulse is the effort to bring indigenous knowledge systems into science. In this book, the analysis presented by Edward Chauca of the career as a doctor and writer of the Peruvian Hermilio Valdizán (1885–1929) reveals how significant integrations of indigenous and European knowledge enable systems of diagnosis and treatment that fit Peru's ethnic and cultural singularity. Such cases recognize the importance of indigenous heritage in the histories of Latin America; however, there is still much work to be done not only in recovering different sites of collaboration and exchange but also in representing responsibly the vastness and complexity of indigenous erudition concerning the natural world. Laurelyn Whitt has documented the legal ramifications of contemporary "biocolonialism" and the continued power imbalances evident in the contact between Western and indigenous bodies of knowledge. She offers the example of the Guajajara in Brazil, who have for a long time prescribed the use of the *Pilocarpus jaborandi* plant to treat glaucoma.[29] Western corporations have moved in to profit from this treatment, subjecting the Guajajara to "debt peonage and slavery" by these businesses.[30] By-products of liberal and neoliberal agendas, such injustices remain a pressing problem in the study of Latin America's position in the world of science.

If Latin American contributions are often erased through a lack of attention to the relationality of knowledge production in linear chronologies of science, conversely, these also often fail to account for the emergence of similar ideas at around the same time in different, apparently unconnected locations. Carlos Fonseca Suárez's contribution to this volume begins with Alexander von Humboldt's encounter with an awe-inspiring electrical machine invented by Carlos del Pozo, a "lonely plainsman" who had had no contact with North Atlantic centers of science and whose reading went not much further than Benjamin Franklin and Sigaud de Lafond. This example, as Fonseca suggests, points to the paradox of a Latin American scientist-inventor who, despite his isolation, also contributes to the global development of science. Pozo's obscurity reveals much about the problematic elision of "universal" and "European" as terms to describe scientific development; it also shows the need to move away from linear chronologies of scientific progress in order to capture the more properly global dimensions of knowledge production.

Science for Emancipation

A number of the chapters in this volume explore how scientific knowledge has been employed to serve political projects with emancipatory ends. In this way, they contrast with more familiar narratives of the use of scientific knowledge to shore up political repression and economic exploitation in the region. One need only think of the roles played by scientific racism in slavery, eugenics, and genocide since the colonial period. Nor are such instances confined to colonial times; scholars have increasingly turned their attention to the ubiquity of eugenic thought in Latin America during the late nineteenth century and first half of the twentieth century, especially as different national governments began developing sanitation practices to control their swelling urban populations. Writing about the Brazilian case in particular, the historian Nancy Leys Stepan notes that the success of vaccination campaigns between 1902 and 1917 "had given great cachet to the 'sanitation sciences' and had stimulated the growth of a scientifically oriented professional and medical class that was increasingly visible and integrated into state and federal policy organizations."[31] During a period in which natural and social sciences remained significantly interconnected, the existence of eugenics in Latin America was especially problematic given the region's racial diversity, as Stepan and others have pointed out. The pursuit of biological and anthropological knowledge in particular has often been carried out with little respect for indigenous culture or even the lives of indigenous people, as has been highlighted in a range of recent documentary and fiction films including *El abrazo de la serpiente* (Ciro Guerra, Colombia, 2015), *El botón de nácar* (Patricio Guzmán, Chile, 2015), and *Damiana Kryygi* (Alejandro Fernández Mouján, Argentina, 2015). Many archaeologists, botanists, geographers, and anthropologists traveled to the south alongside the 1870s military campaigns in Argentina that sought to claim territories inhabited by indigenous people; however, as Sylvia Molloy suggests, they were "less intent on pushing the Indian away from the city than on bringing him back there, albeit in pieces, as an object of study, an item for a collection."[32]

In the context of the widespread abuse of science for the benefit of national elites or foreign powers, it is all the more pressing to document examples of resistance that can undo this practice and the perspectives that underpin it. Several of the chapters in this volume focus on the inspiration science has provided for a number of utopian and emancipatory projects in the region. In her chapter, Lina del Castillo offers new insights into the figure of José María Samper (1828–1888) in Colombia and his development of what del Castillo calls "postcolonial sciences," which, disseminating local solutions to local problems, reversed the center-periphery model of knowledge transfer. Building on the work of Mabel Moraña, Brais Outes-León sees in the work of the Peruvian thinker José Carlos Mariátegui (1894–1930) a blueprint for a new, scientific form of emancipation, one that combines Einstein's theory of relativity with what Outes-León calls "revolutionary mysticism" (300, this volume). In his compelling study of letters sent to the Argentine Technical Secretariat in the 1950s, Hernán Comastri examines the extent to which the popularization of science in that country resulted in the democratization of scientific inventiveness.

The role of science in nation-building in nineteenth-century Latin America is often associated—especially in Argentina, Brazil, and Mexico—with the elites' embrace of positivism and its particular understanding of scientific progress and modernization, which has been charged with leading to the aggravation of social and economic inequalities. However, two chapters here provide different accounts of the role of science in the national imaginary in the nineteenth century. In his discussion of the importance of the sciences of astronomy and botany in the development of national symbols and anthems in post-independence Argentina, Miguel de Asúa offers a nuanced account of the political use of science to encode both neoclassical grandeur and dynamic progress in the development of a patriotic discourse of liberation, demonstrating the close relation of science with republican and Enlightenment ideals. In her chapter, Gabriela Nouzeilles studies counterhegemonic readings of science produced by the renowned Argentine naturalist and paleontologist Florentino Ameghino (1853/1854–1911), whose stories about Patagonia's animal fossils challenge distinctions between fact and fiction, science and indigenous knowledge. Nouzeilles exposes ways in which Ameghino's work throws into relief the geopolitics of science as well as its importance in the formation of a national geographic imaginary in nineteenth-century Argentina.

An Intercultural and Interdisciplinary Science

The choice for many Latin American researchers and institutions between complying with the agendas of "international science" or responding to the real challenges faced at home has been set out in starkly binary terms by many historians of science since the 1950s. Varsavsky railed against scientists who choose to take a prescribed role in the international science market, thereby giving up all concern with the social meaning of their research, disconnecting it from political problems, and surrendering themselves to careers to be pursued according to the norms and values of the international hierarchy. For Varsavsky, "scientificism," which he equates with "indifference" to the potential local benefits of science, reinforces the cultural and economic dependency of undeveloped countries such as Argentina, condemned in this way to become "satellites" of the major world powers.[33] Even those scientists such as geologists and anthropologists whose work is founded on local investigations are in reality providing valuable raw data that will be used to further the careers of scientists in the North, he argues.[34]

Reflections on such global inequalities and on the urgent need for a Latin American science that responds to local needs have constituted some of the region's most significant contributions to the philosophy of science and epistemology more generally. As Pablo Kreimer and Hebe Vessuri affirm, "Latin America has pioneered the study of 'center-periphery' relations in scientific research" and has become an important source of debate on the question of ownership in relation to knowledge, particularly indigenous knowledge.[35] Michael Lemon and Eden Medina argue, similarly, that Latin America is a crucial site for the study of technology (and the same would clearly be true of science, too): "Latin American scholarship is centrally concerned with themes that are frequently absent from existing scholarship in the history of technology, including social justice, indigenous identities and indigenous rights, the causes and effects of economic dependency, neoliberalism, and human rights—themes central to a full understanding of technology and its role in society."[36] They argue that understanding the evolution of technology in Latin America necessarily brings to the fore "international relations, political economy, colonialism, imperialism, hegemony, and the materiality of these relationships as elements of analysis."[37]

An understanding of this complex web of relations has led to a number of innovative interdisciplinary approaches that situate science within a wide range of other forms of knowledge and experience. Some of these provide the focus for chapters in this volume. Jens Andermann explores the emerging genre of environmental writing in 1930s Argentina that sheds light on the endpoint of the capitalist exploitation of the Latin American landscape: the destruction or "unlandscaping" of the region and its eventual uninhabitability. Andermann argues that these narratives develop a critical language in which to evaluate what he calls "a natural history of the Anthropocene" that predates the emergence of narratives of political ecology in Western academia.

Andermann's contribution recognizes that Latin American thinkers have long been interested in problems that have more recently become urgent and extremely relevant within the global scientific community and that they have frequently sought to deploy scientific knowledge to transform local landscapes and practices. An oft-quoted preface to the 1811 almanac by the celebrated New

Granadan scientist Francisco José de Caldas (1768–1816) epitomizes the Creole interest in the patriotic role of science in social and economic development rather than in the production of abstract, universal knowledge. Caldas writes,

> Observar el cielo por observarlo sería una ocupación honesta, pero no pasaría de ser una curiosidad estéril que llenase los momentos del hombre ocioso y acomodado. Este observador sería inútil, y la Patria lo miraría como un consumidor de quien no esperaba nada. Nosotros no queremos representar este papel en la sociedad: queremos que nuestros trabajos astronómicos mejoren nuestra geografía, nuestros caminos y nuestro comercio.[38]

> (To observe the skies for the sake of observing them would be an honest activity, but it would be no more than a sterile curiosity to fill the time of a wealthy man at leisure. Such an observer would be useless, and the Nation would consider him a consumer from whom nothing was expected. We do not want to play that role in society: we want out astronomical work to improve our geography, our roads, and our trade.)

Antonio Lafuente shows that Creole arguments about science in the late eighteenth century were not about the viability of specific models or methodologies but about how science could be made more "relevant . . . , accountable and politically committed."[39]

Interdisciplinary thinking and a hybridization of the history and philosophy of science in Latin America have paved the way for important contributions to epistemology and particularly the epistemology of modernity. Mara Polgovsky Ezcurra, in her contribution to this collection, offers a vivid example of epistemological innovation in her analysis of the work of the Argentine scientist Rolando García (1919–2012). His theory of complex systems offers a nonlinear, nonmechanistic approach to contemporary ecological studies, emphasizing the importance of politics and ethics in such explorations of the natural world. This kind of interdisciplinary thinking marks a significant contribution on the part of Latin American thinkers to the philosophy of science.

It is perhaps the close relation—for good or ill—between science and politics in Latin America that has established a framework for interdisciplinary approaches to science in the region. Marcos Cueto points to "the unique combination of modernity and underdevelopment in Latin America" that "questions the 'natural' frontiers between nature and society determined by politicians and science practitioners, making them appear clearly artificial."[40] Evelyn

Fishburn and Eduardo L. Ortiz even assert that C. P. Snow's famous notion of "two cultures" divided "was never as pertinent in Latin America, where much closer links have always existed between the humanities and the sciences." As examples of such integration they list "the effects of positivism in Argentina, the *científicos* in Mexico, the motto *Ordem e progresso* on the Brazilian flag, the prominence of science in the educational system and the many statesmen and indeed writers who were also scientists."[41] Their overview, however, compiling examples from the nineteenth and early twentieth centuries, does not discriminate between the uncritical co-option of science's universalizing claims to modernity for political purposes in many Latin American countries, on the one hand, and a more positive, integrated thinking that might result from a genuine interdisciplinarity in approaches to understanding the role of science in society.

Science, Writing, and Culture in Latin America

Interdisciplinarity is furthered by the close relation between science and writing in Latin America that is discussed widely across this volume. The significant number of artistic and literary texts that have engaged creatively with scientific ideas in Latin America have often been generated by scholars who have combined the pursuit of scientific knowledge with literary, philosophical, or theological interests. Notable examples would include the seventeenth-century poet and scholar Sor Juana Inés de la Cruz; Eduardo Ladislao Holmberg (1852–1937), one of the leading figures in natural history in Argentina and a writer of the nation's earliest science fiction narratives; and the Nicaraguan poet and priest Ernesto Cardenal (b. 1927). Soledad Quereilhac explains in her discussion of the connections between spiritualism and scientific writing in Argentina how science opened up at the end of the nineteenth and beginning of the twentieth centuries to speculation and fantasy about all manner of mysteries of life and the universe. In her study of the highly eclectic nature of knowledge and culture in the period, Quereilhac links an interest in moving away from the strictures of positivism with an openness to spiritualism, theosophy, and the eroticism and mysticism of literary *modernismo*.

Both Quereilhac and María del Pilar Blanco focus in their chapters on serial publications and periodicals, forms of literary dissemination that enjoyed a boom in the late nineteenth and early twentieth centuries in urban centers such as Mexico City and Buenos Aires. Periodicals, from newspapers to supplements and stand-alone magazines, encompassed diverse fiction and nonfiction narrative genres as well as a dizzying range of forms of knowledge, from science to spiritualism and medicine to mesmerism. These facilitated the interaction of science with other disciplines and forms of writing in ways that threw into relief the social implications of scientific ideas. Looking at *La Ciencia Recreativa*, an understudied serial aimed at children and women readers in 1870s Mexico, Blanco reflects on the emergence of popular scientific writing at a time when the country was beginning to tap into its potential as an actor on the stage of global science.

At many points and most strikingly during the final decades of the nineteenth century and first decades of the twentieth, fiction played a significant role in the exploration of unproven hypotheses and as a platform for discussing different kinds of knowledge that were not yet approved by the academy.[42] Cristina Beatriz Fernández notes that at the end of the nineteenth century in Argentina, medical case histories and medicolegal reports borrowed forms from fictional genres. The discourses of science, history, realism, and fantasy were brought together in "una unidad narrativa simple: el *caso*" (a simple narrative unit: the *case*).[43] Interestingly, Fernández argues that this narrative strategy is closely linked to a shift within medical knowledge from a focus on individual health to the broader social sphere and a concomitant change in the role of the practitioner, who becomes "un intelectual científico" (a scientist-intellectual) with wider social responsibility.[44]

If the "intelectual científico" is a feature of writing of the period, in the increasingly autonomous field of contemporary Latin American literature we find numerous writers for whom new science invites novel forms of thinking about fiction itself. This is the kind of experiment that Joanna Page studies in her chapter, in which she focuses on recent writing by the Argentine Marcelo Cohen and the Mexican Jorge Volpi. She argues that their work stages a new (imagined) reconciliation between science and the arts in which metaphors drawn from evolutionary biology—and particularly the pioneering work of the Chilean biologists Francisco Varela and Humberto Maturana—throw into relief the defining role of fiction in the evolution of human culture and society as well as questioning the relation between science and capitalism.

Many of the chapters in the volume uncover underexplored archives, engaging closely with texts produced by Latin American scientists, philosophers, and artists that have rarely figured in global or even Latin American histories of science and culture. In doing so, they expand a sense of how Latin American science "has contributed to the hegemony of metropolitan Western knowledge" and our understanding of the extent to which "international science" creates and maintains (neo)colonial relations of power.[45] More than that, however, many of the texts provoke a kind of thinking that challenges the norms and values of global science. What would a science truly led by and responding to Latin American concerns look like? Is it possible to imagine a science that would be more egalitarian, more thoroughly mindful of its rootedness in the social and political contexts of its production, that would engage seriously with alternative cosmologies, or that would properly debate the political and ethical implications of its discoveries?

What emerges from the chapters compiled in this volume is the possibility of a powerfully integrative vision. It is one that brings science together with other forms of knowledge, within an understanding of the complex—and often conflictive—intercultural exchanges that have shaped the production of knowledge in and of Latin America. Isabelle Stengers calls for a "slow science" that would break with the model of scientific research developed in the nineteenth century, which held as "a general ideal the fast, cumulative advance of disciplinary knowledge along with a correlative disregard for any question that would slow this advance down."[46] She maintains, "The symbiosis of fast science and industry has privileged disembedded knowledge and disembedding strategies abstracted from the messy complications of this world."[47] In its place, she argues for a scientific practice that would deal in precisely this messiness: "Slowing down means becoming capable of learning again, becoming acquainted with things again, reweaving the bounds of interdependency. It means thinking and imagining, and in the process creating relationships with others that are not those of capture."[48] The chapters of this book give ample evidence of the extent to which Latin American scientists, thinkers, artists, and writers have long been engaged in exactly this work.

The Scope and Approach of This Volume

This book's cover image is of a maquette for the installation *Mente abierta/ Open Mind* (2008–2009) by the contemporary Cuban artist Yoan Capote. At first sight, the white walls bring to mind a labyrinthine art gallery, while the subterranean location suggests an archaeological dig; both are spaces that human agents may enter and explore in order to gain new insights. The shape of the site itself depicts a human brain as a landscape that is easily accessible. As the artist himself has inscribed on a 2008 watercolor sketch for the same installation, it is "el espacio abierto al libre acceso de los transeúntes como metáfora de nuestra mente. Obra que alude a la tolerancia como necesidad psicológica en la contemporaneidad" (a public open space as a metaphor for the mind. A work that alludes to tolerance as a psychological need in the contemporary moment).

But the artist's landscape of the mind might also prompt other, less utopian readings. The simple, clean lines of the maquette suggest a clinical or reductionist approach to mapping the brain that strips it of the more complex and chaotic reality of the human mind. The stances assumed by the anonymous figures inside the brain's galleries denote an act of intense observation. The relation between bodies and entered space provokes a number of questions. Is it the brain's owner who has given access to it, or someone else? Who are the observers, and what is the purpose of their observation? The apparent freedom of the observers to enter and study the mind of another, together with the passivity of the subject, might cause the viewer to question whether the openness of the brain to examination does not, in fact, create an asymmetry of power. On the other hand, the apparent blankness of the site's walls suggests an elusive unreadability, and the complexity of the labyrinthine form makes it a space that is difficult to negotiate and to master. These tensions between subjectivity, power, scientific inquiry, space, place, and landscape inform many of the chapters in this volume.

Unusually, this volume is intended to combine insights from the history of science, science and technology studies, and cultural studies; previous publications have largely focused on one or at most two of these approaches. This breadth is crucial to our aim of demonstrating the integrative, interdisciplinary, intercultural ways in which reflections on science and the scientific imaginary have developed in Latin America. It does bring certain challenges that are the inherent risks of transdisciplinary exchange in this field. Some historians of science reading this collection may miss in some of its chapters a more cohesive, archive-based, or rigorously comparative attempt to account for the development of science and the scientific imaginary in Latin America—which we do not pretend to offer—or some may feel that the engagement in the more literary chapters with European thinkers such as Latour is less appropriate in a Latin American context. On the cultural studies side, there can be an impatience with the reluctance of historians to draw broader conclusions from the specific cases they study. There is often a tendency in both camps to overstate the reach of science and technology studies by assuming that the social construction of science means the demolishing of all referentiality. This can lead to a misuse within cultural studies of theories arising from science and technology studies or to misreadings of the use of such theories on the suspicion that they are being used to further the claim that nothing exists beyond our linguistic, social, and cultural constructions of it.

Such areas of friction or skepticism are familiar to anyone who has engaged in genuinely transdisciplinary exchanges. They offer valuable opportunities to reassess our own disciplinary practices and also, hopefully, to rearticulate the importance of approaching a subject from the widest possible set of standpoints in order to bring them into active dialogue with each other and to balance their respective strengths and weaknesses.

The chapters of this volume have been grouped into thematic sections, each of which is preceded by an introduction; the introductions situate the individual case studies within broader debates in their fields of scholarship, drawing where appropriate on Anglo-American, European, and Latin American sources. What brings the chapters together despite the authors' different approaches is the aim to explore the relations between science, geopolitics, and culture in Latin America in ways that challenge received ideas of the development of science and the scientific imaginary in the region and beyond. The chapters, most of which have been penned by writers hailing from Latin America, contribute to scholarship across disciplines in a number of ways. The central value of some of them is to be found in their presentation of hitherto unacknowledged contributions to knowledge made by Latin American scientists, while others explore the geopolitics of science in the region more broadly and emphasize the importance of a relational rather than a linear-chronological approach to understanding knowledge production. We have also included chapters that demonstrate the sustained tradition in Latin American thought and culture of deploying scientific ideas for the purpose of developing emancipatory and utopian political or cultural projects—and not only exploitative or authoritarian ones—as a way of redressing what we feel to be a current imbalance in the field. A further series of chapters emphasizes the strongly interdisciplinary and intercultural approaches that have long marked epistemological and cultural reflections on science in Latin America; they uncover the productive hybridizations between art and science that have been the norm rather than the exception in the evolution of disciplines in the region. Thinking through how science is and may be disseminated throws light on new, hybrid spaces in which science flourishes alongside other kinds of knowledge and practice and where it shapes and is shaped by diverse subjectivities.

This book evolved out of many presentations and discussions that took place across four events organized as part of an international research network on Science in Text and Culture in Latin America, funded by the Arts and Humanities Research Council in England. At events held in San Juan, Puerto Rico, in Buenos Aires, and in Oxford and Cambridge, England, we were able to learn about the valuable research being undertaken by many more scholars than we have been able to include here. Indeed, this book brings together only a selection of the topics currently under investigation by researchers across the world in a field of burgeoning interest. Our hope is that this collection will spark discussion and further explorations of the role science has played in Latin America and that Latin America has played in science.

Notes

1. Achim, "Science in Translation," 107.
2. Achim, "Science in Translation," 108.
3. Podgorny, "Fossil Dealers," 673.
4. Podgorny, "Fossil Dealers," 674.
5. Kuhn writes, "Individual scientists embrace a new paradigm for all sorts of reasons and usually for several at once. Some of these reasons—for example, the sun worship that helped make Kepler a Copernican—lie outside the apparent sphere of science entirely. Others must depend upon idiosyncrasies of autobiography and personality. Even the nationality or the prior reputation of the innovator and his teachers can sometimes play a significant role" (*Structure of Scientific Revolutions*, 151–152).
6. Vessuri, "Introducción," 16–17.
7. Vessuri, "Introducción," 18–20.
8. Vessuri, "Introducción," 20–22.
9. Varsavsky, *Ciencia, política y cientificismo*, 27.
10. Kreimer, *Ciencia y periferia*, 69, 201, 215.
11. Kreimer, *Ciencia y periferia*, 217.
12. Sigal, *Intelectuales y poder*, 100–101.
13. Cazaux, *Historia de la divulgación científica*, 195.
14. Nobel Prize, "César Milstein."
15. See Medina, da Costa Marques, and Holmes, "Introduction," 8; Kreimer and Vessuri, "Latin American Science."
16. Medina, da Costa Marques, and Holmes, "Introduction," 2.
17. Hurtado, *La ciencia argentina*, 21.
18. Dussel, "Eurocentrism and Modernity," 65.
19. Dussel, "Eurocentrism and Modernity," 67–68.
20. Quijano, "Coloniality of Power," 542.
21. Quijano, "Coloniality of Power," 543.
22. Quijano, "Coloniality of Power," 552.
23. Saldaña, "Introducción," 9.

24. See Anthony Aveni, *Skywatchers of Ancient Mexico* (1980); Johanna Broda and Jorge Báez, *Cosmovisión, ritual e identidad de los pueblos indígenas de México* (2001); Broda, Stanislaw Iwaniszewski, and Lucrecia Maupomé, eds., *Arqueoastronomía y Etnoastronomía en Mesoamérica* (1991); and Jesús Galindo Trejo, *Arqueoastronomía en la América antigua* (1994).

25. See Estrella, *Medicina aborigen*.

26. Cañizares-Esguerra, *Nature, Empire, and Nation,* 8.

27. Burga, "De la patria a la nación," 310.

28. See Manuel Gamio, "Nuestra cultura intelectual," in *Forjando patria,* 165–179.

29. Whitt, *Science, Colonialism,* 4–5.

30. Whitt, *Science, Colonialism,* 5.

31. Stepan, "Eugenics in Brazil, 1917–1940," 113.

32. Molloy, "National Parts," 45.

33. Varsavsky, *Ciencia, política y cientificismo,* 45.

34. Varsavsky, *Ciencia, política y cientificismo,* 46.

35. Kreimer and Vessuri, "Latin American Science." In addition to their own previous publications, Kreimer and Vessuri also cite here Greiff, *A las puertas del universo derrotado.*

36. Lemon and Medina, "Technology in an Expanded Field," 112.

37. Lemon and Medina, "Technology in an Expanded Field," 112.

38. Caldas, "Preliminares para el almanaque de 1811," *Obras completas,* 402.

39. Lafuente, "Enlightenment in an Imperial Context," 172–173.

40. Cueto, Foreword, viii.

41. Fishburn and Ortiz, Introduction, 1.

42. Gasparini, *Espectros de la ciencia,* 22–24, 44.

43. Fernández, *José Ingenieros,* 20.

44. Fernández, *José Ingenieros,* 20.

45. Cueto, Foreword, vii.

46. Stengers, *Another Science Is Possible,* 98.

47. Stengers, *Another Science Is Possible,* 120.

48. Stengers, *Another Science Is Possible,* 81–82.

References

Achim, Miruna. "Science in Translation: The Commerce of Facts and Artifacts in the Transatlantic Spanish World." *Journal of Spanish Cultural Studies* 8, no. 2 (July 2007): 107–115.

Aveni, Anthony. *Skywatchers of Ancient Mexico.* Austin: University of Texas Press, 1980.

Broda, Johanna, and Jorge Báez. *Cosmovisión, ritual e identidad de los pueblos indígenas de México.* Mexico City: Fondo de Cultura Económica, 2001.

Broda, Johanna, Stanislaw Iwaniszewski, and Lucrecia Maupomé, eds. *Arqueoastronomía y Etnoastronomía en Mesoamérica.* Mexico City: Universidad Nacional Autónoma de México, 1991.

Burga, Manuel. "De la patria a la nación/del mundo natural al mundo cultural: La ciencia en el Perú, 1790–1930." In *Las ciencias en la formación de las naciones americanas,* edited by Sandra Carrera and Katja Carrillo Zeiter, 295–316. Madrid: Iberoamericana Vervuert, 2014.

Caldas, Francisco José de. "Preliminares para almanaque de 1811." *Obras completas de Francisco José de Caldas*, 401–410. Bogotá: Universidad Nacional de Colombia, 1966.

Cañizares-Esguerra, Jorge. *Nature, Empire, and Nation: Explorations of the History of Science in the Iberian World.* Stanford: Stanford University Press, 2006.

Cazaux, Diana. *Historia de la divulgación científica en la Argentina.* Buenos Aires: Teseo, Asociación Argentina de Periodismo Científico, 2010.

Cueto, Marcos. Foreword to *Beyond Imported Magic: Essays on Science, Technology, and Society in Latin America*, edited by Eden Medina, Ivan da Costa Marques, and Christina Holmes, vii–ix. Cambridge, MA: MIT Press, 2014.

Dussel, Enrique. "Eurocentrism and Modernity." *boundary 2* 20, no. 3, *The Postmodern Debate in Latin America* (Autumn 1993): 65–76.

Estrella, Eduardo. *Medicina aborigen: La práctica médica aborigen de la Sierra Ecuatoriana.* Quito: Época, 1977.

Fernández, Cristina Beatriz. *José Ingenieros y las escrituras de la vida: Del caso clínico a la biografía ejemplar.* Mar del Plata, Argentina: Eudem (Editorial de la Universidad Nacional de Mar del Plata), 2014.

Fishburn, Evelyn, and Eduardo L. Ortiz. Introduction to *Science and the Creative Imagination in Latin America*, edited by Fishburn and Ortiz, 1–12. London: Institute for the Study of the Americas, 2005.

Gamio, Manuel. *Forjando patria.* Mexico City: Porrúa Hermanos, 1916.

Gasparini, Sandra. *Espectros de la ciencia: Fantasías científicas de la Argentina del siglo XIX.* Buenos Aires: Santiago Arcos, 2012.

Greiff, Alexis. *A las puertas del universo derrotado.* Bogotá: Universidad Nacional de Colombia, 2012.

Hurtado, Diego. *La ciencia argentina. Un proyecto inconcluso: 1930–2000.* Buenos Aires: Edhasa (Editora y Distribuidora Hispano Americana), 2010.

Kreimer, Pablo. *Ciencia y periferia: Nacimiento, muerte y resurrección de la biología molecular en la Argentina.* Buenos Aires: Eudeba (Editorial Universitaria de Buenos Aires), 2010.

Kreimer, Pablo, and Hebe Vessuri. "Latin American Science, Technology, and Society: A Historical and Reflexive Approach." *Tapuya: Latin American Science, Technology, and Society* 1, no. 1 (2018): 17–37.

Kuhn, Thomas. *The Structure of Scientific Revolutions.* Chicago: University of Chicago Press, 2012.

Lafuente, Antonio. "Enlightenment in an Imperial Context: Local Science in the Late-Eighteenth-Century Hispanic World." *Nature and Empire: Science and the Colonial Enterprise*, special issue, *Osiris* 15 (2000): 155–173.

Lemon, Michael, and Eden Medina. "Technology in an Expanded Field: A Review of History of Technology Scholarship on Latin America in Selected English-Language Journals." In *Beyond Imported Magic: Essays on Science, Technology, and Society in Latin America*, edited by Eden Medina, Ivan da Costa Marques, and Christina Holmes, 111–136. Cambridge, MA: MIT Press, 2014.

Medina, Eden, Ivan da Costa Marques, and Christina Holmes. "Introduction: Beyond Imported Magic." In *Beyond Imported Magic: Essays on Science, Technology, and Society in Latin America*, edited by Medina, da Costa Marques, and Holmes, 1–23. Cambridge, MA: MIT Press, 2014.

Molloy, Sylvia. "National Parts and Unnatural Others: A Reflection on Patrimony at the Turn of the Nineteenth Century." In *Science and the Creative Imagination in Latin America,* edited by Evelyn Fishburn and Eduardo L. Ortiz, 44–58. London: Institute for the Study of the Americas, 2005.

Nobel Prize. "César Milstein, Biographical." Accessed March 25, 2018. https://www.nobelprize.org/nobel_prizes/medicine/laureates/1984/milstein-bio.html.

Podgorny, Irina. "Fossil Dealers, the Practices of Comparative Anatomy, and British Diplomacy in Latin America, 1820–1840." *British Journal for the History of Science* 46, no. 4 (December 2013): 647–674.

Quijano, Aníbal. "Coloniality of Power, Eurocentrism, and Latin America." *Nepantla: Views from South* 1, no. 3 (2000): 533–580.

Saldaña, Juan José. "Introducción: Teatro científico americano. Geografía y cultura en la historiografía latinoamericana de la ciencia." In *Historia social de las ciencias en América Latina*, edited by Saldaña, 7–41. Mexico City: Coordinación de Humanidades, UNAM (Universidad Nacional Autónoma de México), 1996.

Sigal, Silvia. *Intelectuales y poder en la década del sesenta*. Buenos Aires: Puntosur, 1991.

Stengers, Isabelle. *Another Science Is Possible: A Manifesto for Slow Science*. Translation by Stephen Muecke. Cambridge, England: Polity, 2018.

Stepan, Nancy L. "Eugenics in Brazil, 1917–1940." In *The Wellborn Science: Eugenics in Germany, France, Brazil, and Russia*, edited by Mark B. Adams, 110–152. New York: Oxford University Press, 1990.

Trejo, Jesús Galindo. *Arqueoastronomía en la América antigua*. Madrid: Equipo Sirius, 1994.

Varsavsky, Oscar. *Ciencia, política y cientificismo y otros textos*. Buenos Aires: Capital Intelectual, 2010.

Vessuri, Hebe. "Introducción: Consideraciones acerca del estudio social de la ciencia." In *La ciencia periférica: Ciencia y sociedad en Venezuela*, edited by Elena Díaz, Yolanda Texera, and Hebe Vessuri, 9–35. Caracas: Monte Ávila, 1983.

Whitt, Laurelyn. *Science, Colonialism, and Indigenous Peoples: The Cultural Politics of Law and Knowledge*. Cambridge, England: Cambridge University Press, 2009.

I

Latin America's Scientific Landscapes

Introduction to Section I

MARÍA DEL PILAR BLANCO AND JOANNA PAGE

In one of the earlier episodes related by Charles Darwin in *The Voyage of the Beagle* (1839), the young Englishman describes the visit to an estate on the Rio Macaé in Brazil. It was "two and a half miles long, and the owner had forgotten how many broad."[1] Remarking that only a small portion of this land had been put to use, Darwin notes that "considering the enormous area of Brazil, the proportion of cultivated ground can scarcely be considered as anything, compared to that which is left in the state of nature." The observation leads him to muse that "at some future age, how vast a population it will support!" (33). The estate on the Macaé River, while producing in the budding naturalist a sense of wonderment at the vast expanse of untampered South American nature, is also the scene for a strikingly different observation, this time of the economic substructure that sustained Brazil into the final years of the nineteenth century. For it is also there that Darwin becomes a witness to the facets of life in a "slave country": families being driven apart and a black man who, upon thinking the Englishman was about to strike him, crumbled in submissive fear (33–34). "This man," Darwin writes, "had been trained to a degradation lower than the slavery of the most helpless animal" (34).

This brief diary entry in one of the best-known scientific travel accounts in the Western tradition reveals much about perceptions of the Latin American landscape during the nineteenth century. Following in the footsteps of Alexander von Humboldt and Aimé Bonpland, Darwin takes part in a second age of exploration, one in which a scientific imperialism had come to replace the religious and extractive missions of previous centuries. The enthusiasm shared by Humboldt and Darwin when faced with the landscapes south of the equator stems in part from a renewed ability to enter into spaces that had been sealed off to non-Iberian Westerners during the centuries of Spanish rule. Their expeditions, transpiring during the decades immediately preceding and following the wars of independence across Latin America, can be read both as inspirations for emancipation, in that they reveal to the Latin American reader the significant natural wealth of their region, and as rehearsals of a well-known scene in the history of Western science, that is, Latin America as open ground of exploration by foreigners—particularly those from Britain, France, Germany, and the United States—in whose hands the future of scientific development rests.[2] Darwin's retelling of his encounter with the owner of the estate on the Macaé rehearses an opposition between European liberalism and scientific exactitude on the one hand and *criollo* inefficiency and cruelty on the other. In adhering so steadfastly to a slave economy, *criollo* society remains barbaric in the eyes of the European scientist.

Such episodes in Darwin's account reveal a scene of incompatibility between the European scientist and the Latin American subject. The clear binarisms that underpin accounts of encounters like this one need to be revised, however, in the light of a greater understanding of the complex interchanges that mark scientific knowledge production and dissemination in the Latin American region throughout the colonial and postcolonial periods, in which diverse agents are involved. The historian of science Antonio Lafuente has opposed models of understanding scientific exchange that advocate the "automatic identification of science with emancipation, or radical contrasts such as 'Creole versus metropolitans' or 'ancient versus modern,'" as these only sustain long-held conceptions about science as a universal standard rather than a set of socially embedded practices.[3] Lafuente instead suggests reading for historical complexity and against homogeneous categorizations, to seek "a less mechanical but more organic view, a less geometric but more historical view of the movements and metamorphoses experienced and described by science in the course of its chronological and spatial development."[4]

Lafuente proposes an understanding of the globalization of science and innovation as a process of "mutual renewal."[5] As a corrective to established modes of thinking about unilateral center-periphery transmissions of scientific knowledge, recent work in the history of science situates canonical figures within webs of mobile intellectual interaction. Writing about Humboldt, Laura Dassow Walls describes the work of the naturalist as "deliberately comparative," leading him to understand civilization as having "multiple centers of origin" and "many cradles."[6] For Humboldt, travel and contact were antidotes to forms of imperial thinking that judged the New World unfavorably from afar. Emerging from this contact, Humboldt's "environmental discourse," Walls argues, considers humanity and nature locked in constant conversation such that "the ways various societies construct their views of nature were crucial to understanding their physical environment."[7] What Walls perceives as Humboldt's paradigmatic shift in perceptions of the so-called New World also represents a different way of reading how and where scientific knowledge happens, for it is not that Humboldt enacted change in this part of the world but that his contact with its populations and their uses of nature transformed his way of thinking science. Such a rationalization of the scientific enterprise enables a different outlook on the roles Latin America, with its diverse populations and environmental realities, may play in the history and development of global science.

Exploring perceptions of Mexican and Argentine environments throughout the nineteenth and twentieth centuries, the chapters in this section actively and critically engage with a transformation of attitudes toward the flows of information described above. In different ways, the cases presented here demonstrate how scientific reflection grows from moments of situated observation to become part of complex systems of exchange, some of which, as Lafuente's examples show, are not without tension and controversy. Examining the narrativizations of science at turning points in Latin American history, each of the chapters offers a response to the North Atlantic claim to authority over the region's natural configurations as well as its scientific potential. In their contributions, Gabriela Nouzeilles, Jens Andermann, and María del Pilar Blanco explore how Argentina and Mexico sketched out their national personalities in a modern world in which spaces and forms of scientific dissemination were becoming ciphers of biopolitical and cultural power. The development of such personalities entailed, in turn, diverse levels of rapprochement as well as disengagement with metropolitan scientific theories and methodologies. Nouzeilles and Blanco note that such positionings within the geopolitical maps of the late nineteenth century and beyond are deeply invested in the representation of national landscapes that endure throughout the length of prehistory and in the depiction of a continual adaptability of those landscapes to technological modernity. In his chapter Andermann offers a fascinating perspective on how Latin American environmentalists' writings from the mid-twentieth century trace the devastating underside of the fantasy of a landscape that is constantly exploited.

In her contribution to this section, Blanco studies experimental publication genres in the República Restaurada (1868–1876) in Mexico—the popular science magazine as well as the *novelita científica*, the scientific novel published for young readers—as adaptations of European literary forms employed to represent the country's environmental exceptionality. Rather than offering examples of the periphery rehearsing the activities of the center, these publications engage in a form of avid local exegesis that reflects the inner workings of scientific work more generally: how it relies on continuous inscription, comparison, and corroboration of such inscriptions by a wide population of science workers. Nouzeilles, in her chapter on the paleontological imagination in Argentina at the end of the nineteenth century, explores the role of Patagonian fossil discoveries in the legitimation of the nation's position as a major player in Western modernity. Focusing on the Argentine naturalist Florentino Ameghino (1853/1854–1911)—a figure whose career, as Irina Podgorny argues, raises important questions for contemporary historians of science[8]—Nouzeilles explores his "geopolitical readings of the past" (39, this volume), delineating how universally established scientific ideas are contested at the local level in response to institutional, political, and cultural pressures on national and international levels. Moving northeastward in Argentina to Santiago del Estero during the period between the 1930s and 1950s, Andermann examines a body of remarkable regionalist writings that spell out the unraveling of the capitalist exploitation of Santiago's landscape. Reframing such regionalism as radically modern, Andermann explores how these publications trace the "natural history of the Anthropocene" in unprecedented ways.

Together, the chapters in this opening section help reframe commonly held assumptions of science's direction of travel into Latin America from elsewhere, demonstrating the complexity of encounters between the local and the global. They reveal the extent to which scientific ideas originated in or were appropriated for social, cultural, and economic approaches to the specificities of land, geography, and space within national contexts as well as how Latin American writers and scientists have deployed such ideas to author the natural histories of their country and region and to imagine futures to come.

Notes

1. Darwin, *Voyage of the* Beagle, 33.
2. For a discussion of the marginalization of Mediterranean science within the broader

European network, see Papanelopoulou, Nieto-Galan, and Perdiguero, eds., *Popularizing Science*.

    3. Lafuente, "Enlightenment in an Imperial Context," 157.

    4. Lafuente, "Enlightenment in an Imperial Context," 157.

    5. Lafuente, "Enlightenment in an Imperial Context," 156.

    6. Walls, *Passage to Cosmos*, 6.

    7. Walls, *Passage to Cosmos*, 8.

    8. Podgorny, "Human Origins in the New World?" 69.

References

Darwin, Charles. *The Voyage of the* Beagle. New York: P. F. Collier and Son, 1909.

Lafuente, Antonio. "Enlightenment in an Imperial Context: Local Science in the Late-Eighteenth-Century Hispanic World." *Osiris* 15 (2000): 155–173.

Papanelopoulou, Faidra, Agustí Nieto-Galan, and Enrique Perdiguero, eds. *Popularizing Science and Technology in the European Periphery, 1800–2000*. Aldershot, England: Ashgate, 2009.

Podgorny, Irina. "Human Origins in the New World? Florentino Ameghino and the Emergence of Prehistoric Archaeology in the Americas (1875–1912)." *PaleoAmerica* 1, no. 1 (2015): 68–80.

Walls, Laura Dassow. *The Passage to Cosmos: Alexander von Humboldt and the Shaping of America*. Chicago: University of Chicago Press, 2013.

1

Bone Tales

Patagonian Monsters and the Paleontological Imagination

GABRIELA NOUZEILLES

Perhaps, however, someone will mount the converse argument,
and say that the ancients not only knew as many large animals as we
do—as we have just shown—but that they described several that we
have not; that we are too hasty in regarding these animals as fabulous;
that we ought to go on looking for them before we believe we have
exhausted the history of the existing creation; and finally that among
these putatively fabulous animals—when we know them better—will
perhaps be found the originals of our bones of unknown species.[1]

Georges Cuvier, Recherches sous les ossemens fossiles, *1799*

Georges Cuvier, the founder of modern paleontology, rejected the use of the
imagination in science, but he hinted at the possibility that giants, dragons, and
monsters of lore could actually be cultural allegories of empirical evidence provided in nature, thereby suggesting that there might be some truth concealed
behind fantasy and myth. Cuvier's interest in fictional accounts of monsters
is important for understanding the early formation and subsequent development of a visual rhetoric of paleontology and its ideological undercurrents
in the nineteenth century. The association of extinct creatures with the world
of fantasy became a "stereotypical way of presenting extinct ecosystems," and

"dangerous-looking monsters and other fiends were systematically reproduced in the visual rhetoric and iconography of antediluvian worlds" that eventually dominated scientific illustrations, public exhibitions, and museum displays.[2] The Crystal Palace dinosaurs designed and sculpted by Benjamin Waterhouse Hawkins under the scientific direction of Sir Richard Owen and unveiled in London in 1854 are a manifestation of this iconographic trend.[3]

The modern paleontological imagination had a strong narrative component. Cuvier's comparative method suggested that every fossilized bone or skeleton could be made to tell a story that ultimately revealed the identity of an extinct, unknown animal. Influenced by Charles Lyell's vision of geological change as the result of the cumulative power of incremental transformation in geologic or deep time,[4] in *On the Origin of Species* (1859) Charles Darwin presents the history of the earth as a sensational narrative shattered into pieces and locked up in geological layers, waiting to be read by a naturalist:

> For my part, following out Lyell's metaphor, I look at the natural geological record, as a history of the world imperfectly kept, and written in a changing dialect; of this history we possess the last volume alone, relating only to two or three countries. Of this volume, only here and there a short chapter has been preserved; and of each page, only here and there a few lines. Each word of the slowly-changing language, in which the history is supposed to be written, being more or less different in the interrupted succession of chapters, may represent the apparently abruptly changed forms of life, entombed in our consecutive, but widely separated, formations.[5]

Darwin sees the geological record as an incomplete history book written in a language with a changing grammar for which there is no dictionary. In order to reconstruct the missing parts of the geological macro narrative, the naturalist needs to look for clues in the only chapter available. This conjectural approach corresponds to what Carlo Ginzburg calls the "evidential paradigm" that shaped the methods of inquiry of several disciplines toward the end of the nineteenth century.[6] According to this paradigm, the naturalist operates as a sort of detective seeking to reconstruct a past event by looking at the scattered evidence of a crime scene.

But, as is the case in many detective stories, the causality of the past can be interpreted in many different ways. Even though the notion of the extreme age of the earth was eventually accepted in all scientific circles, there was no clear agreement on the narrative logic behind geological time and life evolution. In the second half of the nineteenth century, Darwin's sensational history of terrestrial life was subjected to several rewritings in order to adapt it to different and at times conflicting visions of the natural world.

In this essay I focus on the paleontological rhetoric around Patagonian fossils in late nineteenth-century Argentina and the ways in which local science responded and contributed to imperial arrangements regarding narrative power and the symbolic and material production of scientific knowledge. On the one hand, as both Irina Podgorny and Jens Andermann have argued, local scientific practices and institutions actively channeled the demands of primitive accumulation sanctioned by international capital by providing a physical foundation and a visual grammar (in the form of catalogs, drawings, sculptural contraptions, public exhibitions, and architectural structures of display) for the hegemonic interpellation enacted through the idea of "national" nature on behalf of the state idea.[7] On the other hand, local naturalists and paleontologists also participated in the development of a global society of the spectacle and an international market of collectible artifacts that put forward contradictory claims regarding the meaning of deep time and the value of progress and modernity.

My argument goes beyond a critique of science based on the sway of historical ideological formations on scientific thinking put forward by Latour,[8] in order to stress the epistemological instability of paleontological objects and theories as the result of the impact of competing and conflicting epistemological, institutional, political, and affective claims on local scientific practices. I also seek to complicate the view of local science as a mere branch of state power and institutional racism by hinting at the impact of cultural and social dynamics, including aesthetic primitivism and masculinist definitions of scientific authority. In this I follow Stephen Gould when he states that scientists are social beings "immersed in culture, and struggling with all the curious tools of inference that mind permits—from metaphor and analogy to . . . flights of fruitful imagination."[9]

Among the many elusive objects that fed the Patagonian paleontological imagination, the legendary and overdetermined figure of a prehistoric animal of gigantic dimensions has been a central element in the history of the region's modern representations. If, as Emily Apter and William Pietz have argued, one of the defining characteristics of fetishism is that it destabilizes the base of representations even as it helps them function as monolithic symbols of a culture, the figure of the Patagonian monster can be aptly read "as the site both of the formation and revelation of ideology and value-consciousness."[10] Using the figure's inherent instability as a means of access, I propose an archaeology of some of the political and cultural metamorphoses through which the Patagonian fossil fetish has gone since its emergence in the late nineteenth century, when scientific debates on the geological interpretation of the planetary past were the public arena for openly political interpretations of the global order of nations.[11] The historical triggers that stimulated the production of fetishistic representations of the geological past were fundamentally two: the creation of the Argentine modern state not only as a sum of institutions and practices but most importantly as an idea, and the sweeping modernization of the country through massive urbanization, biopolitical policies, and technological change. Complementing the military control and the ideological reinvention of Patagonia as a promissory land that characterized the utopian visions of the ruling class, the task of locating, collecting, identifying, and dating fossils began to contribute to the formation of a geographic national imagery.

Argentina's paleontological imagination was constituted in an oppositional dialogue with metropolitan imperial views of deep time. Like the United States, Argentina located its symbolic foundations in nature. Lacking the ancient and illustrious civilizations of Europe and Mexico, the nation's power would come from the wealth and antiquity of the natural constitution of its plants, animals, soil, and fossil sites. Once the humid and fertile pampas became the central stage for the country's modern agrarian economy and urban development, the Patagonian expanses, both emptied and constituted by state violence, became natural blank pages for ideological constructs. While the militarization of the region helped consolidate the national state in 1879, the paleontological imagination turned Patagonia into the new nation's *political unconscious*—a heterotopic site on which to unleash genocidal violence and counterhegemonic desires. Patagonia as heterotopic space, however, was the repository of political and cultural fantasies that exceeded the national and appealed to an infectious modern fascination with the geological traces of deep time, that is, with a temporality embedded in nature that went beyond the human and defied established religious and scientific views.

Central to this epistemological process were the activities and contributions of the naturalist Florentino Ameghino (1854–1911), the founding father of Argentine paleontology and one of the most renowned Latin American scientists of his time. Ameghino's geopolitical interpretation of the meaning of Patagonia's animal fossils offers a counterhegemonic rereading of dominant scientific narratives of the geological past. At the same time, his enthusiastic participation in local and international debates regarding the existence of natural lost worlds inhabited by living prehistoric monsters points to a paleontological poetics that questions and complicates the divisions between evidence and fiction and between scientific and indigenous knowledge. Ameghino's stories, poised between the factual and the fictional, were a sign of scientific progress and a means to negotiate social, cultural, and political tensions.

Rewriting the Book of Nature

For most of his professional life, Florentino Ameghino was passionately devoted to the study of fossil mammals from Patagonia, with the steady support of his brother, Carlos (1865–1936). Between 1887 and 1902, Carlos made fourteen trips to the region, where he collected numerous fossil faunas and made important stratigraphic observations on behalf of his brother. In 1906 Florentino Ameghino published his most important work on the region, "Les formations sedimentaires du Crétacé Supérieur et du Tertiaire de Patagonie" (Sedimentary formations of the Upper Cretaceous and Tertiary Periods in Patagonia), in which he modified his classification system of the strata from Patagonia and the Pampas and presented his interpretation of the ancient connections between South America, Australia, Africa, and North America during the Cretaceous and early Eocene. In this book, Patagonia became not only the center of origin and distribution of mammals but also the actual scenario of Ameghino's phylogenetic tree, according to which the genus Homo and all Old World human fossil specimens were offspring of the small bipedal Homunculus from the Early Tertiary of Patagonia.[12]

Ameghino's voluminous and erudite publications on the geological composition of Patagonia and its massive archives of fossils, as well as his heated debates with the international scientific community about how to interpret the history of the planet from a different geopolitical perspective, turned him into a national and international celebrity. But what most likely put Ameghino among Argentina's founding fathers were the political undertones of his scientific fictions. His paleontological theories can be read as foundational fictions based not on an erotics of reconciliation, as was the case with hegemonic romances of state love as the critic Doris Sommer has argued, but on an open war between fossil "tribes"—gangs of Leviathans struggling for control of the planet.[13] It is important to note, however, that even though Ameghino contributed to the production of national fictions, he made an effort to remain independent from the state apparatus. In this he may be differentiated from other Argentine naturalists such as Francisco P. Moreno.

In the texts Ameghino dedicated to Patagonia, the recodification of the Patagonian fossils as national wealth works in at least three ways. First, Ameghino bases the fossils' epistemological value on the amount, accessibility, and rarity of specimens. The scattered fossils in the Patagonian desert were enormous archives of scientific evidence unavailable in other parts of the world and therefore indispensable to the reconstruction of the rings and branches of the evolutionary tree. It was this evidence that, according to Ameghino, had elicited Darwin's "Eureka" upon his discovery of the universal laws of life:

> Todos vosotros sabéis sin duda que Darwin puede considerarse como uno de nuestros sabios, pues el descubrimiento de su teoría está ligado a la historia de nuestro progreso científico, por ser aquí, entre nosotros, donde recogió los materiales de ella y tuvo su primera idea.[14]

> (All of you are surely aware that Darwin can be considered one of our brilliant minds. After all, the development of his main theory is intimately related to the history of our scientific progress, since it was here, among us, where he collected the primary data and got his first hunch.)

It had been there, in the pampas and in Patagonia, where he had found unmistakable proof of the laws of evolution. Whereas Europe's natural past looked like a book from which four-fifths of its pages had been torn, Ameghino argues, local nature was an open book that naturalists could read from beginning to end almost without interruption.

Ameghino's second way of endowing Patagonia with value was by superimposing a lush prehistoric landscape over the Patagonian dry and bare desert. This rhetorical strategy complements the futuristic visions of progress that at the time imagined an apparently useless space radically transformed by modernity. While the proleptic imagination of the state engineers saw fertility as the promise of technology, Ameghino unveiled the region's potential by evoking the hyperactivity and abundance of a quasitropical geological past. Reminiscent of Hawkins's illustrations of the tertiary fauna circulating in the second half of the nineteenth century in Europe and beyond, Ameghino's virtual landscapes filled the barren expanses of Patagonia with palm trees, conifers, and enormous beasts such as *Megatherium*. Ameghino's most radical scientific rereading of the landscape, however, was in the field of scientific theory. In his view, the dominant versions of global prehistory and animal dispersal needed to be revised and amended. Filling in the blanks in the prehistoric map was not enough; Ameghino wanted to turn the map upside down. This new perspective understood the deep history of the globe to have originated in Patagonia and by extension, in a prehistoric Argentina. Ameghino's radical politics of representation were from the outset structurally limited by the nature of science itself as an apparatus of power and by the imperial ambitions of Argentine nationalism.

The first contested terrain was that of classification. The meticulous catalogs of national fossils were seen as acts of appropriation of local nature and of contestation against the imperial machine of nomination.[15] The epistemological battle between local and metropolitan scientists also inspired Ameghino's hierarchical descriptions of Patagonia's prehistoric fauna. Anticipating the phallic allegories that would later cover the walls and atriums of many museums of natural history, usually focused on a fight to the death between two prehistoric animal titans, Ameghino would forever endow Patagonian paleontological fictions with a sense of postcolonial resentment. The Patagonian fossils were equal or superior to those of the Northern Hemisphere in strength and size. In the group of dinosaurs, although the most remarkable specimens had been found in the deserts of the United States, Argentina did not lack for superior types. Among the herbivorous dinosaurs, Ameghino argued, Patagonia offered extremely competitive examples. The genus *Argyrosaurus*, similar to the North American *Brontosaurus* and *Atlantosaurus*, exceeded both of them in size. Even the local *Titarosaurus*, of more moderate dimensions, could easily measure up to Cretaceous representatives from India, England, and Madagascar.[16]

But it was in the lost fauna of Patagonian mammals that Ameghino mounted his third and strongest vindication of Patagonian fossils. His argument combined strategies of inversion and transvaluation.[17] First, he inverted the hierarchical order by size that portrayed the fossil animals from the south as lesser versions of those of the north. According to Ameghino, the geological and fossil evidence showed that when the northern mammals still consisted of "a few simple and feeble marsupials," the southern mammals had already reached gigantic proportions. Second, and this was his most daring line of argument, Ameghino replaced the geographic axis that explained the distribution and evolution of mammals as a movement north-south with one that went south-north. As a result of this change, modern mammals (including humans) had originated in Patagonia, from where they had slowly moved northward until after many transformations they had occupied the whole planet. In this account, Patagonia's fossil sites were the monumental ruins of a glorious past of biological supremacy. Ameghino vehemently maintained that the Northern Hemisphere had not been su patria de origen, porque cuando atravesaron en peregrinación hacia el Norte la línea ecuatorial, miles de siglos hacía que pisaban las tierras australes, donde habíanse desarrollado y diversificado en faunas sucesivas con numerosísimas formas.[18]

(their fatherland; when they finally crossed the Equator *en route* to the North, they had lived for thousands of centuries in the South, where they already had developed and diversified into succeeding faunae adopting many different shapes.)

In open contradiction to the dominant paleontological theories of geographic distribution, Ameghino's chain of biological metamorphosis located the genesis of the Quaternary Era in Patagonia.[19] From there, the ancestors of all the mammals had dispersed throughout the world, making use of the continental bridges that connected South America to Africa and Europe at the end of the Cretaceous Period. The *Pyrotheridae*, whose fossilized remains had only been found in Patagonia, offered a model case of this narrative fiction of migratory transformation through many continents and geological eras:

Estos animales constituyen indiscutiblemente el tronco de los Proboscidios que en el antiguo continente no aparecen sino a partir del Mioceno bajo la forma de *Dinotherium*. Los *Pyrotheria* debieron pasar al África hacia fines del Cretáceo o principios del Terciario y se transformaron gradualmente en *Dinotherium* y esa es la forma bajo la cual aparecen en Asia y en Europa durante el Mioceno medio. El *Dinotheirum*, o una forma cercana, se transforma en *Mastodon* y en *Elephas*, géneros que son hallados en todo el Antiguo continente, y pasaron también a América del Norte: el *Mastodon* hacia fines del Mioceno y el *Elephas* a principios del Plioceno.[20]

(These animals are unquestionably a branch of the *Proboscidii*, which in the Old Continent did not appear until the end of the Miocene in the form of *Dinotheria*. The *Pyrotheridae* must have crossed over Africa toward the end of the Cretaceous or at the beginning of the Tertiary, and they gradually turned into *Dinotheria*—the form in which they showed up in Asia and Europe during the mid-Miocene era. The *Phinotetherium*, or a similar form, transformed into the *Mastodon* and the *Elephas*, genera that are found throughout the Old Continent. These also moved on to North America: the *Mastodon* by the end of the Miocene and the *Elephas* at the beginning of the Pliocene.)

Like the *Pyrotheridae* in their journey from South America to Asia, the *Archaeohyracoidea* had turned into the *Hyracoidea*, the *Notohippidea* into horses, the *Condylarthra* into *Artiodactyles* and *Perissodactyles*, the *Sparassodonta* into *Creodontes* and *Carnivorae*, and the Patagonian apes into *Homunculidae*. From Asia, Ameghino concludes, the slow invasion of the "Argentine" mammals had conquered Europe in order to move on to North America, where their specialization would reach the most "bizarre and fantastic" forms.[21]

Although the naturalists and researchers working for the Museum of Natural Sciences of La Plata questioned Ameghino's geopolitical readings of the past, the museum's structural organization gave architectural and iconographic expression to the interpretation of the local fossil record as the natural foundation of the nation. In the museum, conceived as the site of national nature, the state's impure sacred body revealed itself in the spectral sculptures of biological wonders such as the *Megatherium*, *Mylodon*, and *Gliptodon*. In them the state idea is presented as a spectacle of animal resurrection in contrast to the rows of forced fossilization of Amerindian corpses displayed in the anthropological rooms, obliterated from coeval history by state violence and scientific necropolitics.

The virtual space of totemic nature needed to be large and dense. In a letter to Ameghino written in 1886, the director of the museum, Francisco P. Moreno, insists on the need to find large specimens for the museum's displays:

> Insisto en la conveniencia de las grandes fieras, y allí podría encontrarse algo. . . . Necesitamos apurarnos para tener con qué llenar esos cientos de metros. En caso que no lo hagamos, mucho me temo serias amonestaciones por el gasto, inútil por ahora, pero indispensable para nosotros. . . . Mucho deseo que le vaya bien por ahí. ¡Recuerde la necesidad de piezas grandes! ¡No se ría![22]

> (I insist on the convenience of having great predators, and something could be found there. . . . We need to obtain something to fill these hundreds of meters as soon as possible. If we fail, I fear serious questions about expenses, unnecessary for now but indispensable to us later. . . . I very much hope it goes well for you down there. Remember the necessity of large items! Don't laugh!)

The emphasis on scale and size points to the transformation of large fossils into a modern spectacle for mass consumption. In 1887, after visiting the museum, the American paleontologist Henry A. Ward compared the effects that such a spectacle had on him to the mesmerizing power of dreams and fantasies:

En ninguno de los museos públicos ó privados de los Estados Unidos hoy, ni en museo alguno de las capitales de Europa en la última ocasión cuando yo los visité, durante el año 1885, existen colecciones tan numerosas de grandes fósiles armados, de ningún orden de mamíferos, como las que hay aquí en el Museo de La Plata. Tan sorprendido estuve de cuanto vi en él, que mi primer visita me parecía un ensueño en el que me había entregado a saborear las delicias de fantásticas visiones. Solo después de repetidas visitas pude convencerme de que todo aquello era, en efecto, una realidad.[23]

(In none of the public or private museums of the United States today, nor in any museum in the capitals of Europe when I last visited them in 1885, exist as many collections of large fossil animal reconstructions, of the order of mammals, as those that one finds in the Museum of Natural Sciences of La Plata. I was so surprised during my first visit to the museum that it felt like a dream I had abandoned myself to, while savoring the pleasures of fantastic visions. Only after several visits was I able to convince myself that everything in the museum was indeed real.)

The vacillation between fiction and reality produced by the vision of fossil animals at museums points to the instability of scientific objects and the hybrid nature of the paleontological imagination.

The Mysterious Mammal

> Can you draw out Leviathan with a fish hook, or press down his tongue with a cord? Can you put a rope in his nose, or pierce his jaw with a hook? Will he make many supplications to you? Will he speak to you soft words? Will he make a covenant with you to take him for your servant forever? Will you play with him as with a bird, or will you put him on a leash for your maidens?
>
> *Book of Job, 41:1–5*

With their rituals of resurrection, paleontologists opened channels of reciprocal communication between worlds. Just as the present invaded the past in search of political symbols, the geological past intruded into the present with its wonders and monsters. Around 1898, rumors began to circulate in the daily press and in specialized scientific publications that there were living specimens of prehistoric fauna roaming Patagonia, giving rise to "the case of the mysterious mammal." These rumors had been encouraged by a discovery made by the rancher Hermann Eberhard in 1895. In the cavern of Última Esperanza, at the southern end of Patagonia (figure 1.1), he had found a strange piece of mammal hide with hair and osteoderms attached to it. After Eberhard's finding, new hide remains were collected by other explorers, including O. Nordenskjöld in 1896, Francisco P. Moreno in 1897, and Rodolfo Hauthal in 1899.[24] The skin, extraordinarily well preserved, did not belong to any known living animal. The remains were attributed to a Mylodontid ground sloth, named *Neomylodon listai* by Ameghino in 1899. The finding promoted several publications and a strong competition for sorting the baffling paleontological enigma, especially between Moreno and Ameghino.

The enigmatic fragments of skin and bones triggered a noteworthy subseries of resurrections and interpretations of fossil remains that captured the imaginations of local and international scientists as well as that of the general public. Within the web of meanings in which it was inserted, the skin functioned as a floating signifier without a fixed content—a fetish in circulation through which different political and cultural fictions of Patagonia were articulated. Paleontology, anthropology, the aesthetics of the rare, *criollista* primitivism, and Amerindian lore came together to form a dense discursive arrangement of scientific theories and tall tales.

From this network, different configurations of the mysterious mammal would emerge as hypothetical puzzles from a small number of loose pieces. Among this multitude of possible images, two stood out: the ferocious and invincible beast that, like the biblical Leviathan, had "bones of bronze" and "muscles of steel" and the tamed giant on a leash, obediently following his conqueror. Both animal configurations were related to hunting accounts and male prowess. The hunt was either understood as a symbolic capture through scientific classification and naming or in a more literal sense as the hunting expeditions arranged by the brave scientists and adventurers who rushed to Patagonia with the goal of catching the mysterious mammal—as cadaver or bone booty—and bringing it back as trophy.

The notoriety attained by the natural enigma was partly due to the prodigious modernization that the central regions of the country went through in the last two decades of the nineteenth century. The transformations produced by sweeping urbanization in Buenos Aires and the eastern coastal provinces also affected Patagonia, although on a lesser scale. Over the span of a few decades, towns and livestock settlements multiplied at an unprecedented rate.

Figure 1.1. Interior of the larger cave of Last Hope in the Patagonian Andes. "El mamífero misterioso de la Patagonia," *Revista del Museo de la Plata* 9 (1899): 474.

The inauguration of the railroad between Buenos Aires and the province of Neuquén in northern Patagonia in 1899 prefigured the end of the proverbial isolation of the southern region and its definitive incorporation into the territory and national economy. The swift transformation of nature in its pure state generated a paradoxical nostalgia that manifested as speculation about the existence of "lost worlds" in which the authenticity of the primordial was still intact.[25] Having escaped from one of these worlds in suspension, the mysterious mammal roamed the Patagonian desert as a ghostly remnant of a nature in retreat.

It was Ameghino who again left his professional peers bewildered by affirming that the shadowy beast that had been sighted was a survivor from the remote past that, favored by the geological irregularity of Patagonia, had remained imprisoned in a temporal island in one of the last unexplored regions of the globe. This new, scandalous hypothesis reached the public through two short articles. In the first, entitled "Primera noticia acerca del *Neomylodon listai*: Representante vivo de los antiguos edentados gravógrados fósiles de la Argentina" (First report on the *Neomylodon listae*: A living specimen of the ancient

edentati gravigradi fossils of Argentina, 1898), Ameghino gives scientific support to the indigenous legends and the personal accounts of a few explorers who claimed to have seen a mysterious quadruped of large size and ferocity lurking on the banks of Patagonian rivers. According to Ameghino, the mythical animal was not only alive but possibly the last representative of an animal species that was believed to be totally extinct, a toothless Gravigradus of the *Megatherium* family, related to the *Mylodon* but from a previous era. It was an unknown specimen of the renowned lost fauna native to southern Argentina from which, in Ameghino's opinion, all earth mammals had evolved. Ameghino named the new specimen *Neomylodon listai* after the explorer Ramón Lista (1856–1897), founder of the Argentine Geographical Society, infamous killer of Selknam people in Tierra del Fuego, and the only "civilized" man who had allegedly glimpsed the fleeting contour of the mysterious beast.[26]

In the second article, "Un sobreviviente actual de los megaterios de la pampa" (A living survivor of the pampean *Megatheria*), published the same year, Ameghino constructs his representation of the mysterious mammal by juxtaposing the scientific image of the *Megatheria*, whose skeletons were on display in museums, with the image of the "autochthonous" monster that the Tehuelche called "Iemisch." This unusual combination of science and so-called primitive knowledge was based on the belief that the indigenous peoples were not coeval; that is, they inhabited, like the mysterious mammal, a different temporality. Such lack of synchronicity, Ameghino argues, allowed the native tribes to have a more direct, unmediated relation with nature. By blending two epistemological traditions, Ameghino shifts the West's hierarchical distinction between modern rationality and *penseé sauvage* (primitive thinking) by presenting Amerindian lore as encrypted empirical evidence, as Cuvier had recommended. Once put into contact, Western and so-called traditional forms of knowledge contaminated each other. Thus, at the same time that the empirical logic of science sought to demystify indigenous "superstitions" by finding a rational explanation for them, the signifying power of the native Iemisch was projected back upon the scientific reconstructions of the prehistoric animals, bringing their fetish condition to the surface.

The totemic implications of Ameghino's reconstruction of the last living *Megatherium* are hard to overlook. In order to visualize its portentous body, Ameghino recalls the spectacular displays of the modern museums of natural history. By 1899, he hints, everybody had seen in one way or another the fossil remains of the giant *Megatherium*—the heaviest, strongest, and most massive mammal that had ever existed in the planet. The vision was unforgettable:

La primera vista del esqueleto produce la impresión de una andamiada de una casa en construcción; el propietario de esa armazón podía alcanzar en vida un largo de más de siete metros por dos y medio de alto y una corpulencia extraordinaria. . . . La cadera tiene un ancho y desarrollo extraordinario y la cola un grueso inusitado; las vértebras podrían servir de asientos y las costillas semejan grandes garrotes. Los miembros están sostenidos por huesos cortos y macizos y armados de garras gigantescas de más de un pie de largo; el fémur tiene un metro de circunferencia y es tres veces más grueso que el del elefante. En las quijadas, que son extraordinariamente altas, implántanse en profundos alvéolos grandes dientes parecidos a largos pilares cuadrangulares . . . que . . . constituían un aparato de masticación que debió permitirle triturar hasta las mismas piedras. Las formas toscas de todos sus huesos, cubiertos de apófisis y fuertes rugosidades destinadas a la inserción de tendones y músculos formidables, denotan un animal dotado de fuerza colosal.[27]

(The first time we see the skeleton it produces in us the impression of the scaffolding of an edifice under construction. When it was alive, the owner of such a monumental structure reached a length of more than seven meters, was two and half meters tall and of amazing corpulence. . . . The hips are very wide and extraordinarily developed; the tail is unusually long. Its vertebrae could be used as seats and its ribs look like enormous clubs. Its extremities are supported by short, substantial bones that end in enormous claws, a foot long each. The femur bone is one meter in diameter—three times as thick as that of an elephant. The jaws, extraordinarily high, are set into deep alveoli, with big teeth similar to long quadrangular pillars. . . . These teeth . . . once worked as a chewing machine that allowed the beast to crush even rocks. The rough shape of all its bones, covered by the apophysis and strong ridges where formidable tendons and muscles were once inserted, suggests an animal of colossal strength.)

The *Megatherium*'s skeleton is the hollow structure of a fiction of power that fills the lack of evidence with ideological material. From this imprecise journey through its bones, Ameghino was interested in emphasizing the *Megatherium*'s enormous size and the ferociousness and power that could be read in its jaws, which were able "to crush even rocks." In addition, the ridges on its bones, designed for the insertion of almighty muscles, were characteristic of the most massive animal in the world. This formidable animal, offered up for admiring contemplation in museums, was not only authentically "American" but, more importantly for Ameghino, "Argentine"; its bones and skin were the remains of a national ancestor in which the ineffable power of the natural state revealed itself.

The appearance of the indigenous Iemisch was no less terrifying. Ameghino had received news of the Tehuelche monster through his brother, Carlos, who on one of his exploratory trips through Patagonia had come across a Tehuelche man in possession of yet another piece of skin, the third in the series of fragments of the mysterious mammal. Along with the skin, Carlos had sent to Buenos Aires a summary of the indigenous tales that circulated about the animal. In these stories, the frightful Iemisch is represented as an amphibious animal that walked with the same ease as it swam in the water. Now confined to the most remote corners of Patagonia, Florentino Ameghino argued, it had once ranged as far as the southern border of the province of Buenos Aires. Traits of different modern animals coexisted in the Iemisch's hybrid body:

Es de cabeza corta, con grandes colmillos y orejas sin pabellón o con pabellón rudimentario: pies cortos y aplastados (plantígrados) con tres dedos en los anteriores y cuatro en los posteriores, unidos por membrana natatoria, a la vez que armados de formidables garras. La cola es larga, deprimida y prensil. Su cuerpo está cubierto con pelo corto, duro y rígido de color bayo uniforme. Dicen que su talla es mayor que la de un puma, pero que es de piernas más cortas y mucho más grueso de cuello.[28]

(Its head is short, with big fangs and rudimentary or no external ears. Short and flat feet [plantigrade feet] with three toes in the anterior legs and four in the posterior, united by a natatorial membrane, and armed with formidable claws. The tail is long, flat, and prehensile. Its body is covered by short, hard, coarse hair, yellowish white in color. They say that it is larger than an American lion or puma, but with shorter legs and a much thicker neck.)

The discrepancies between Ameghino's contradictory descriptions of the mysterious mammal make us wonder about his reasons for merging the two animals. It is surprising that one of the most prestigious paleontologists on the continent, known for his scientific knowledge and classificatory skills, found points of support for such an assimilation. One possible explanation resides in the aforementioned view that Patagonia constituted the last refuge of the primitive and therefore of nature at its purest. From this perspective, the Tehuelche Indians were figures equivalent to the mysterious mammal since they also, according to science and modern racism, were the survivors of a world outside history. If their direct contact with the monster corroborated the mysterious animal's existence, this evidence could only be acceptable once it was translated from the language of so-called superstition into the precise language of science. The disagreement between the two descriptions could be explained by the difference in the quality of the representations. However, by using the skin as material proof of the reality of the mythical beast and the mythical beast to prove the existence of a fossilized animal, Ameghino unsettled the established hierarchical relation between rational and primitive thought. The result of his epistemic transgression was the contamination of the representations of the two animals and of their politico-cultural meanings.[29]

Ameghino's essays caused a stir in scientific and journalistic circles. In 1898 the British journals *Nature* and *Natural Science* published a few pieces on the discovery of a living *Mylodon*. Francisco Moreno sent Nicolás Illín, the ex-librarian of the Museum of La Plata, to Patagonia in search of more remains of the mysterious animal; in 1899 Moreno took a piece of skin that Illín had found to London and presented his findings to the British Zoological Society. In an interview that appeared in the Chilean newspaper *El Magallanes* the same year, Ameghino stated that seven expeditions were already in the area looking for the animal.

The Patagonian marvel mesmerized the minds of the most sophisticated and the simplest of spirits. Wanting to utilize such interest for their benefit, representatives of the daily press financed expeditions to gather information on the elusive beast. Far from solving the mystery, accounts produced by the explorers involved only deepened it. In its edition of June 28, 1899, the newspaper *La Prensa* of Buenos Aires was obliged to inform its readers that even though its reporters had not managed to sight the mysterious mammal, they had nevertheless found fresh traces of it. Interest overseas, particularly in England, was considerable. The owner of the *Daily Press* of London financed the expedition to Patagonia led by H. Hesketh Pritchard with the purpose of investigating the puzzling phenomenon. At the end of his travel log, Pritchard's ambiguous conclusions reveal his resistance to abandoning the idea of the beast's existence. Although their research had not found any evidence of the animal's existence, Pritchard says, he did not dismiss the possibility that the rumors had a certain validity: "I would not offer my opinion as an ultimate answer to the problem. In addition to the regions visited by our Expedition, there are, as I have said, hundreds and hundreds of square miles about, and on both sides of the Andes, still unpenetrated by man . . . and it would be presumptuous to say that in some hidden valley far beyond the present ken of man some prehistoric animal may not still exist."[30]

In the modern circles of Buenos Aires, some viewed in the cultural hybridity of the mysterious mammal a sign of fin-de-siècle fascination with all that was rare and primitive. In a sarcastic piece in the magazine *Caras y Caretas* published on May 13, 1899, entitled "Argentine Monsters," the author, F. De Basaldúa, takes Ameghino's syncretism to its extreme and puts the *Neomylodon listai* in the company of all the native monsters that had appeared in the legends passed down by indigenous storytellers and gauchos alike since colonial times, including the Yaguaroy of the Guaraní, the Iemisch of the Tehuelche, and the Sukara of the Puelche. Far from contesting the fabrications of the imagination, the author continues, modern science allowed for the most incredible theories and all kinds of exotic beings. Because of its radical nature, at the very limits of civilization, Patagonia offered fertile ground for extraordinary things since truth and fiction were often confused there. The living Mylodon was "una verdadera maravilla *fin de siècle*" (a true fin-de-siècle wonder), an exquisite commodity that could surely delight any collector in search of rare objects.[31]

Not everyone was seduced by the enigmatic aura of the nomadic *Neomylodon*. A month later, *Caras y Caretas* returned to the topic but this time to talk about more real and dangerous monstruosities, such as the large debts that the national government had contracted with England to subsidize its program of modernization and that, in view of the author, threatened the country with a return to colonial times. Pointing with irony to the paradoxes of uneven modernity, the cover of the issue from June 3, 1899, portrays the mysterious mammal against a Patagonian Andean landscape, its back displaying the word "empréstito" (loan) as it takes hold of a British flag (figure 1.2). With a mocking attitude, the animal looks askance at two hunters in pursuit: one who wears the attire of British explorers in their imperial travels with a riffle engraved with the word "monopoly" and another on his knees, dressed as a gaucho and with an expression of fear, perhaps on the verge of fleeing; the latter figure's resemblance to Argentine President Julio A. Roca, the general who commanded the military expedition that occupied Patagonia on behalf of the state, would surely not have been missed by readers.

In local scientific circles, the general reaction to the rumors about the wonder of Patagonia was one of incredulity and skepticism.[32] In contrast to what the journalist Basaldúa suggested, scientists contended that it was not science that was responsible for the creation of fantastic animals, but Ameghino himself. Guilty of a fertile imagination and a tendency to devise heterodox theories, Ameghino was using his undeniable talent for reconstructing complete specimens from fossilized remains to fabricate fictions that were in and of themselves monstrous and quickly diffused throughout society by a press eager for shocking news. "Siempre sucede así" (It is always like this), comments Santiago Roth, the Museum of La Plata chief geologist, with irritated resignation. "Cuanto menos datos positivos se tienen tanto más se desarrolla la fantasía, y cuanto más misterioso se presenta el asunto, mayor interés general despierta" (The less positive data we have, the more fantasy develops, and the more mysterious the matter, the greater the interest it awakens).[33]

Figure 1.2. Cover of the popular Buenos Aires magazine *Caras y Caretas*, June 3, 1899.

Although most scientists rejected the possibility of a living specimen of an extinct species roaming freely in the Patagonian wilderness, there was little agreement on the identity of the specimen to which the skin belonged. Francisco Moreno thought that it corresponded to a toothless species similar to the Mylodon. At first, Roth was inclined to believe that it belonged to some unknown species of aquatic mammal from the southern seas. Robert Lehmann-Nitsche, the chief anthropologist of the Museum of La Plata, suggested that it was possible that the animal was a natural aberration. In order to solve the enigma, in 1899 the Museo de La Plata commissioned the German geologist and paleontologist Rodolfo Hauthal to visit and explore the cave of Last Hope one more time. There he found a fourth piece of skin a meter in length by ninety-three centimeters in width, in addition to a large number of animal bones embedded in a dense and millennial coating of excrement (figure 1.3), in which there were also signs that so-called troglodytes, primitive groups of humans, had shared the cave.[34] From the analysis of this paleontological booty, Santiago Roth concluded that the skin had belonged to a completely extinct animal, to which he gave the name of *Grypotherium domesticum*. The word "domestic" referred to Hauthal's theory that the beast had been kept in captivity in a sort of corral thousands of years ago.[35]

The *Grypotherium* rapidly became the central figure in yet another foundational scientific fiction. The cave of Last Hope had maintained in a suspended state the primordial scene of Argentine society. In Hauthal's view, his findings had finally provided the first scientific proof of the beginning of culture, when primitive "Argentine" men had begun to exert control over nature. In the shift from the logic of war to the logic of the *oikos*, the formidable Leviathan of Ameghino's totemic vision became a pet and yard animal that had been sacrificed for the benefit of the Argentine primitive community and its first proto-*asado*:

> No se puede saber con seguridad, por los restos, de qué manera fue muerto el animal. *Este ser indefenso y pesado, con sus molares inofensivos, probablemente fue matado a golpes de maza en la cabeza.* Una vez sacado el cuero, el cadáver ha sido desmembrado. Las partes mayores fueron cortadas en pedazos pequeños y comidas *con placer.* No dejaron nada más que las inserciones de los músculos y los tendones duros. . . . En el festín, la carne ha sido arrancada de los huesos con los dientes o quizás ayudándose de un cuchillito. No es seguro que haya sido asada; su sabor ha sido igual al de un herbívoro; las astillas de huesos de animales jóvenes nos demuestran que aquellos glotones supieron apreciar muy bien la carne tierna.[36]

> (We cannot know from the remains how the animal died. *This inoffensive and heavy being, with its harmless molars, was likely beaten to death on the head with a club.* Once its fur was skinned off, the animal was dismembered. The larger chunks were cut into smaller pieces and *eaten with pleasure.* Only the ridges of the muscles and the hard tendons were left

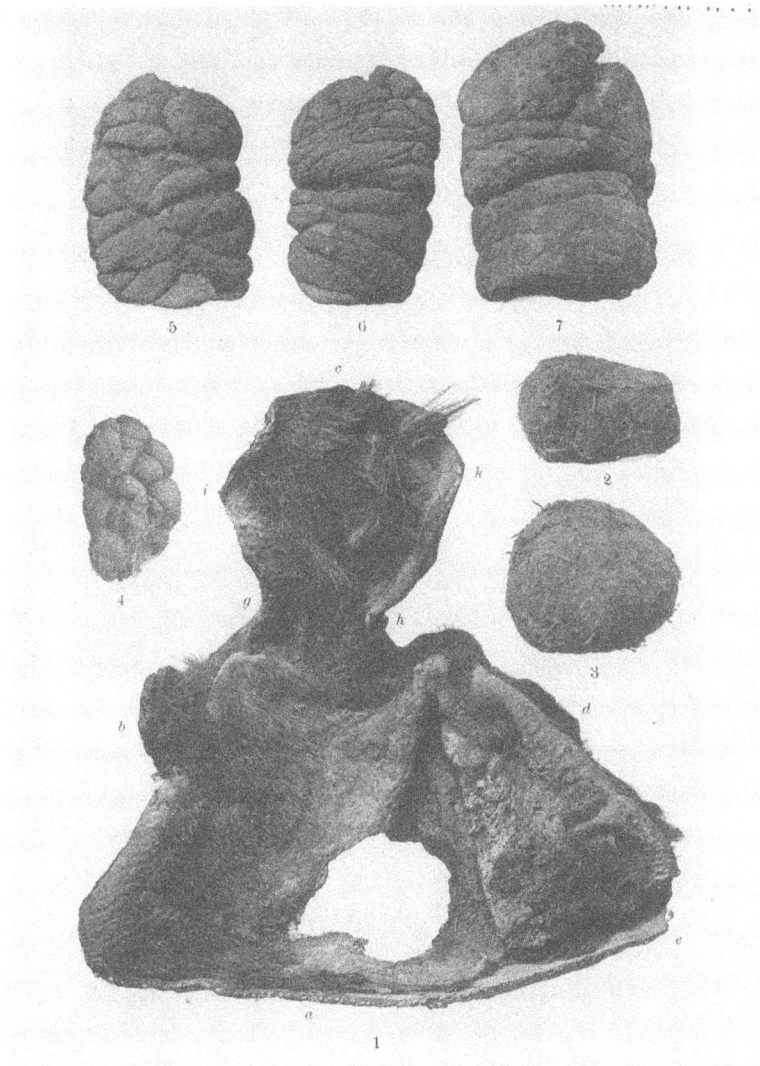

Figure 1.3. Piece of leather (1, *bottom*) and three fecal balls (5, 6, 7, *top*) found in 1899 in the cave of Last Hope and thought to be from a *Grypotherium domesticum*. The arrangement includes feces from an elephant (2, 3) and a horse (4) for comparison.

untouched. . . . *During the feast,* the meat was pulled off the bones with the teeth or perhaps with the help of a little knife. It's not clear if it was cooked; its flavor must have been like that of an herbivore. *The chips of bones from young animals demonstrate that those gluttons knew very well how to appreciate tender meat.*)

With this reconstruction of a communal banquet during which the *Gripotherium* specimen found in the cave had been consumed, the anthropologist Robert Lehmann-Nitsche introduced yet another way of reading the skin and bones. In his interpretation, what was important was not so much the identity of the animal in question but the clues provided by the remains (the small knife, the discerning bite that the men had inscribed upon it in the primordial stage of national evolution). What was at the origin was not animal violence but rather the subordination of nature by men and the shared experience of the pleasure of eating. In this foundational fiction, the moment of separation from nature preceded the distinction between the raw and the cooked, with the caveat that there were no clear indications that the meat had been grilled. In Hauthal's scientific fiction, the development of a rudimentary domestic economy and the ordered disposal of the corpse of a domestic animal marked the onset of a national imagined community.

In the series of representations I have traced, the ghostly mammal keeps transfiguring itself according to different scientific, cultural, and political codes. Not only do the fragments of skin and bones multiply, but the scientists also keep assembling them in different ways, generating divergent political fictions. What returns repeatedly in the cultural genealogy is the shadow of the extractivist capitalist state and its ideological masks as they manifest in a heterotopical geography, simultaneously inside and outside politics. But this is not the only narrative thread in the thick web of scientific, cultural, and mythical knowledge that fed the paleontological imagination of the time. Intersecting with the political lineage are the indigenous stories about bones and monsters and the extraordinary beasts of modern nostalgia. Ameghino saw in indigenous legends empirical proof of the *Neomylodon*'s existence. Conversely, Roth and Lehmann-Nitsche insisted that indigenous testimonies lacked credibility, considering them to be metaphorical elaborations of fear with no scientific value.[37] The terrifying monsters of indigenous myths, they argued, were just inadequate explanations of natural phenomena that the savage mind was unable to explain rationally. Despite their differences, all these scientists considered indigenous communities to be living remnants of a buried past, asynchronic growths of an anachronic world.

We could speculate, however, that if the Patagonian indigenous myths about extraordinary animals were indeed elaborations of fear, they pointed to the Amerindians' coevalness and involvement with history rather than to their alleged prehistoric condition. Indigenous monsters could be read as metaphorical political fictionalizations of real encounters with state violence, as seen from the other side of hegemonic articulation. Read from the side of the subaltern, the

Tehuelche Iemisch and its variants are signs in the language of war. While one face of the monster makes visible the overwhelming violence of the state, the other reflects the angry gesticulations of a beast in retreat that, as the nomadic tribes decimated by the military campaigns of the late nineteenth century did, sought refuge in the lost worlds of Patagonia.

In a kind of contrapuntal dialogue, indigenous myths were the unexpected supplement of paleontological knowledge and state fetishism, interrupting their epistemological legitimacy by introducing possible subaltern elaborations of modern prehistory. Either as monster or tamed beast, the mysterious mammal and its ghostly shadows keep pointing to the unresolved contradictions of scientific authority, Western racism, and postcolonial modernity.

The Tail of the Series

The episode of the mysterious mammal was perhaps the strongest articulation of the series of fossils as the primordial writing of Patagonia. Although there would be a succession of archeological discoveries in various parts of the region throughout the twentieth century, accompanying the steady advance of modernization and the construction of dams and oil plants, many decades would pass before the lost fauna of the south regained the central role in the national geographic imagination that it had occupied earlier in Ameghino's paleontological fictions. Its symbolic appeal continued to diminish until it was limited to professional circles. In 1964 a minor theoretical resurrection took place. In that year the book *Estudios icnológicos: Problemas y métodos de la icnología con aplicación al estudio de pisadas mesozoicas (reptilia, mammalia) de la Patagonia* (*Ichnological Studies: Problems and Methods in Ichnology Applied to the Study of Mesozoic [Reptilia, Mammalia] Footprints in Patagonia*) appeared. There the Patagonian paleontologist Rodolfo Casamiquela proposed a classification of the prehistoric Patagonian fauna that would be based not on the study of bones but of fossilized footprints. In Casamiquela's opinion the phylogenetic analysis of the tracks left by ancient animals permitted not only the reconstruction of the mechanical characteristics of their gait and ultimately of the complete structure of their bodies but also the identification of previously overlooked specimens whose study until then had been impeded by the fragmentary condition of fossil evidence. Like a hunter in search of virtual prey, Casamiquela sought among the numerous fossilized tracks scattered throughout Patagonia the traces of a new species of mammal, which he named *Ameghinichnus* in memory of the father of Argentine paleontology, Florentino Ameghino.[38] Because of their Mesozoic origins, the faded tracks left by the gait of this small animal's five-toe feet on the Patagonian surface, in Casamiquela's opinion, vindicated Ameghino and refuted the metropolitan theory that the mammals of the Southern Hemisphere were mere derivatives of those of the Northern Hemisphere. Voicing Ameghino's nationalist arguments and the political accusations against imperialism made by the new Latin American Left in the 1970s, Casamiquela argues that the new paleontological findings in Patagonia provided a less hierarchical and more inclusive image of life on earth, pointing to "un desarrollo cosmopolita, panecuménico . . . de la vida" (a cosmopolitan and pan-ecumenical development of life.)[39]

Although quite eccentric, Casamiquela's political reading of fossil tracks remained an anecdote in the history of national theories on Patagonian fossils. Not until the discovery of "Argentine" dinosaurs in the underground deposits of the Neuquén province in the late twentieth century could Patagonian fossils begin to regain the spectacular presence that the *Megatherium* had nearly one hundred years earlier. Once again the primordial masculinity of the gigantic fossils returned to lend its power to the idea of the state. An article published in *Science News* on September 23, 1995, clearly records the nationalistic associations incited by the new discoveries as well as their repercussions in the political imaginary of paleontology:

> In the wilds of northwestern Patagonia, they grow their dinosaurs big. Six years ago, Argentine paleontologists discovered remains of what may be the largest dinosaur ever known, a plant-eating behemoth named, appropriately, *Argentinosaurus*. Now, they have found the perfect counterpart: the world largest meat-eating dinosaur.[40]

Highlighting the megalomaniac enthusiasm that the appearance of this epitome of the national Argentine diet produced, the new monster was baptized with the name *Giganotosaurus carolinii* (figure 1.4). In an interview, Rodolfo Coria, an Argentine paleontologist involved in the recovery, dating, and classification of the fossil remains, remarked with competitive eagerness that the specimen, at a length of thirteen meters and a weight of approximately nine tons, surpassed the famous American *Tyrannosaurus rex* specimen in size and volume by one meter and three tons.[41] In February 2001, the exhuming of the remains of yet another herbivorous dinosaur of even more colossal dimensions in Loma de la Lata intensified even further the phallic triumphalism already caused by the *Giganotosaurus*.[42]

Figure 1.4. The paleontologist Jorge Calvo with a life-size model of the head of the *Giganotosaurus carolinii*, 1995. Image (E446/0206), reprinted with permission from SciencePhoto.com.

Paradoxically, the raucous entrance of these giants of imperial lineage into the national imagination occurred when, debilitated by the economic policies of neoliberalism, the Argentine state was falling apart and the dreams of domination and economic and political competition with the United States were definitely out of reach. That the new archeological excavations in search of primitive worlds in the Neuquén province replaced the exploratory drillings of Yacimientos Petrolíferos Fiscales, the state oil company that for almost a century fed the regional economy, is a telling indicator of the transformations in the political economy of fossilized bones. In search of new sources of work and wealth, the citizens of Plaza Huincul, one of the communities most devastated by the dismantling of the oil company, decided to take advantage of the interest awakened by the series of archeological findings in the area and sell the soil that they could no longer exploit as a commercial spectacle. Surrounded by the ruins of peripheral development and using metal scavenged from old oil rigs,

Figure 1.5. Remaining metal structure of the replica of *Argentinosaurus huinculensis*, Plaza Huincul, Neuquén, Argentina. Photograph courtesy of Fernando Leonel Ranni.

the town built a life-size replica of the *Argentinosaurus huinculensis* (figure 1.5) as a totem to be offered as a virtual sacrifice to the thousands of tourists who continue to go to Patagonia in search of lost worlds, outside modernity. Soon after its inauguration in 2001, the replica caught fire, and only its metal structure remains today. In this late modern catastrophic metamorphosis,[43] the colossal body of the dinosaur became the hollow mask of a peripheral state on the edge of extinction.

Resurrected over and over from its dusty ashes, the Patagonian fetish keeps coming back, ever mightier and more colossal. In 2017 Diego Pol of the Egidio Feruglio paleontology museum in Argentina unearthed yet another Patagonian giant, the *Patagotitan mayorum*, the new largest dinosaur on earth. As was the case in Ameghino's time, the paleontological imagination continues to produce bone tales about monsters whose overbearing masculinity alternately embodies national pride, postcolonial resentment, and uncontrollable nature. Within this conceptual frame, scientists continue to be the heroic explorers and bone hunters of a geological past ever receding into the future.

Notes

1. "Mais peut-être quelqu'un fera-t-il un argument inverse, et dira que non-seulement les anciens, comme nous venons de le prouver, ont connu autant de grands animaux que nous, mais qu'ils en ont décrit plusiers que nous n'avons pas; que nous nous hâtons trop de regarder ces animaux comme fabuleux; que nous devons les chercher encore avant de croire avoir épuisé l'histoire de la creation existante; enfin que parmi ces animaux prétendus fabuleux se trouverons peut-être, lorsqu'on les connaîtra mieux, les originaux de nos ossemens d'espèces inconnues" (Cuvier, *Recherches sous les ossemens fossils,* xxxv).

2. Talairach-Vielmas, "Shaping the Beast," 270.

3. Visual representations of prehistoric life were ways of making the unthinkable span of geological time more accessible. It was also a way for naturalists to gain access to a wider public. See Rudwick, *Scenes of Deep Time.*

4. The concept of "deep time" was first described in 1788 by the Scottish geologist James Hutton (1726–1797) but only coined as a term two hundred years later by the American writer John McPhee in his book *Basin and Range* (1981). Hutton posited that geological features were shaped by cycles of sedimentation and erosion, a process of lifting up and then grinding down of rocks that required timescales much grander than those of prevailing biblical narratives, according to which the earth was just six thousand years old. This epistemological shift threw both God and man into question.

5. Darwin, *On the Origin of Species,* 251.

6. Ginzburg, *Clues, Myths,* 106. According to Ginzburg, only heuristic disciplines such as medicine, law, art criticism, philology, and psychoanalysis use the evidential paradigm to produce knowledge about the past and sometimes to predict the future. Ginzburg argues that sciences are excluded from this trend because the empirical method implies the quantification and repetition of phenomena, but I contend that paleontology, like the aforementioned forms of interpretation, is an evidential and conjectural discipline.

7. See Podgorny's introduction to *El argentino despertar de las faunas y de las gentes prehistóricas* and Andermann's chapter 1 in *Optic of the State.*

8. See in particular the section entitled "Irréductions" in Latour's *Les Microbes.*

9. Gould, *Time's Arrow,* 6–7.

10. Apter, Introduction, 3.

11. See Podgorny, "Human Origins," 68–80.

12. Podgorny, "Human Origins," 74–75.

13. Sommer, *Foundational Fictions.*

14. Ameghino, "Un recuerdo," 43. On the variants of evolutionism and Darwinism in the scientific and philosophical work of Ameghino, see Oriola Rojas, "Florentino Ameghino."

15. Ameghino, "Paleontología argentina," 111. Ameghino himself eventually broke all the records of classification. On his death, at the age of fifty-seven, he had classified more than a thousand species of extinct mammals.

16. Ameghino, "Paleontología argentina," 115–116.

17. For an analysis of the use of these strategies by subaltern subjects, see Harding, "Appropriating the Idioms of Science."

18. Ameghino, "Paleontología argentina," 106.

19. Although his was a minoritarian position, Florentino Ameghino was not alone in his views; many scientists at the time postulated that the territories of the Southern Hemisphere had been a center of animal distribution. See Simpson, *Discoverers of the Lost World*, and Podgorny, "Bones and Devices," 251.

20. Ameghino, "Mamíferos cretáceos de la Argentina," in *Obras completas* 12:321–322.

21. Ameghino, "South America as the Source of the Tertiary Mammals," in *Obras completas*, 12:295.

22. Moreno, "Carta a Florentino Ameghino."

23. Ward, *Los museos argentinos*, 149. The letter was published in Spanish in the newspaper *El Censor*, Buenos Aires, October 7, 1897, and then reproduced in the *Revista del Museo de La Plata*.

24. The hide found by Eberhard was taken to the Museo de La Plata in 1898 and then to London in the beginning of 1899. Currently it is housed at the Natural History Museum of London. The remains found by Hauthal's expedition in 1899 are still in the Museo de La Plata. In 1914 Moreno gave three fragments of the hide to the American president Theodor Roosevelt during his visit to Argentina. Roosevelt later donated one of the pieces to the Museum of Natural History in New York (Pérez et al., "Los restos tegumentarios").

25. The notion of "lost world" as refuge of primitivism is another manifestation of what Rosaldo calls "imperial nostalgia" and constitutes the core of a long cultural genealogy in which the local palaeontological imagination participated. In general, the notion refers to certain areas of the Third World and at least since *The Lost World* (1912) by Arthur Conan Doyle to South America.

26. Ameghino, "Primera noticia del *Neomylodon listai*," in *Obras completas* 12:477–483. This article appeared as an independent booklet in La Plata in 1898. In the November issue of that year, the British magazine *Natural Science* included a summary of the text.

27. Ameghino, "Un sobreviviente actual de los megaterios de la pampa," in *Obras completas*, 12:755. This article appeared for the first time in two parts in the journal *La Pirámide* 1 (June 15, 1899): 51–54 and (July 1, 1899): 82–84.

28. Ameghino, "Un sobreviviente actual," 12:757.

29. A Tehuelche man had the skin and had possibly hoarded it as a fetish, suggesting that the figure of the paleontologist is just a modern version of the primitive fetishist.

30. Pritchard, *Through the Heart of Patagonia*, xiv.

31. Basaldúa, "Monstruos argentinos."

32. Among the exceptions, besides Florentino Ameghino and his brother, Carlos, is his friend the paleontologist André Tournouer, who in a report to the Academy of Paris said that he himself had seen the Neomylodon alive in the water (Tournouer, "Sur le Néomylodon").

33. Roth, "Descripción de los restos," 422.

34. Hauthal, "Reseña de los hallazgos," 411–420.

35. Roth, "Descripción de los restos," 433.

36. Lehmann-Nitsche, "Coexistencia del hombre," 464, my emphasis.

37. In "La pretendida existencia actual del Grypotherium" Lehmann-Nitsche offers the two Araucanian myths as proof that the Tehuelche Iemish was a completely imagi-

nary animal, based on the combination of features from two real animals, the lutra and the American lion, the puma.

38. Casamiquela, *Estudios icnológicos,* 89–90.

39. Casamiquela, *Estudios icnológicos,* 164, emphasis in original.

40. Monastersky, "New Beast Usurps T Rex."

41. Monastersky, "New Beast Usurps T Rex."

42. Argentine journals devoted many articles to the new discovery. See Ortiz, "Hallaron en Neuquén restos"; Iglesias, "Creen que el dinosaurio."

43. The Plaza Huincul sculpture is only one among many scientific commercial resurrections of Argentine dinosaurs in recent years. Since the end of the twentieth century, there have been many exhibits displaying fossils of Patagonian monsters in Argentina and abroad. In August 1998 the Buenos Aires zoo exhibited the skeleton of the *Argentinosaurus huinculensis.* In June 2001 there was a very large exhibition at the Museum of Natural Sciences Bernardino Rivadavia in Buenos Aires that displayed more than ten reconstructions of "great Argentine dinosaurs." Among them were *Giganotosaurus carolinii* and *Argentinosaurus huinculensis.* During the administration of Cristina Fernández de Kirchner, the government sponsored an itinerant paleontological exhibition entitled *Dinosaurs: Argentine Giants*, which traveled to Germany in 2010 and Hungary in 2013.

References

Ameghino, Florentino. "Les formations sedimentaires du Crétacé Supérieur et du Tertiaire de Patagonie." *Anales del Museo Nacional de Buenos Aires* 3, no. 8. Buenos Aires: Imprenta de Juan A. Alsina, 1906.

———. "Mamíferos cretáceos de la Argentina: Segunda contribución al conocimiento de la fauna mastológica de las capas con restos de *Pyrotherium.*" In *Obras completas y correspondencia científica de Florentino Ameghino*, vol. 12. La Plata, Argentina: Talleres de Impresos Oficiales, 1921, 301–461.

———. "Paleontología argentina: Relaciones filogenéticas y geográficas." In *Doctrinas y descubrimientos*, edited and revised by Alfredo J. Torcelli. Buenos Aires: La Cultura Argentina, 1915, 99–166.

———. "Primera noticia del *Neomylodon listai*: Representante vivo de los antiguos edentados gravógrados fósiles de la Argentina." In *Obras completas y correspondencia científica de Florentino Ameghino*, vol. 12. La Plata, Argentina: Talleres de Impresos Oficiales, 1921, 477–483.

———. "South America as the Source of the Tertiary Mammals." In *Obras completas y correspondencia científica de Florentino Ameghino*, vol. 12. La Plata, Argentina: Talleres de Impresos Oficiales, 1921, 290–297.

———. "Un recuerdo a la memoria de Darwin: El transformismo considerado como ciencia exacta." In *Obras completas y correspondencia científica de Florentino Ameghino*, vol. 4. La Plata, Argentina: Taller de Impresiones Oficiales, 1915, 45–55.

———. "Un sobreviviente actual de los megaterios de la pampa." In *Obras completas y correspondencia científica de Florentino Ameghino*, vol. 12. La Plata, Argentina: Talleres de Impresos Oficiales, 1921, 753–760.

Andermann, Jens. *The Optic of the State: Visuality and Power in Argentina and Brazil.* Pittsburgh, PA: University of Pittsburgh Press, 2007.

Apter, Emily. Introduction to *Fetishism as Cultural Discourse,* edited by Emily Apter and William Pietz, 1–9. Ithaca, NY: Cornell University Press, 1996.

Basaldúa, F. de, "Monstruos argentinos." *Caras y Caretas* 2, no. 32 (June 3, 1899): n.p.

Casamiquela, Rodolfo. *Estudios icnológicos: Problemas y métodos de la icnología con aplicación al estudio de pisadas mesozoicas (reptilia, mammalia) de la Patagonia.* Buenos Aires: Librart, 1964.

Cuvier, Georges. *Recherches sous les ossemens fossils.* Discours preliminaire. Paris: Chez G. Dufour et E. D'Ocagne, Libraries, 1821.

Darwin, Charles. *On the Origin of Species by Means of Natural Selection, or the Preservation of Favoured Races in the Struggle for Life.* Oxford: Oxford University Press, 1998 [1859].

Doyle, Arthur Conan. *The Lost World.* New York: A. L. Burt, 1912.

Ginzburg, Carlo. *Clues, Myths, and the Historical Method.* Baltimore, MD: Johns Hopkins University Press, 1986.

Gould, Stephen. *Time's Arrow, Time's Cycle: Myth and Metaphor in the Discovery of Geological Time.* Cambridge, MA: Harvard University Press, 1987.

Harding, Sandra. "Appropriating the Idioms of Science." In *The Racial Economy of Science,* edited by Harding, 170–193. Bloomington: Indiana University Press, 1993.

Hauthal, Rodolfo. "Reseña de los hallazgos en las cavernas de Última Esperanza (Patagonia Austral)." *Revista del Museo de La Plata* 9 (1899): 411–420. La Plata, Argentina: Talleres de Publicaciones del Museo.

Iglesias, Mariana. "Creen que el dinosaurio descubierto en Loma de la Lata es el más grande del mundo." *Clarín,* February 25, 2001.

Latour, Bruno. *Les Microbes: Guerre and paix, suivi par irréductions.* Paris: A. M. Métailier, 1984.

Lehmann-Nitsche, Robert. "Coexistencia del hombre con un gran desdentado y equino en las cavernas patagónicas." *Revista del Museo de la Plata* 9 (1899): 460–478. La Plata, Argentina: Talleres de Publicaciones del Museo.

———. "La pretendida existencia actual del Grypotherium. Supersticiones araucanas referentes a la lutra y al tigre." In *Revista del Museo de la Plata* 10 (1902): 271–281. La Plata, Argentina: Talleres de Publicaciones del Museo.

McPhee, John. *Basin and Range.* New York: Farrar, Straus, and Giroux, 1981.

Monastersky, R. "New Beast Usurps T Rex as King Carnivore." *Science News* 148, no. 13 (September 23, 1995): 199.

Moreno, Francisco P. "Carta a Florentino Ameghino." In Ameghino, *Obras completas y correspondencia científica de Florentino Ameghino,* vol. 20. La Plata, Argentina: Talleres de Impresos Oficiales, 1934, 407–410.

Oriola Rojas, Margarita. "Florentino Ameghino." In *El movimiento positivista argentino,* edited by Hugo E. Biagini, 399–409. Buenos Aires: Belgrano, 1985.

Ortiz, Osvaldo. "Hallaron en Neuquén restos de un gigantesco dinosaurio." *Clarín,* February 22, 2001.

Pérez, Leandro M., Néstor Toledo, Sergio F. Vizcaíno, and M. Susana Bargo. "Los restos tegumentarios de perezosos terrestres (Xenarthra, Folivora) de Última Esperanza (Chile). Cronología de los reportes, origen y ubicación actual." *Publicación Electrónica de la Asociación Paleontológica Argentina* 18, no. 1 (2018): 1–21.

Podgorny, Irina. *El argentino despertar de las faunas y de las gentes prehistóricas. Colec-*

cionistas, museos y estudiosos en la Argentina entre 1880 y 1910. Buenos Aires: Eudeba, Libros del Rojas, 2000.

———. "Bones and Devices in the Constitution of Paleontology in Argentina at the End of the Nineteenth Century." *Science in Context* 18, no. 2 (2005): 249–283.

———. "Human Origins in the New World? Florentino Ameghino and the Emergence of Prehistoric Archaeology in the Americas (1875–1912)." *PaleoAmerica* 1, no. 1 (2015): 68–80.

Pritchard, H. Hesketh. *Through the Heart of Patagonia.* New York: Appleton, 1902.

Roth, Santiago. "Descripción de los restos encontrados en la caverna de Última Esperanza." *Revista del Museo de La Plata* 9 (1899): 421–453. La Plata: Talleres de Publicaciones del Museo.

Rudwick, Martin J. S. *Scenes of Deep Time. Early Pictorial Representations of the Prehistoric World.* Chicago: University of Chicago Press, 1992.

Simpson, George Gaylord. *Attending Marvels: A Patagonian Journal.* New York: Time Reading Program Special Edition, 1965.

———. *Discoverers of the Lost World: An Account of Some of Those Who Brought Back to Life South American Mammals Long Buried in the Abyss of Time.* New Haven, CT: Yale University Press, 1984.

Sommer, Doris. *Foundational Fictions: The National Romances of Latin America.* Berkeley: University of California Press, 1993.

Talairach-Vielmas, Laurence. "Shaping the Beast." *European Journal of English Studies* 17, no. 3 (2013): 269–282.

Tournouer, André. "Sur le Néomylodon et l'animal misterieux de la Patagonie." *Comptes rendus des séances de l' Academie des Sciences de Paris* 32, no. 2 (January 14, 1901): 96–97. Paris: Gauthier-Villars, Impremeur-Libraire.

Ward, Henry A. "Los museos argentinos." Carta del Señor Henry A. Ward [1887]. *Revista del Museo de la Plata* 1 (1890–1891): 145–151. La Plata: Talleres de Publicaciones del Museo.

2

Nation as Laboratory

Rethinking Science Writing in Mexico's
República Restaurada, 1868–1876

MARÍA DEL PILAR BLANCO

The 1870s represent an important decade in Mexico's relation with science. Known as the time of the República Restaurada, a political shift spurred by the defeat and execution of Maximilian I in 1867, these years led to the beginning of Porfirio Díaz's rule in the country in late 1876. As a period of "restoration," it was marked by relative stability after decades of deep political and military unrest; it was, as Carlos Monsiváis has called it, "ese breve período de creencia en la unidad profunda y venidera de México" (that brief period of belief in the deep and forthcoming unity of Mexico).[1] The years of the República Restaurada are notable for their pervasive utopianism. In this period the country's *letrados* actively pursued ways of building the nation's cultural and educational infrastructure. Alfonso Reyes, many years later, describes the projects along these lines:

> Bajo la marejada imperial, la república queda reducida a las proporciones de la carroza en que emigraba Benito Juárez. Pero, revertida la onda, triunfa para siempre la república. El país había quedado en ruinas, era menester rehacerlo todo. Las medidas políticas ofrecían alivios inmediatos. Sólo la cultura, sólo la Escuela, pueden vincular alivios a larga duración.[2]

(After the imperial swell, the republic was reduced to the dimensions of the carriage in which Benito Juárez fled. But, the waves having turned, the republic triumphed permanently. The country had been left in ruins; everything had to be redone. Political measures offered immediate relief. Only culture, only the School [Escuela Nacional Preparatoria], could extend long-term relief.)

A few months after Maximilian's death, the philosopher Gabino Barreda, who had studied under Auguste Comte in France, read out his "Oración cívica" in Guanajuato. In this famous speech, he expresses Mexico's post-independence need for a triple "emancipación científica, emancipación religiosa, emancipación política" (scientific, religious, and political emancipation).[3] He went on to found the Escuela Nacional Preparatoria, mentioned by Reyes, which the latter calls the "'alma mater' de tantas generaciones" (the alma mater of so many generations).[4]

Science, as can be intimated from his proclamation, was seen as a way out of the shadows of clerical and political backwardness and a way to partake in an established sense of modern progress. Science was not defined along specific "hard" or "soft" lines in those decades; instead, it was a broad concept that encapsulated different areas of organized knowledge now more commonly called social and natural sciences. This explains, at least partly, why "science" (*ciencia*) is so prevalent a term in political discourse across Latin America in the late nineteenth century; many of the writers expounding on science at the time were really delineating a *social* science when they aligned matters of nation-building with this ample field of knowledge. At the same time, however, the very malleability of the term "science" led to its misinterpretation and misuse, particularly if one takes into account the many dangerous justifications of social and ethnic coercion in the name of scientific advancement and progress during that time.[5] As an open-ended concept, science writing from the period, particularly in the popular press, moved fluidly from specific appreciations of natural phenomena to broader discussions of its applicability to different levels of national organization.

Taking these definitional challenges into account, in this chapter I explore publishing projects from that short but significant period and their engagement with the kind of "emancipación científica" hailed by Barreda in 1867. Focusing principally on a set of science primers written by José Joaquín Arriaga, entitled *La ciencia recreativa* (1871), and on Santiago Sierra's popular science magazine *El Mundo Científico* (1877–1878), I explore how these publications with distinct purposes imagine science through a curiously diversified contemplation of Mexico's autochthonous landscape. In the spirit of total renewal that the República Restaurada represented, for the authors and editors of these publications the Mexican landscape appeared like a slate that invited varied forms of writing, all of which were produced as autochthonous contributions to science. The figures studied in this essay perceived the country as a large laboratory ready to yield discoveries and advancements; they also projected desires for those discoveries to bear importance on both national and international levels. The publications dramatize the need felt by the Mexican literate class to stake a claim in the field of Western science, which was by all means a sign of distinction and modernization.

I employ the concept and space of the laboratory as a way to think about the state of science during the República Restaurada. In *Laboratory Life* (1979)—a work that begins to spell out the imbrication of science within the social and political fields, thus dispelling the myth of its autonomy—Bruno Latour and Steve Woolgar contemplate the relation between experimentation and text that occurs within this particular space. Instead of positing discovery as the main activity that happens in laboratories, the predominant actions in this site are, for Latour and Woolgar, a set of processes they term "literary inscriptions," the constant exegesis of literature through which the persuasion of readers ensures that a given idea is accepted as fact.[6] In a further development of this argument, Latour returns to the relation between scientific experiment and writing when he remarks on the "little difference" that exists between "observation" and "experiment": "An observation is an experiment where the body of the scientist is used as instrument, complete with its writing device."[7] That writing is, in turn, caught up in the process of exegesis of scientific texts. In the Mexican context in the late 1860s and 1870s, I argue, science primarily manifested as a system of inscription, with a difference: if advanced laboratories were in short supply, the observable landscape itself was perceived as a laboratory, a space that was read in the light of and in relation to other external inscriptions, which in turn invites continuous writing, cataloging, and measurement of local phenomena and conditions that can eventually be construed as what Latour and Woolgar call "fact."

The last three decades of the nineteenth century were crucial in opening Mexicans' eyes to the abundance of pre-Columbian ruins in the country and how that abundance translated into national scientific development. Indeed, as Justo Sierra observes, "es lo único que caracteriza la personalidad de México ante el mundo científico: todo lo demás es lo mismo que existe en otras partes y está realizado aquí por extranjeros" (it is the only thing that describes Mexico's personality within the world of science; everything else is the same thing that exists elsewhere and is done here by foreigners).[8] Sierra's quite prophetic statement explains why so much contemporary scholarship on nineteenth-century Mexican science turns to the development of archaeology as a discipline that gained political importance in this period and particularly during Porfirio Díaz's regime. Much less attention has been placed, however, on the activity on and about science in the years that precede the Porfiriato; looking more closely at that brief period is one of my main aims in this chapter. If interest in unearthing pre-Columbian artifacts is barely evident in the press in the decades that preceded the República Restaurada, newspapers in the late 1870s began to print an increasing number of articles on (proto)archaeological matters. Those who reported on science in newspapers during this time puzzled over the ruins, auguring a future in which they would finally be legible. Reporting on the move of a Chac-Mool statue to Mérida in Yucatán in early 1877, one journalist for *El Monitor Republicano* newspaper writes,

> Este precioso objeto de la antigüedad, cuyo origen se pierde en la noche de los tiempos, es digno del estudio de los hombres pensadores. La historia y la arqueología, en sus graves y profundas investigaciones, descubrirán tal vez algún día el arcano que encierra, como todos esos preciosos monumentos que riegan la extensión de nuestro rico suelo, prueba evidente de la antigua civilización de los mayas.[9]

> (This beautiful object from ancient times, whose origin is lost in the darkness of the ages, is worthy of study by learned men. History and archaeology, in their serious and deep investigations, could one day discover the important secret that it keeps, as do all those other beautiful monuments scattered across the extent of our rich terrain, vivid proof of the Mayans' ancient civilization.)

This excerpt reveals much about the state of scientific exploration in the 1870s. For one, the unnamed author sets the moment of scientific discovery in a future in which research will be both possible and effective. The author's incomprehension and wonder at the Chac-Mool statue is a common feature in articles devoted to archaeological topics. It is also interesting to note how, in this particular piece, the concepts of ruin and natural terrain deeply intertwine. Ancient Mayan artifacts, it would seem, are seen to be as plentiful and regular within the Mexican landscape as the elements that make up the country's "rich terrain" (*rico suelo*). They are, then, other fruits that can be unearthed and count as evidence of Mexico's archaeological and scientific wealth. Writing about the development of archaeology during Porfirio Díaz's regime, Christina Bueno explains the necessity—shared among the country's elites—to develop a discourse of the nation as heaving with archaeological wealth. Mexico, she writes,

> was a nation deemed inferior by the dominant Eurocentric racist thinking of the day, a nation that Europeans and Americans saw as backward and uncivilized. Recasting this image became a Porfirian concern, one that scholars attribute to the ruling class's desire to attract foreign immigration and investment, but which was also a matter of sheer pride. Through antiquity, elites aimed to present Mexico as a unified nation with ancient and prestigious roots.[10]

Considering Bueno's argument alongside the article from *El Monitor Republicano*, one notices how these very elites were keen to promote an idea of a nation whose soil bore organic and historical fruit, both of which equipped the nation for competition and equal exchange with other North Atlantic countries like France, England, and the United States. Still in formation in the late nineteenth century, archaeology, like metallurgy and botany, was one of the many wings of science that could unearth a national wealth visible to and exchangeable with key players in the Western world.[11]

If science writers perceived Mexico's landscape as mysterious and enticing in the science writing of the period, they also imagined it as an ideal platform for experimentation. This perspective moved beyond the archaeological and into other areas of science such as geography and meteorology. Arriaga's *La ciencia recreativa*, written for children and women (several "novelitas" open with an address to "lectoras"), invites readers to learn how different areas of science relate to an observation of the local environment. Arriaga, an engineer by training, was far from being a supporter of Benito Juárez's republicanism. His disagreements with the liberal establishment of the time led him to found a number of Catholic periodicals, including *La Revista Universal* in 1867, which he directed until 1869. Regardless of his staunch conservatism, Arriaga was a vocal figure in the drive to nationalize science and to promote science as one of the goals of a modernized Mexico. According to a 1900 biographical essay on Arriaga by Santiago Ramírez, in August 1868 the municipality of Tulyehualco wrote to D. Ramón J. Alcáraz, then director of the Museo Nacional in Mexico City, to say that á inmediaciones de aquel pueblo se habían descubierto los restos de una población antigua, sepultada bajo las lavas que había arrojado un volcán inmediato, cuya población fué designada con el pomposo nombre de el Pompeya Mexicano.[12]

(the remains of an ancient population had been discovered near that town, buried under the lava that a neighboring volcano had expelled, and whose population had been anointed with the pompous name of the Mexican Pompeii.)

Arriaga was one of four scientists promptly sent to explore the site. The remains of the ancient population never materialized, but Arriaga did take the opportunity to produce a report about the flora and fauna of the Valley of Mexico that included a translation of an ancient manuscript by Faustino Galicia Chimalpopoca, one of the few authorities on Nahuatl language in that period.[13] This resulted in Arriaga being inducted into the Sociedad de Historia Natural.

Arriaga's excursion into Tulyehualco sheds light on a number of features of scientific life in the country in the República Restaurada. In Latour's epistemology of laboratory life one can recognize Arriaga's shift from the hope of discovery to the necessity for inscribing and performing the act of scientific experimentation through the production of literature. It is also important to consider the actors and material conditions of this expedition, the men who were sent to the site came from different backgrounds (engineering, medicine) and had no formal training in archaeology. Also interesting here is the urgency with which the matter was treated. International interest in Central and South American pre-Columbian artifacts had been reaching an all-time high in the mid-nineteenth century. The impulse to retain these objects for the purposes of national patrimony was one important feature of the restabilization of the Mexican nation in the years following the liberal defeat of Maximilian's imperial rule. Shelley Garrigan has noted that this defeat was quickly followed by the transfer from the emperor's residence of "all documents pertaining to the archaeological monuments and collections of private interested parties (including Boturini, León de Gama, Dupaix, and Humboldt)" to a different site where they could be studied.[14] Arriaga's participation within the failed expedition that yielded the report is evidence of the interest in reclaiming national science by producing autochthonous accounts of the country's biosphere and therefore handing over the act of inscription and exegesis to Mexican subjects.

Arriaga's proselytization of science did not end with the report on Tulyehualco;

it also led him to experiment with a new genre within Mexican publishing. Popular science publications for children had been in existence in Europe since the eighteenth century. Bernard Lightman explains that it was John Newbery who inaugurated this form in 1761 with his Tom Telescope series *Newtonian System of Philosophy, adapted to the capacities of young gentlemen and ladies.* This particular publication combined science teaching alongside instruction on matters of morality and religion. Later, in the first half of the nineteenth century, female science writers played key roles in developing the genre. Notable examples, according to Lightman, are Jane Marcet and Mary Sommerville, who arranged their narratives in the form of conversations between curious subjects. In this "familiar format," the discussion became the dramatic grounds to stage doubts, ideas, knowledge, and ultimately "consensus."[15] Arriaga's text, to the best of my knowledge, is the first to adapt the genre to the Mexican context; *La ciencia recreativa* also follows the familiar format closely.

Arriaga's interest in what he called "vulgarizing" science came from his realization that such attempts had been successful in other, industrially and scientifically advanced societies. In the prospectus that appeared in *El Siglo Diez y Nueve* on April 2, 1871, he writes,

Mirad . . . á las naciones en las que la ciencia, por medio de mil ingeniosos artificios se vulgariza; allí todos sus ramos se fortifican y producen abundantes frutos: las artes se embellecen y se ilustran, porque descansan sobre el fundamento científico; la industria marcha á pasos agigantados impulsada por la voz autorizada del saber.[16]

(Look at the nations in which science, by means of a thousand ingenious devices, is popularized [se vulgariza]; there, all of its branches are strengthened and produce abundant fruits; arts are made more beautiful and are illustrated because they rest on a scientific foundation; industry takes enormous steps, propelled by the authoritative voice of knowledge.)

In the author's well-orchestrated advertising campaign he looks to models of vulgarization of science from elsewhere that, to his mind, strengthen the reach of science across a nation but also across different strata of society. Seeing the successes of popularized science in western Europe, especially in France and the United Kingdom, Arriaga therefore promotes one possible social reform for the Mexican population, that is, through the democratizing force of science. He affirms later in the article that science is a fair entity, able to reach into even "las clases más oscuras de la sociedad" (the darkest social classes) to find its future geniuses "que la han enaltecido con sus obras" (who have extolled her with their inventions).[17] It is important to note here that such efforts were dependent upon widespread literacy, a problem that put Mexico—with a literacy rate that fell below 15 percent in the final decades of the nineteenth century—in a very different position than England, which had a literacy rate of 90 percent in the 1870s–1880s.

Arriaga effects an approximation of literature and science by proposing the novel as a genre to be exploited in pedagogy: "El artificio de la novela se ha apoderado ya de la historia para hacer agradable su estudio" (The artifice of the novel has already taken hold of history to make the study of the latter more amenable).[18] Thanks to the novelization of French history, he writes, his fellow Mexicans knew more about that European nation's events than they did of their own. Such an observation leads Arriaga to propose the "novela científica," which would "generalizar los conocimientos científicos, embelleciéndolos con el artificio de la novela" (broaden scientific knowledge, beautifying it with the novel's artifice).[19] The novel, critics from Ian Watt to Franco Moretti have argued, is assumed to be the genre most linked to both a material and ideological understanding of modernity. Within this outlook, not only is the genre's spread across the globe possible thanks to technological improvements that go hand in hand with the spread of Western modernity, but novelistic narratives also succeeded in promoting a bourgeois way of life—its tastes, preferred infrastructures, and manners—as the ideal model for progressive societies.

For an author with rather limited literary abilities who pursued an ambitious combination of purposes and genres in this publishing venture, Arriaga's novelizing often results in clumsy and sudden shifts in plot and register, veering as it does from comedy to fantasy to didacticism. In *La primera semilla*, the first volume of *La ciencia recreativa*, an unnamed narrator retells what happened one evening while he was sitting in contemplation in his humble home. The fatigue and the cold of a winter's night make for an especially apt moment in which to sit in his armchair and allow his mind and spirit to roam into the realms of memory. He happily recalls his past voyages to foreign lands where he witnessed what to him was a beautiful form of progress. This advancement came in the shape of improved agricultural methods, factories, schools, and "establecimientos científicos" (scientific establishments).[20] His mind then turns to something altogether different: the "dulces impresiones" (sweet impressions) of distant busy industrial landscapes begin to slowly make way for other visions, this time of a beautiful, untouched natural landscape that is ravaged by war and therefore does not show signs of industry:

Ni en aquellos bosques, ni en aquellos campos, se escuchaba el canto alegre del colono que volvía de su trabajo. Ni en las comarcas solitarias percibíase el murmullo cadencioso de las máquinas, que es la voz de la actividad y del progreso. (6)

(In those fields and forests one could not hear the merry song of the laborer returning home after a day's work. Neither could one hear in those regions the cadenced murmur of machines, which are the voice of activity and progress.)

To add to this apparent horror, which is linked to profound disappointment, the virgin fields were also the stage for a bloody civil war. Horror gives way to a "narcotic" stupor, and then a moment of catharsis strikes the narrator as he realizes that morbid landscape was his native Mexico: "¡Era mi patria la que se había presentado ante mis ojos, adornada con sus inimitables galas y abatida por inmensa desventura!" (7) (It was my homeland that I had been seeing, adorned as it was with its unparalleled gifts and battered by an immense misfortune!). He is shocked by this realization, and a voice tells him, "Espera, esta es la ciencia del hombre" (Wait, this is the science of man). The voice comes from a winged young woman who has appeared in his room out of nowhere. They begin to converse about his terrible visions, which she admits to having placed within the deepest recesses of his mind:

Yo fuí la que evoqué en tu imaginacion los primeros recuerdos que tanto te deleitaron, y la que te espantó al presentarte algunos rasgos de la triste situacion que guarda este paraiso que es tu patria. (9)

(It was I who evoked in your imagination those first memories that you found so pleasurable and who frightened you by presenting some of the traits of the sad situation that awaits this paradise that is your homeland.)

Up to this point, Arriaga's narrative follows not only the familiar format of popular science publications described by Lightman but also echoes what Kenneth Roemer has called the "dialogic" form of utopian literature.

The first number of the series focuses on a dialogue between the narrator and science, and subsequent numbers also follow this format. Many of the volumes, which appeared between 1871 and 1874, relate a conversation between a scientist named Petit, a reference to any number of French scientists from the eighteenth and nineteenth centuries, and a character named Tío Pablo, who repeatedly moves from an ignorance of science to an understanding of its processes. Many critics contend that much of nineteenth-century utopian literature from the European and Anglo-American traditions follows the model, as Roemer describes it, of "didactic guide-visitor narratives that are heavy on long socio-economic dialogues, lightened by touches of romance and travel-adventure episodes and firmly grounded in cooperative or socialistic ideologies."[21]

Notably, the utopia that Arriaga's narrator envisions offers a radical reinterpretation of the pastoral idyll: the untampered natural landscape, as the character Science would have it, gives way to an industrial future where workers are nevertheless working merrily. In *La primera semilla*, an unmediated dialogue between a human being and science constitutes a particular fantasy in which the divine is entangled with a post-Enlightenment picture of progress. As Roemer also notes, in order for a utopian literature to be convincing, "the readers' culture has to provide perceptual tools in the forms of shared worldviews, ideologies and values that invite readers to 'see' utopia as an important and even inspirational guide to the past, present and future."[22] In his deification of science, Arriaga implements a shared worldview of a global model of progress and underdevelopment that was a staple of nineteenth-century thought in Mexico and elsewhere.

Like many fantastic narratives from the nineteenth century, *La primera semilla* has an overactive plot line. At the conclusion of this episode, the character Science takes the narrator on a final journey, this time to the Popocatepetl volcano that he passed earlier in his flight. Once they are there, the mischievous spirit turns him into a "*Buprestis gigangteus*" beetle, and he finds himself in a crystal jar in the cabinet of an old sage named Dr. Ramiro at the foot of the volcano, where a "reunión científica" (scientific meeting) is about to take place.[23] The narrator listens in on the conversation between peers in which Dr. Ramiro promises to plant that "first seed" (*primera semilla*) of science into the minds of his interlocutors, who come from different social strata:

> Estudiad con afan los riquísimos productos de vuestra patria; en los lechos de sus cristalinos rios brillan con deslumbrantes reflejos, las lentejas de oro y las piedras preciosas; en sus bosques de cedros y de pinos seculares se abrigan las plantas de flores seductoras. (32)

> (You must study the products of your nation; specks of gold and precious stones lie on the beds of its crystalline rivers; seductive flower plants take shelter in its forests of oak and cedar.)

The doctor projects an industrialized future in which his audience has become specialized: "Cuando ya seais mecánicos inteligentes, hábiles agricultores ó químicos industriales" (31) (When you have already become intelligent mechanics, able farmers or industrial chemists), he tells them, they will be able to take full advantage of the country's riches. Then, turning his attention to the beetle specimen he has in his cabinet, the doctor proceeds to remove the transmogrified narrator with a pair of tweezers. As he is about to spear the insect with a needle, the narrator wakes up and finds himself back in his room. The dream creeps into reality when the friend who wakes the narrator mentions he has received an invitation to a scientific expedition from none other than Dr. Ramiro.

Ramiro's laboratory at the base of Popocatepetl—a location that leads the narrator to ask whether the place would one day suffer the same fate as Pompeii or Herculaneum (24)—situates the scientific conversation within the folds of the autochthonous natural landscape, while the narrator is (temporarily) transformed into a Linnaean specimen. This final setting for the "novelita científica" recalls Arriaga's own expedition to Xochimilco, where at the base of a volcano he was hoping to make an archaeological discovery and performed instead a literary act. The volcano acts as a symbol that potentially connects Mexican history and geography with other histories across the world. In the absence of bustling factories, it metonymically represents a natural, organic site of ebullience and production and marks the site for a future eruption of scientific activity. Put differently, Arriaga strategically translates the volcanic site into a metaphor that services national self-projection. Moreover, everything in the story—the meandering plotline, the visions of a dystopian present and a future industrial idyll—leads to this locale, in which the many forms of work of the laboratory can be initiated.

While the image of the national volcano is imagined, in Arriaga's work, as a symbolic site for new beginnings, it also becomes a space for actual experimentation in popular scientific literature of the time. One of the longest articles by a Mexican author to appear in Santiago Sierra's *El Mundo Científico* is "Ascención al Ajusco" (Ascension to Ajusco [volcano]) by Vicente Reyes, a noted civil engineer. Published in the October 20, 1877, edition of the biweekly publication, it is an account of an exploratory expedition to the volcano south of Mexico City on September 30 of that year. According to Reyes in the opening of the article, elevated areas, "rodeadas de grandes misterios" (shrouded in great mystery), invite the observer to "averiguar su constitucion y penetrar sus fenómenos" (discover their constitution and penetrate their phenomena).[24] In addition, the volcano can reveal infinite amounts of knowledge to the scientific mind, for it can encerrar en el menor espacio posible y presentar en el mas breve tiempo la fisonomía de diferentes regiones y los fenómenos de climas diversos . . . alimentando con profusion esa avidez de sentir y de conocer, que incesantemente crece y se desarrolla en el espíritu humano. (33)

(enclose in the smallest space possible and present in the briefest time the physiognomy of different regions and the phenomena of diverse climates . . . thus feeding that desire for feeling and learning that incessantly grows and develops in the human spirit.)

Reyes's piece opens with a metaphysical reflection on nature that brings to mind the style of Arriaga's *Primera semilla* and then turns to quite a different mode of inscription. The details of the expedition start with measurements of atmospheric pressure with various instruments including "un barómetro de sifon de la fábrica de Secretan" (a syphon barometer made in the [French] Sécretan factory), "un aneroide compensado de la fábrica Negretti y Zambra, número 6481, cuya escala circular tiene un diámetro de 37.5mm" (an aneroid barometer, number 6481, from the [British] Negretti and Zambra factory, whose circular scale has a diameter of 37.5 mm), "un hipsómetro de Regnault, de la fábrica de Troughton y Simms" (a Regnault hypsometer, made by [British] Troughton and Simms), and so on.[25] The writing becomes numerical, as equations take over the prose (figure 2.1). Alongside numerical expressions are comparisons to other measurements of the Ajusco, including one by Alexander von Humboldt, who visited Mexico City in 1803–1804.

This particular article exemplifies what Latour addresses as the "experimental," the act of observation that is intertwined with the act of writing and dissemination. Reyes incorporates his own (native) inscriptions into the ongoing experiment of measuring the atmosphere in that region of Mexico; by doing so, he contributes to and annotates an existing body of work, turning his personal and physical act of exploration into a scientific fact.

These connected readings of volcanic appearances in popular Mexican texts point to the circular processes of scientific progression as they were perceived by writers during the period of the República Restaurada. While José Joaquín Arriaga's text is a dreamlike projection of future scientific ebullience in Mexico, Vicente Reyes's article makes manifest, through detailed inscriptions, a series of processes inherent in that development. As experiments in new forms and genres of writing—the "novelita científica" and the popular science magazine—they are literary accounts of science as well as creative accounts of scientific

rmómetros. La cima es árida y entre los inters-
s de un hacinamiento de lajas de pórfido, so-
recen algunas gramíneas: mas abajo se encuen-
pequeños ocotes [pinus communis,] que cuen-
algunos años de existencia y que apenas han ad-
do una altura de 1.m50.
uestra primera diligencia, despues de echar una
da al magnífico panorama, fué instalar los instru-
tos que habíamos conducido sin accidente algu-
para hacer las observaciones simultáneas con las
á las 10.h A. M. ejecutaba en México el Sr. Col-
, ingeniero auxiliar del Observatorio Meteoroló-

anéroide (6481) marcaba una presion de......
m50; el [7442], 477.mm80, siendo la temperatu-
el termómetro libre á la sombra, y á un metro de
ra sobre el suelo, 11.° El termómetro húmedo
alaba 7.°7 y en consecuencia el enfriamiento
lucido por la evaporacion, estaba representado
3.°3. Con estos elementos y empleando las fór-
as de August, modificadas por Regnault, hemos
ulado la tension del vapor contenida en el aire
grado de humedad relativa, habiendo encontra-
para el valor de la primera 7.mm22 y para la se-
da 0.09. En el mismo instante, la temperatura re-
rada en el Observatorio ascendia á 17.°9, siendo
ension del vapor 8.mm56 y la humedad relativa
aire 0.53. Soportábamos, pues, una temperatu-
erca de 7° mas baja que la que experimentaban
México y respirábamos un aire notablemente mas
nedo. En el Observatorio soplaba el viento del
y en la cima del Ajusco la corriente traia la
eccion del N. E. Sobre el vasto horizonte que
iamos, la cantidad total de nubes cubria ménos de
dos décimas partes del cielo y las formas domi-
ites observadas sobre las crestas de las lejanas
dilleras, tenían el aspecto de los cúmulos. En el
it se presentaba una nube, formada de diversos
mentos de cirrus, que parecian animados de un
vimiento ascendente y de otro giratorio en el sen-
horizontal, pues se les percibian distintas direc-
nes. El mismo hidrometeoro, visto desde el Ob-
vatorio, tenia la forma y caractéres de los cú-
lus.
usimos en práctica el aparato de ebullicion, y el
mómetro subió á 87°55.—La presion correspon-
nte á esta temperatura es 478mm26, segun las ta-
s de Regnault; y 481mm48, conforme á las tablas
Williamson.
Al ir á observar el barómetro, notamos con pena
el polo del menisco en la rama menor del sifon,
ba mas arriba del límite de la escala, quedando
lto en el estuche del tubo barométrico. Esta cir-
stancia nos impedia registrar una de las lecturas
ponentes de la presion, y nos resolvimos á tomar
base para nuestros cálculos las indicaciones de
aneroides, que habian marchado satisfactoria-
te. Nos era conocida la correccion de ambos
trumentos bajo la presion que se experimenta en
xico, pero era indispensable establecerla para
a presion próxima á la que tiene lugar en el Ajus-
. Esta correccion se dedujo de las observaciones
chas al pié del cerro del Aguila, que dieron los re-
ltados siguientes:

rómetro de Sifon á 0°.. 492mm84
eroide (6146)........ 493 - 80..Correc.—0mm96
eroide (7442)........ 489 - 30..Correc.+3 - 54

Con estos datos y tomando el promedio de las lec-
ras de ambos aneroides, se tiene para las observa-

ciones practicadas en la cúspide de la montaña á las
10h15 Å. M:

Presion reducida y corregida............ 481mm94
Term. libre............................. 11° - 0

Repetimos la observacion á las 11, y obtuvimos:

Aneroide (6146). 482mm60 Aneroide (7442). 476mm50
Correccion..... 0.96— Correccion..... 3·54+

Presion corregida 481.64 Presion corregida 480.04

Promedio.. $\frac{481.64+480.04}{2}$ =480mm84 Term. libre. 8°5

Hipsómetro 87°53...... Presion correspondien-
te: 477mm86, por las tablas de Regnault y 481mm08,
por las de Williamson.

Term. seco.................. 8°5
,, húmedo................. 5.7
Enfriamiento................ 2.8
Tension del vapor.......... 6mm36
Hum. relativa.............. 0.71

El viento habia cambiado de direccion, soplando
del N. con bastante intensidad: de ahí el abatimien-
to de la temperatura y el aumento de la humedad.
Persistia la misma cantidad de nubes, presentando
los mismos aspectos.
Las observaciones correspondientes en México,
fueron:

Presion á 0°.... 586mm63 Tension del vapor. 8mm47
Term. libre..... 19°0 Hum. relativa.... 0 . 49
Viento del E.; cúm. en el horizonte.

Si aplicamos la fórmula

$$n = A\,D\,(\log.\ B - \log b)\left(1 + \frac{2\,r + n}{R}\right)$$

tomando los promedios respectivos de las dos pares
de observaciones simultáneas, tendremos:

DATOS.

B..586mm75 T..18°5
b...481 . 39 t... 9.7 } Latitud media.. 19°20' N.

DESARROLLO.

log. B...... 2.7684531   log. A...... 4.26522
log. b...... 2.6824971   log. D...... 0.02383

log. B.—log b 0.0859560............ 2.03428

                                    3.22333
Correccion........ 0.00011

            log n............ 3.22344
            n=1672m8

Y entonces:

Altura del Ajusco sobre el observatorio.... 1672m8
Altitud del observatorio................. 2283.4

Altura del Ajusco sobre el nivel del mar... 3956.2

Diversos observadores han calculado la altura de
esta montaña, asignándole los valores que á conti-
nuacion se expresan:

Observadores.	Alturas.
Iberri....................	3859m
Dr. Maire.................	3921,,
Humboldt..................	3674,,
Burkart...................	3677,,

Figure 2.1. Page 35 of Vicente Reyes, "Asención al Ajusco," *El Mundo Científico*, October 20, 1877 (Mexico City). Photograph courtesy of University of New Mexico Libraries.

observation. Moving across the heights and depths of the Mexican landscape, they invite readers to view national space through the accrued methods of science. There is much at stake in this invitation, for the authors and editors of those publications also persuade readers to translate minute observation into an abstraction of future national progress.

While the novel genre becomes Arriaga's solution to move across the two interpretative planes, encapsulating as it does a sense of Western modernity in a nineteenth-century context, Reyes operates within the format of the popular science magazine, whose general aim is to democratize science. As a period of high hopes, the República Restaurada deserves to be studied in terms of the many literary experiments it witnessed. Scientific experimentation in these texts connects Mexico to an expanding intellectual tradition, one that involves observing the particularities of the physical universe in order to speak *of* and *to* global communities. The correspondences between science and literature and between scientific knowledge and geopolitical aspirations in Mexico during this period are revelatory of science's broader operations. Mexico is perceived as a large-scale laboratory in that period, representing an expanded platform on which experimentation and inscription are caught in a continuous cycle of translation, in a manner that recalls Latour's theorizations. The Mexican example, rather than a solitary story from the periphery, is revelatory of science's nonautonomy, its embeddedness in the social fabric of public life.

Notes

1. Monsiváis, "Manuel Gutiérrez Nájera," 27.
2. Reyes, "Pasado inmediato," 12.
3. Barreda, "Oración cívica," 8.
4. Reyes, "Pasado inmediato," 13.
5. Stepan, *"Hour of Eugenics."*
6. Latour and Woolgar, *Laboratory Life*, 76.
7. Latour, "Force and the Reason of Experiment," 56.
8. Sierra, *Epistolario y papeles privados*, 290.
9. "La estatua de Chac-Mool," *El Monitor Republicano*, March 30, 1877, quoted in Lombardo de Ruiz, *El Monitor Republicano*, 52–53.
10. Bueno, *"Forjando Patrimonio,"* 219.
11. In her essay Bueno discusses the lack of expertise of men who participated in archaeological digs in the Porfiriato: "Training in the discipline [of archaeology] would not exist on a regular basis until 1906, when classes began in the [national] museum. Most of the archaeologists instead came from other established fields, especially history, medicine, and law. . . . This raises the question of what constituted the expert. The expert

was essentially someone who studied antiquity, who published on it, and who frequently had some sort of connection to the state, quite often as an employee of the museum" ("Forjando Patrimonio," 233).

12. Ramírez, "Estudio biográfico," 17.
13. Ramírez, "Estudio biográfico," 18.
14. Garrigan, *Collecting Mexico*, 77.
15. Lightman, *Victorian Popularizers*, 20–21.
16. Arriaga, prospectus for *La ciencia recreativa* in "Editorial."
17. Arriaga, "Editorial."
18. Arriaga, "Editorial."
19. Arriaga, "Editorial."
20. Arriaga, *La primera semilla*, 5–6.
21. Roemer, "Paradise Transformed," 80.
22. Roemer, "Paradise Transformed," 81.
23. Arriaga, *La primera semillia*, 25.
24. Reyes, "Ascención al Ajusco," 33.
25. Reyes, "Ascención al Ajusco," 33.

References

Arriaga, José Joaquín. "Editorial." *El Siglo Diez y Nueve* (Mexico City), April 4, 1871, 2.

———. *La primera semilla*. Vol. 1 of *La ciencia recreativa: Publicación dedicada a los niños y a las clases obreras*. Mexico City: J. M. Aguilar Ortiz, 1873.

Barreda, Gabino. "Oración cívica." *Latinoamérica: Cuadernos de cultura latinoamericana*, vol. 72. Mexico City: Universidad Nacional Autónoma de México, 1979, 5–19.

Bueno, Christina. "Forjando Patrimonio: The Making of Archaeological Patrimony in Porfirian Mexico." *Hispanic American Historical Review* 90, no. 2 (2010): 215–245.

Garrigan, Shelley. *Collecting Mexico: Museums, Monuments, and the Creation of National Identity*. Minneapolis: University of Minnesota Press, 2012.

Latour, Bruno. "The Force and the Reason of Experiment." In *Experimental Inquiries: Historical, Philosophical, and Social Studies of Experimentation in Science,* edited by Homer E. Le Grand, 49–80. Dordrecht, Netherlands: Reidel, 1990.

Latour, Bruno, and Steve Woolgar. *Laboratory Life: The Construction of Scientific Facts*. Princeton, NJ: Princeton University Press, 1986.

Lightman, Bernard. *Victorian Popularizers of Science: Designing Nature for New Audiences*. Chicago: University of Chicago Press, 2010.

Lombardo de Ruiz, Sonia. *El Monitor Republicano (1877–1896)*. Vol. 1 of *El pasado prehispánico en la cultura nacional (Memoria hemerográfica, 1877–1911)*. Mexico City: Instituto Nacional de Antropología e Historia, 1994.

Monsiváis, Carlos. "Manuel Gutiérrez Nájera: La crónica como utopia." *Literatura Mexicana* 6, no. 1 (1995): 27–43.

Ramírez, Santiago. "Estudio biográfico del Sr. Ingeniero D. José Joaquín Arriaga." Mexico City: Oficina Tipográfica de la Secretaría de Fomento, 1900.

Reyes, Alfonso. "Pasado inmediato." In Alfonso Reyes, *Pasado inmediato y otros ensayos.* Mexico City: Colegio de México, 1941, 3–64.

Reyes, Vicente. "Ascención al Ajusco." *El Mundo Científico*, October 20, 1877, 33–36.

Roemer, Kenneth. "Paradise Transformed: Varieties of Nineteenth-Century Utopias." In *The Cambridge Companion to Utopian Literature,* edited by Gregory Claeys, 79–106. Cambridge, England: Cambridge University Press, 2010.

Sierra, Justo. *Epistolario y papeles privados.* Vol. 14 of *Obras completas.* Mexico City: Universidad Nacional Autónoma de México, 1984.

Stepan, Nancy L. *"The Hour of Eugenics": Race, Gender, and Nation in Latin America.* Ithaca, NY: Cornell University Press, 1991.

3

Natural Histories of the Anthropocene

Santiago del Estero, Argentina, in the 1930s

JENS ANDERMANN

In the first half of the twentieth century, on Latin America's advancing frontiers of predatory extraction of natural resources, a new mode of essayistic reflection emerges that attempts to tackle the social, ecological, economic, and cultural aspects of environmental destruction. What these writings set out to name is, in fact, nothing less than a natural history of the Anthropocene. The term "Anthropocene," first introduced in 2000 by atmospheric chemist Paul J. Crutzen and biologist Eugene Stoermer in the International Geosphere-Biosphere Programme's *Global Change Newsletter*, refers to the geological era succeeding the interglacial state of the Holocene. It is associated with the rapidly incrementing impact of human activities on the earth's biota that "have become so pervasive and profound that they rival the great forces of Nature and are pushing the Earth into planetary *terra incognita*."[1] In Latin America, I suggest, the wave of regionalist essays written between the 1930s and the 1950s offers us a first glimpse of the ground zero, the belt of eco-hazards that accompanies an expanding capitalist modernity. Moreover, it does so from the vantage point of critical languages that have often themselves been superseded, made anachronical, by this same movement of expansion. Yet this very untimeliness, I argue, also allowed the regionalist essayists to perceive, long before the emergence of "political ecology" in Western academia,[2] the interrelatedness between the crisis of biodiversity and that of certain modes of conviviality and of linguistic, musical, and material culture in which the regionalist movement sought to ground its claims for the particularity and cultural richness of local societies.

Revisited from a biopolitical perspective, these writings were not just negotiating their resistance or cooptation with regard to a metropolitan, hegemonic modernity. Rather, and more importantly, they also attempted to force their traditionalist languages and knowledges to reveal something as yet unnamed: the radical novelty of an all-encompassing destruction.[3] Even when its prime subject might have been an oral or artisan tradition on the verge of disappearance, this regionalism of the zone of emergency was radically modern. Its very enunciation stemmed from a sense of rift, from a radical break between past and present. Indeed, it was an attempt to forge nothing less than a fractured way of thinking, one that would be the bearer of a local memory at the same time as it recognized the irredeemable loss of what this memory aimed to hold onto. Regionalist writing produced an essayism of flight, of exodus. At the same time as this body of writing sought to preserve in words certain human and nonhuman geographies whose destruction had already taken place, it also already looked forward to the uncertain horizon of life after abandonment and toward a region now bereft of its landscape (*despaisada*) and thus of the very bond that had previously underwritten its regionality.

The Argentine Northeast is an eloquent example of this constellation. Already in 1900, the local newspaper *El Liberal* celebrated the logging of *quebracho* forests as the main source of work and income in the province of Santiago del Estero, calculating the total exports that year at 1 million logs, 900,000 boards, 600,000 tons of firewood, and 25,000 tons of charcoal.[4] In 1915, according to the forest inspector and future provincial governor Antenor Álvarez, 137 *obrajes* (logging stations) were operating in Santiago del Estero, employing more than 15,000 lumberjacks, and turning out an estimated total of 20,700,000 planks since 1906, equivalent to 1,600 kilometers of railroad track.[5] The logging stations, frequently established on public lands sold off at ridiculous prices, with technology and investment kept to a minimum, were a cruelly efficient mechanism of capital exportation. The pillaging of primary resources was accompanied by a system of *proveedurías* (company-owned general stores) where subcontracted laborers were forced to buy their provisions at prices more than double the market rate. In encouraging seasonal labor migration, the logging stations accelerated the depopulation of rural areas, which entailed the abandonment of maintenance tasks such as irrigating, plowing, and clearing of mud from wells and springs. The railway-logging assemblage refashioned regional space in accordance with metropolitan requirements regardless of the destruction wrought upon local ecosystems.

The year 1937 marks a point of culmination of the combined effects of socioecological devastation in the Argentine Northeast as an extreme drought lasting several years caused widespread famine and rural exodus, with Santiago del Estero suffering the most severe effects. That same year, Bernardo Canal Feijóo—a *santiagueño* sociologist, poet, and literary critic—published *Ensayo sobre la expresión popular artística en Santiago* (Essay on popular artistic expression in Santiago), in which he aimed to give a comprehensive overview of the multiple interrelations between the local environment and its creative expressions, the "juego integral de paisaje, costumbres, tonada, locales" (integral pattern of landscape, customs, accent, localities) that Canal conceived of as the region's very numen.[6] Santiago, the first colonial city to be founded in Argentina, Canal argued, could claim "una pequeña superioridad" (a small superiority) over the rest of the country inhabited by more recent immigrants, "dimana[da] precisamente de cierta capacidad de conservación" (emanating precisely from a certain conservative capacity) as manifest in popular forms of expression and in the persistence of the local indigenous language, Quichua Rinasumi.[7] This in turn spoke to a convivial rather than merely contemplative relation with the environment. The *santiagueño* landscape, Canal argues, was a constitutive part of this "integral pattern," since it actively denied locals the objectifying detachment that characterized a colonial, extractive relation with the earth and, on the contrary, invited an attitude of immersion, of self-absorption ("ensimismamiento"):

> Para muchos sé que no existe como paisaje, pues no es ni pampa ni montaña. Es bosque, broza, maleza, salina. Mientras los otros paisajes están diseñados en distancia, en fuga, en infinitud, en masa, éste sólo se dibuja en rincones, en ocultos detalles casuales. No es para ser visto desde el tren, o desde el aeroplano. En cierto modo, pide la convivencia del sujeto humano; no su simple éxtasis. El hombre está *ante* la pampa, *ante* la montaña, desde el punto de vista del sentimiento del paisaje; desde el mismo punto de vista nunca podría estar "ante" el bosque: precisa estar *en* él, envuelto, inmerso en él.[8]

(For many, I know, it doesn't exist as a landscape, for it is neither plain nor mountain. It is forest, scrub, undergrowth, salt flat. Whereas the other landscapes are formed in distance, in flight, in infinity, as a whole, this one only takes shape in small corners, in dark, casual details. It is not made to be seen from the train or from an airplane. In a certain way, it demands the cohabitation of the human subject; not simply its extasis. Man is *faced with* the plain, *faced with* the mountain, from the point of view of his affective relationship with the landscape; from this same point of view he could never be "faced with" the forest: he needs to be *in* it, surrounded, immersed in it.)

This intense attachment, argues the sociologist, also conferred on the cultural idiosyncrasy of the province a particular density, at the same time as it made this same nature-culture ensemble extremely vulnerable in the face of external forces that, since they could not be incorporated into the "integral pattern" of native society, were fatally bound to turn into pure destruction. As an economic system that remains "fríamente extraño" (coolly external) to the reasons of local historical reality, Canal continues, the logging of Santiago's hardwood forests may have produced "fabulosa riqueza pecuniaria" (fabulous pecuniary riches). However, these never actually belonged to the province: "En ingentes chorros se trasladaba directamente de la fuente a otra parte: de la Provincia afuera, a Buenos Aires, a Londres, a Bruselas" (In mighty streams, they were transported directly from the source to other parts: away from the Province, to Buenos Aires, to London, to Brussels).[9] None other, he concludes, has been the basic system of provincial history over the last fifty years, the impact of which, "reflejado en el alma nativa" (reflected in the native soul), has resulted in "la categoría más patética de una destrucción del paisaje" (the more pathetic category of a destruction of the landscape).[10]

Canal Feijóo's reflections emerged in the context of a vibrant cultural and literary scene fueled not least by the active presence of a combative trade unionist movement over the first decades of the twentieth century. Both of these, intriguingly, were the product of the very same logging boom, the disastrous human and environmental effects that Canal was denouncing. It was during this same period that cultural initiatives like the Archaeological Museum and the People's University of Santiago del Estero sprang to life, popular libraries were being set up, and various newspapers were being published throughout the province. The Asociación Cultural La Brasa (The Ember Cultural Association), founded in 1925, aimed to gather these cultural and literary ventures under a common umbrella despite their ideological discrepancies.[11]

It is important to bear in mind the context of thriving intellectual activity, as much the effect of an accelerated modernization as of the destruction the latter wrought on the very foundations of local society, in order to understand the extreme tension that runs through the literary and cultural production of Santiago del Estero, present already in early forerunners such as Ricardo Rojas's *El país de la selva* (The forest country, 1907). However, the peculiar challenge confronting the writers and artists of La Brasa and of Santiago's literary and artistic scene more generally was to find an expression for the destructive and fallacious nature of a modernity that at the same time provided them with the conceptual and aesthetic languages to name this very process. Indeed, the regionalist essay was intended to force the philosophical and literary arsenal of modernity to address what that modernity had already destroyed: the "hecho ancestral" (ancestral fact) of an enduring conviviality between material and symbolic regimes of production and their environment, "el fenómeno permanente de incidencia" (the phenomenon of permanent repercussion) of nature within culture and vice versa.[12]

Orestes Di Lullo, from whose work the latter quote is taken, was a medical doctor and founding member of La Brasa and the author of *El bosque sin leyenda: Ensayo económico-social* (The forest without legend: Economic-social essay), released by a local *santiagueño* printing house in 1937, the same year Canal's *Ensayo sobre la expresión popular* was published in Buenos Aires. Like Canal, Di Lullo in his work points to the close interaction between the devastation of the natural environment and the social, cultural, and political process at large: "Ahora, la tierra rapada es otra, y nosotros también," he writes. "La industria foresta ha destruído el paisaje" (56) (Today, the razed earth has become something else, and so have we. The logging industry has destroyed the landscape).

A truly exceptional work not only for its acute analytical insights but also for its powerful images and epic composition, *El bosque sin leyenda* occupies a threshold in Di Lullo's oeuvre and career. Having trained as a medical doctor specializing in regional pandemics such as Chagas disease, the skin infection known locally as *paj* or *quebracho* illness (the focus of his doctoral thesis), Di Lullo gradually developed his lifelong interest in traditional medical knowledge and the curative properties of plants into a more comprehensive engagement with questions of popular nutrition and access to health care and other social services. Following studies on *La medicina popular en Santiago del Estero* (Popular medicine in Santiago del Estero, 1929) and *La alimentación popular* (Popular nutrition, 1935), between 1943 and 1944 Di Lullo completed the multivolume *El folclore de Santiago del Estero* (Folklore of Santiago del Estero), building on research carried out during the organization in 1941 of the provincial Museo Histórico of which he was the driving force.

Set midway between medicine and cultural history, then, *El bosque sin leyenda* combines both types of knowledge in an unprecedented attempt to understand the mutual interrelations between social, economic, and cultural processes and nature understood from a kind of Catholic-pantheistic perspective informed by Di Lullo's parasitological and epidemiological research as an organic totality, "porque la creación, es la organización de lo eterno, o mejor, perpetuación de la vida que se suplanta a sí misma, constantemente, de modo a ofrecer una sola fisonomía, indeformable, a través de los tiempos" (11) (because Creation is the organization of the eternal, or better, the perpetuation of life, constantly renewing itself, such that it offers a single physiognomy impossible to deform, which persists through time).

Combining a highly evocative literary language with a holistic scientific vision, Di Lullo's text divides its attention between the botanical and ecological effects of deforestation and a kind of anatomy of the *obraje*, the logging station, described as a tentacular apparatus—half organic, half techno-economic. The book is divided into three parts in a movement of progressive abstraction that takes the reader from naturalist description toward conceptual and historical syntheses. The text begins with a kind of flow chart introducing in short narrative vignettes the main actors and productive intervals of the logging machinery, from the lumberjacks' departure from their villages ("El éxodo"—Exodus) to their solitary work felling trees in the forest ("La hachada"—Axing); the storage, loading, and transport of the logs ("La rodeada," "La cargada," "La acarreada"—Assembling, loading, moving); and the tasks of working the logs and carbonizing firewood at the mill ("La labrada," "La quemada"—Working, burning), as well as, finally, the lumberjacks' return ("El regreso"—The return), their bodies and spirits broken, the men frequently on the run from the *capangas*, the stations' militialike overseers.

This impressionistic catalog of human types and their functions in the apparatus is followed by a more systematic historical and social analysis of the mechanisms of workers' enslavement, such as the exclusive sale of provisions through the station-owned stores where the lumberjacks would inevitably run up heavy debts ("La Proveeduría"—General store) and the state's complicity in granting logging concessions and building railroads for the benefit of the industry ("El ferrocarril"—The railway). The survey concludes with an economic, social, and political assessment ("Resultados de la explotación forestal"—Results of forest exploitation) and an ecological evaluation ("La madre tierra"—Mother Earth) before moving on to an apocalyptic vision of the burned-down forest ("El incendio del bosque"—The forest on fire). Here, however, in a sudden change of course, Di Lullo sees emerging from the ashes the faint hope of a "Tregua" (Truce), calling on an interventionist state to make common cause, finally, with science rather than capital and to reorganize the exploitation and preservation of nature ("Parques nacionales"—National parks) and employ the marvels of technology for the benefit of workers and the environment alike ("El obraje de mañana"—The logging station of tomorrow).

For Di Lullo the apparatus of the logging industry entailed two major problems: on the one hand, it corroded the social texture of the province; on the other hand, it interfered with the area's natural rhythms. The terms carry a strong moral undertone in which *El bosque sin leyenda* alerts readers to the loss of attachment to the native soil on the part of subsistent farmers transformed into nomadic day workers:

> El peón era la energía, en potencia, del campo. La industria forestal, la succionó, la malgastó. . . . No más granjerías de la tierra bendita, ni cosechas rebosantes, aunque sufridas. No más tranquila paz de los campos germinados. . . . Desertaban de la vida. Y con el toque de atención del bosque se fueron a la muerte. No fue tanto el engaño ni la esclavitud lo que más daño le hicieron. Lo peor fué la pérdida de su vocación agrícola-pastoril. Cincuenta años de industria forestal han destruido la tradición de un pueblo criado sobre el arado y en pos de los rebaños. (34)

> (The farmhand had been the energy reserve of the land. The logging industry sucked it up, threw it away. . . . No more orchards of blessed earth, no more overflowing, although hard-won, harvests. Gone is the peace and tranquility of germinating fields. . . . They had deserted life. And on being called by the forest, they went to their deaths. It wasn't even betrayal or enslavement that caused the most damage. The worst was losing their vocation for agriculture and herding. Fifty years of logging industry have destroyed the tradition of a people reared to work a plow and attend to flocks.)

What is more, by interfering with the evaporation and rainfall cycles and diminishing the native fauna and flora, deforestation had left behind a lifeless, barren soil. Di Lullo describes with a naturalist's precision the zoobotanical process of the forest's transformation into steppes and swamps:

El humus del suelo, estratificado por infinitos y lentos años de exfoliación y humedad, se desecó bajo el intenso sol y se pulverizó en brazos del viento huracanado que ya recorría la llanura con su cortejo de polvaredas. La gordura de la tierra, antes protegida por espesas frondas, fué sepultada por las arenas de los vendavales, y su milagrosa potencia germinativa, henchida y lujuriante, se sofocó en la inmersión de otras tierras estériles, que los remolinos transportaron a distancia (62).

(The humus of the soil, slowly stratified through infinite years of exfoliation and moisture, dried up under the intense sunlight, to be pulverized in the arms of the hurricanes crossing the plains with their court of dust clouds in tail. The fat of the land, once protected by dense canopies, has been buried under gales of sand, and its miraculous, bursting, and luxuriant germinative power has been suffocated on becoming immersed in other, more sterile soils, which the whirlwinds carry away with them into the distance.)

The logging of the native forests, in short, gave way to a "vegetación de postguerra—guerra cruel del hombre contra el bosque—es una vegetación que se defiende, que vive en contínua lucha de defensa" (62) (a postwar vegetation—an all-out war that's being waged by Man against the forest—it's a vegetation that defends itself, that lives as if it were under constant attack).

This shared degradation of men and their environment harks back, for Di Lullo, to a common cause: the perverse transformation, on behalf of a predatory capitalism, of human labor into pure negativity, pure destructiveness, into "el trabajo de todo el ser para matarse" (14) (the labor of the entire being for the purpose of killing itself). The logging companies, Di Lullo argues, acted in Santiago del Estero as mechanisms of dehumanization. At the logging stations, men cease to be human, "trabajo para vivir esclavizado. Trabajo para no morir" (33) (working to live as a slave. Working so as not to die). Logging, he writes, is a systematic activity that leaves the earth "despoblada de vida" (15) (depleted of all life). The culmination of this destructive frenzy arrives through the apocalyptic image of trees piled up and ready to be burned and reduced to charcoal, an industrialized form of death that is described here in open allusion to the Christian martyrological tradition. At the same time, it is in this tragic and final moment of suffering, immediately before it is reduced to ashes, that the forest community gathers once more in its richness and diversity:

Por fin, el horno, listo, levanta su giba inerte. Yacen en su entraña, el aromo y el chañar, la tusca florida y la brea resinosa, el mistol y el algarrobo, la cina-cina y el espinillo, el guayacán, el quebracho, el molle, y el itin, toda la riqueza de la tierra exprimida de jugos. Y de pronto el quemador, con la tea encendida prende fuego a la pira. Se ha consumado el sacrificio. . . . El penacho de humo que ensombrece la selva, escapado a borbotones por la boca del horno, tiene otro significado que el humo de la fábrica. No es humo que redime, sino humo estéril, humo de destrucción. (21)

(At last, the furnace, ready once more, raises its lifeless hump. In its entrails, side-by-side, lie the myrrh tree, the chañar, the flowering acacia and the resinous brea, the mistol tree and the oak, the cina-cina and the espinillo, the guayacán, the quebracho, the pepper tree, and the itín, all the wealth of the earth, its juices squeezed from it. And suddenly the wood burner, flaming torch in hand, sets fire to the pile. The sacrifice has been consummated. . . . The plume of smoke shadowing the forest, rising in thick bubbles from the oven's mouth, has a different meaning from the smoke that rises from a factory chimney. It is not smoke that redeems; it is sterile smoke, the smoke of destruction.)

In his attempt to identify the structural causes underlying the clash between environmental and historical forces, however, Di Lullo's essay becomes caught up in a tension its author only partially acknowledges, expressed in the two poetics the text mobilizes to narrate this natural history: the pastoral and the epic. In the idealized vision of the first, the world prior to the onset of deforestation resurges full of Edenic associations of bread-yielding fields and fertile plains (13), while from the point of view of the second, the trails departing in the direction of the logging stations are the tragic heroes of a vast rebellion against the regime of the latifundio (81), the extreme concentration of scarce fertile lands in a few hands. The tension remains unresolved since Di Lullo seeks at once to rescue some kind of stable colonial kernel as the very hallmark of *santiagueño* identity and to denounce the way the logging industry has betrayed and thrown away the energies it received from a peasantry longing for "buenos campos y . . . mejores tiempos, y . . . una sabia política de colonización" (36) (good lands and better times, and for a sensible politics of colonization).

In other words, rural populations were expecting from the logging companies the kind of sensible colonization they had been denied for centuries by the colonial organization of landownership. They had been hoping, in the end, for nothing other than progress such that, with the aid of modern technology, their plight would finally yield the means for modest survival. There were still some woodlands left, Di Lullo writes, perhaps even enough for the "logging station of tomorrow" to redeem its monstrous forerunner and to finally make good on its promise of bringing "el criterio científico" (scientific judgment) to bear on the exploitation of natural resources (79).

Suddenly the tone shifts toward what Adrián Gorelik, reflecting on Canal Feijóo and the regionalist essay, describes as "un pasaje típicamente voluntarista, que propone resolver a través de un salto tecnológico una fractura que es en realidad política y cultural" (a typically voluntaristic passage that aims to repair by way of a technological leap a rift that is really political and cultural in kind).[13] In Di Lullo's book, the rift that this technovoluntaristic leap is meant to overcome is one that opens up between two different forms of thinking about the relation between culture and the environment. While in the first of these the idea of a residual and atemporal equilibrium prevails, in the second this stable relation gives way to a mutable, contingent constellation of both "natural" and "historical" agencies in mutual engagement. The achievement of a relatively stable relation, the "juego integral" (integral pattern) forged by traditional modes of farming and cattle-raising, is what Di Lullo and his fellow La Brasa members call a "landscape," and it is for this reason that the term plays such a key role in their conceptualization of a distinctive *santiagueño* identity and that the impact of deforestation far exceeds the ecological havoc caused by rapacious and indiscriminate logging.

In Canal Feijóo's writings, the difficulty experienced by Di Lullo in reconciling an ahistorical environmental determinism with a vision of multiple historical and natural agencies reappears precisely in association with the concept of landscape and its interpretation. In "El paisaje y el alma" (The Landscape and the Soul), the opening essay of *Ñan* ("path" in Quichua), the journal he edited from 1932 to 1934 and of which he was the sole author, Canal questions the meaning of this intangible, perhaps even already extinct entity that is the landscape of Santiago del Estero. As a "suceso psico-geográfico" (psychogeographical event), the landscape neither is purely a natural fact nor exists solely in the perception of an observer who projects onto nature his own aesthetic intentions (10). Landscape, on the contrary, describes the interplay, the "juego" (pattern) in which both of these become involved, resulting in mutual affections and transformations. How, then—Canal goes on—can we conceive in such terms of an apparently un- or even de-formed spatial ensemble, devoid of any ordering principle, if not as a non-landscape that encourages uprootedness and forges a mentality of passionate extremes; a culture, in short, of resistance and rebellion against its own environment?

In one of the subchapters of his essay called "El paisaje santiagueño" (The landscape of Santiago), Canal reflects on the singular forms of popular aesthetic expressions in terms of their "translation"—indeed of a "rebote en el alma" (backlash inside the soul)—of this peculiar environment, experienced not as a landscape but through an attitude of rebuff and resignation towards "una constricción telúrica . . . un mundo inacogedor" (a tellurical constriction . . . an unwelcoming world).[14]

The expressions of popular culture are, for Canal, almost instinctive attempts to transcend this inhospitable environment:

> El alma santiagueña sufre la secular esclavización de una naturaleza para ella sin paisaje. Todavía sin recursos para el señorío inteligente del mundo, confía su redención a la tangente musical. En ella se descarta; por ella se evade. Pero en el trance, su inconfundible acento, traduce precisamente el drama de la existencia vernácula.[15]

> (The *santiagueño* soul suffers from a secular enslavement to a nature devoid of landscape. Lacking the means, as yet, of intelligent domination of the world, it entrusts redemption to the musical realm. In music, it excuses itself; through music, it takes flight.)

It is the feebleness rather than the solidity of people's ties with the earth that accounts for the singularity of the "fenómeno santiagueño": the peasant

> no se "encuentra en el paisaje" dentro de la naturaleza vernácula. Encontrarse en paisaje es sentirse señor de la Naturaleza, o huésped grato. Ni lo uno, ni lo otro el santiagueño, en su propia tierra, se siente ajeno, o a lo más, prisionero.[16]

> (does not find himself "inside the landscape." Being inside the landscape is to feel dominion over nature, or to be its welcome guest. Neither one nor the other is true for the *santiagueño* who feels like a stranger in his own land, at best its prisoner.)

It is only in the subsequent issue of *Ñan*, published two years later, that this psychological and phenomenological landscape will appear as shot through with a historical temporality. In the long essay "Imagen de Santiago: Reconocimiento de una provincia desconocida" (Image of Santiago: Survey of an unknown province), Canal changes focus and zeroes in on the fateful triad of "el rapto del ferrocarril"[17] (abduction by railway), "el asalto de la selva"[18] (the assault of the forest), and "la destrucción del paisaje"[19] (the destruction of the landscape). The bone-dry land where in the previous essay Canal seeks the key for understanding the native "soul" is now revealed as the product of a long-standing process of decay in which both environment and culture have been equally victimized. If the railway started to induce the uprooting, decentering, and depopulation of the countryside, ushering in nomadism, the logging station that arrived on its heels and at the same time fueled its insatiable need for timber and charcoal was "una formidable trinchera" (a formidable trench) in the war of position waged against the forest; in it "sierras cantarinas performan una estricta anatomía industrial"[20] (the song of the saws perform a detailed industrial autopsy). Together, they were bringing about the "despaisamiento" (unlandscaping) of the province.

Even more than deforestation and rural exodus and even more than the cumulative effects of both factors together, in his neologism—"despaisamiento"—Canal attempts to grasp the withdrawal of any other kind of relation between man and nature beyond that of a radical destructiveness. Unlandscaping, in other words, is the figure Canal forges in order to conceive of a natural history of the Anthropocene. The lumberjacks of the logging stations, he writes,

> se encontraron más pobres que al nacer, pues hasta habían perdido su paisaje. ¿Qué otro argentino podría quejarse de una tragedia tan enorme como la de este santiagueño, condenado a servir a la destrucción lisa y llana de su propio paisaje? ¿Y qué había sacado de aquello? . . . Un día se halló súbitamente solo y desguarnecido. Con la última jornada se había ido su paisaje, y el abra de aquél día era ya su destierro. Fué como un súbito despaisamiento. Y en el desconcierto de este trance, el alma se orienta por la brújula hipnótica del tren.[21]

(found themselves more miserable than at birth, since they had lost even their landscape. What other Argentine could weep for a tragedy as enormous as the one endured by this *santiagueño*, condemned to carry out the destruction, pure and simple, of his own landscape? And what had he got out of all this? . . . One day, he suddenly found himself alone and unprotected. With the last working day, his landscape had gone, and that day's clearing was already his exile. It was as if unlandscaping had arrived all of a sudden. And in the uncertainty of this trance, the soul starts following the railway's hypnotic compass.)

In linguistic, cultural, social, and political terms, Canal continues, unlandscaping entails the loss of absolutely all active forces of autonomy and, as a consequence, of the very "integral pattern" that had underwritten the singularity of the province. Even if labor migration had already been a staple of provincial regimes of work and income before the rise of the logging industry, only the combined effects of the railway and the logging station transformed this longstanding practice into a permanent flight, an emigration of between fifty and sixty thousand peasants per year, as he estimates in "Los éxodos rurales" (Rural exoduses), an essay originally written in 1938 and later published as part of *De la estructura mediterránea argentina* (On the structure of inland Argentina).[22]

Part of an unfinished project for a "Sociología mediterránea argentina" (Sociology of inland Argentina), *De la estructura mediterránea argentina* was to focus specifically on the phenomena of the crisis and disintegration of rural communities in the interior (15), a geographical notion that, as Canal goes on to explain, in Argentina is commonly counterposed to Buenos Aires and the coastal region and labeled the cause for all the country's real or perceived ills. The culprit "de las dificultades, de las reviniencias, de las demoras" (of all difficulties, relapses and delays) would always be "'el Interior,' con los nombres, a veces, de campañas, de Provincias, de masas" (12) (the "interior," meaning, at different times, hinterland, province, or masses).

In relation to the material realities of human sociability and its environment, "un espíritu de evasión" (a spirit of evasion) is at work, he suggests, in the "miraje pampeano" (pampean mirage) that has cast a spell on the nation, a disregard for the real that has been cruelly punished:

> La Naturaleza ha perdido generosidad y dulzura en el ciclo de los últimos cincuenta años—acaso sea la indignada respuesta a la insensatez con que el hombre la ha venido tratando—; los términos medios pluviales han disminuído. . . . A las insuficiencias líquidas siguen, mecánicamente, las erosiones, la esterilización de la superficie accesible a los recursos del trabajo elemental, la proliferación de las plagas zoológicas. . . . Cualquiera que fuese la relación con esos eventos climáticos, lo cierto es que el estado de sanidad de nuestras campañas es hoy espeluznante. Suele nombrarse, en primer lugar, al paludismo; pero no ocupan menos lugar, ni son menos graves dentro del marco general de "inasistencia" en que se desenvuelven todas las pestes rurales, el bocio montañés, el tracoma, la leichmaniosis, la brucellosis, . . . las venéreas, la sífilis.[23]

(Over the past fifty years, Nature has lost its generosity and sweetness—possibly in angry response to the foolishness with which it has been treated by man. Rainfall averages have diminished. . . . Insufficiency of liquids is being followed, mechanically, by erosion, sterility of the surfaces accessible to subsistence work, and the proliferation of zoological plagues. . . . Whichever relation there may be with these climatic events, what is beyond doubt is that sanitary conditions in our countryside today are horrifying. We are used to hearing about malaria in particular, but just as bad and widespread is the lack of assistance thanks to which all other rural plagues are spreading: mountain goiter, trachoma, Leishmaniosis, brucellosis, . . . venereal diseases, syphilis.)

In the 1948 essay, then, Canal was already reflecting on a climate of history under the impact of which the environment (climatic cycles, bacteriological and viral processes) *responds*, on a shared plane of historical agency, to the "insensatez" (foolishness) of human actions. At the same time, he regrets "la pérdida de . . . ese profundo señorío que ha ido siempre del brazo con el dominio directo de la tierra" (81) (the loss of . . . a kind of profound dominion that has always gone hand in hand with working the earth). Yet, insofar as these ancestral and convivial forms of knowledge have become lost in the same measure as the soils to which they were once applied have become eroded and sterilized, only a new, disillusioned, and realistic rationality might still be able to contain this vicious circle. Here, as in the coda of Di Lullo's book, the very framework and categories that previously made it possible to conceive of the "integral pattern" of provincial life (in which "landscape" played a fundamental role) need to be sacrificed in order to overcome "el viejo y ya quizás apenas cartográfico concepto de Provincia, por el más vivo, razonable y realista de Región" (the ancient and possibly today not much more than cartographic concept of Province for the more vital, reasonable, and realistic one of Region).[24]

On becoming politicized—in parallel with the emergence of a developmentalist state in which, finally, its proponents believe they have found a sympathetic interlocutor—this new regionalism also had to let go of some of the categories central to its previous iteration in the 1920s and 1930s. As the regionalists of the 1940s and 1950s hedged their bets on a dream of technological damage containment, they also had to acknowledge that the very unlandscaping it was meant to counteract through a self-consciously Cyclopean effort had already run full circle. The "planificación integral" (integral planning), in Canal's 1948 essay,[25] took the place of the "juego integral" (integral pattern) on which the study of popular expressions in Santiago del Estero had been predicated more than a decade earlier. This movement of gradual abstraction and detachment from a material ground that had become little more than a "cartographic abstraction" and in which the author himself was "losing his foothold" is thus not unlike that of the subject of his analysis, the rural migrant.

Indeed, what Canal Feijóo ends up promoting as "un remedio" (a remedy) for the "mal del cuerpo" (142) (sickness of the body) that has befallen the region and its inhabitants is what Roberto Esposito, in his study of modern biopolitics, calls an "immunity shield."[26] Against the betrayal of the promises of modernity and progress, that shield reinstates the very oppositions on which those promises had been grounded, only now underwritten by national development and the state: reason/materiality, man/nature, and so forth. Just as it was for Di Lullo, for Canal Feijóo to let go of the promise of an intelligent colonization capable of providing mankind with a stable and sustainable immunity was in the end unimaginable. Here, I suggest, lies the limit that even the dissident modernity of twentieth-century Latin American regionalism hardly ever dares to cross.

Notes

1. Steffen, Crutzen, and McNeill "The Anthropocene," 614.
2. Escobar, "After Nature," 2–4.
3. For a recent, fascinating rereading of the cartographies of Argentina's national literary canon, see Demaría, *Buenos Aires y las provincias*. Ericka Beckman, in *Capital Fictions*, studies the interrelations between the region's absorption into capitalist modernity and the literary process in a Latin American perspective. Jennifer French has explored the critical dialogue Latin American literary regionalism maintains with British colonial fictions (Kipling, Conrad, Rider Haggard), one that responds to the real subsumption of the Latin American extractive frontiers into the financial networks of British imperialism (French, *Nature, Neo-Colonialism*).
4. Cited in Lascano, *El obraje*, 81.
5. Álvarez quoted in Brailovsky and Foguelman, *Memoria verde*, 180.
6. Canal Feijóo, *Ensayo sobre la expresión popular artística*, 9–11.
7. Canal Feijóo, *Ensayo sobre la expresión popular artística*, 11.
8. Canal Feijóo, *Ensayo sobre la expresión popular artística*, 11.
9. Canal Feijóo, *Ensayo sobre la expresión popular artística*, 16.
10. Canal Feijóo, *Ensayo sobre la expresión popular artística*, 14.
11. Ocampo, *La nación interior*, 74–75; Martínez, "La Brasa," 110–112.
12. Di Lullo, "Grandeza y decadencia de Santiago del Estero," 3–4.

13. Gorelik, "Mapas de identidad," 67.

14. Canal Feijóo, "El paisaje y el alma," 20.

15. Canal Feijóo, "El paisaje y el alma," 24.

16. Canal Feijóo, "El paisaje y el alma," 22.

17. Canal Feijóo, "El rapto del ferrocarril," 58.

18. Canal Feijóo, "El asalto a la selva," 60.

19. Canal Feijóo, "El asalto a la selva," 61.

20. Canal Feijóo, "El asalto a la selva," 60.

21. Canal Feijóo, "El asalto a la selva," 61–62.

22. Canal Feijóo, *De la estructura mediterránea argentina*, 20.

23. Canal Feijóo, *De la estructura mediterránea argentina*, 113.

24. Canal Feijóo, *De la estructura mediterránea argentina*, 142–143.

25. Canal Feijóo, *De la estructura mediterránea argentina*, 117.

26. Esposito, *Bíos*, 50.

References

Beckman, Ericka. *Capital Fictions: The Literature of Latin America's Export Age*. Minneapolis: University of Minnesota Press, 2012.

Brailovsky, Antonio E., and Dina Foguelman. *Memoria verde: Historia ecológica de la Argentina*. Buenos Aires: Sudamericana, 1991.

Canal Feijóo, Bernardo. "El asalto a la selva." *Ñan: Revista de Santiago* 2 (1934): 60–76.

———. *De la estructura mediterránea argentina*. Buenos Aires: López, 1948.

———. "El paisaje y el alma." *Ñan: Revista de Santiago* 1 (1932): 9–31.

———. "El rapto del ferrocarril," *Ñan: Revista de Santiago* 2 (1934): 58–60.

———. *Ensayo sobre la expresión popular artística en Santiago*. Buenos Aires: Impresora, 1937.

Crutzen, Paul J. and Eugene F. Stoermer. 2000. "Have We Entered the 'Anthropocene'?" *IGBP Global Change Newsletter* 41 (2000). http://www.igbp.net/news/opinion/opinion/haveweenteredtheanthropocene.5.d8b4c3c12bf3be638a8000578.html.

Demaría, Laura. *Buenos Aires y las provincias. Relatos para desarmar*. Rosario, Argentina: Beatriz Viterbo, 2014.

Di Lullo, Orestes. *El bosque sin leyenda: Ensayo económico-social*. Santiago del Estero, Argentina: Tipografía Arcuri y Caro, 1937.

———. "Grandeza y decadencia de Santiago del Estero." *Boletín del Museo de la Provincia* 10 (1959): 3–30.

Escobar, Arturo. "After Nature: Steps to an Antiessentialist Political Ecology." *Current Anthropology* 40, no. 1 (1999): 1–30.

Esposito, Roberto. *Bíos: Biopolitics and Philosophy*. Translation by Timothy Campbell. Minneapolis: University of Minnesota Press, 2008.

French, Jennifer. *Nature, Neo-Colonialism, and the Spanish American Regional Writers*. Lebanon, NH: Dartmouth College Press, 2005.

Gorelik, Adrián. "Mapas de identidad. La imaginación territorial en el ensayo de interpretación nacional: De Ezequiel Martínez Estrada a Bernardo Canal Feijóo." In *Miradas sobre Buenos Aires: Historia cultural y crítica urbana*, 17–68. Buenos Aires: Ariel, 2004.

Lascano, Raúl Alen. *El obraje*. Buenos Aires: Centro Editor de América Latina, 1972.

Martínez, Ana Teresa. "'La Brasa,' un 'precipitado del ambiente': Leer, escribir, publicar, entre la provincia y el pago." *Políticas de la Memoria* 14 (Summer 2013–2014): 110–117.

Ocampo, Beatriz. *La nación interior: Canal Feijóo, Di Lullo y los hermanos Wagner*. Buenos Aires: Antropofagía, 2004.

Rojas, Ricardo. *El país de la selva*. Buenos Aires: Taurus, 2001 [1907].

Steffen, Will, Paul J. Crutzen, and John McNeill. "The Anthropocene: Are Humans Now Overwhelming the Great Forces of Nature?" *Ambio* 36, no. 8 (2007): 614–621.

II

Latin America as a Site of Knowledge Production

Introduction to Section II

MARÍA DEL PILAR BLANCO AND JOANNA PAGE

The "contact zone" of the American hemisphere, to use Mary Louise Pratt's terminology, was both the space that witnessed the violence of colonization and also one of organic and cultural *mestizaje* in which European, indigenous, African, and Asian worldviews met and produced new, syncretic forms of interpreting natural phenomena, of treating bodies, and of understanding society. In his *Historia de las Indias*, Bartolomé de las Casas (1484–1566) gives an early glimpse of how the first Iberian colonizers translated indigenous curative practices. Describing the careers of the "Piachas," the "witches" or "doctors" who were "muy expertos en el arte mágica" (well versed in the magic arts), he notes,

> Escogen de los muchachos de 10 y 12 años, los que, por conjeturas que tienen, les parece que son por naturaleza inclinados y dispuestos para ser instruidos en el arte mágica, de la manera que nosotros conjeturamos por señales algunas ser nuestros muchachos hábiles, más que otros, para que estudien gramática y otras ciencias.[1]

> (They choose boys aged between ten and twelve who, in their judgment, seem to be naturally inclined and willing to be instructed in the magic arts, in the same way that we judge some of our boys to be more capable of studying grammar and other sciences than others.)

In this episode from *Historia de las Indias*, two versions of science—understood here as forms of knowledge that make sense and enact control of the natural world—face each other in an interesting comparative tension. In many ways, this section from de las Casas's text begins to dramatize the potential for productive relations between ethnic groups, but it also demonstrates the challenges in reaching mutual understanding that arise from encounters in the contact zone. Centuries later, Xavier Lozoya has suggested quite forcefully that there are two versions of the region's scientific history:

> The first follows, step by step, the history of scientific events in dominant Western cultures, primarily the European, to then research and prove that [in the Americas] . . . also existed wise and studious men who drank from the same fountain, were familiar with the same notes, or were inspired by the same muse; they were, however, always solitary, surrounded by heathens, idolaters, ignorant men, uncultured men, or empiricists, according to the terminology of the age. . . .
>
> The other version has yet to be written. . . . It is a version that begins by rescuing and disseminating indigenous cultural baggage and believes the findings of indigenous Latin American thought, both current and ancient, to be scientific knowledge.[2]

The first version of history as presented by Lozoya bears witness to deep-seated ethnic and cultural impasses, while the second seeks to uncover the hybrid intellectual heritage of the Latin American region. This heritage is one in which a *criollo* elite (with ties to Europe), indigenous people, and those hailing from Africa and Asia, in observing the particularities of their locations within the Americas, have for centuries developed myriad ways of understanding and managing the natural world surrounding them. The version of history to come, according to Lozoya, suspends questions of hegemony in order to make room for an ampler vision of scientific work.

What has so often been missing are ways of telling history syncretically and relationally, ways that might take account of the racial, economic, religious, and cultural differences of the subjects who have inhabited the Americas before and since Columbus's landfall. This reformed view of science has started to emerge in recent decades. As Miruna Achim observes, the search for "independent geniuses" has been supplanted with "more inclusive definitions of scientists and scientific practice, making room for sailors, bureaucrats, travelers, publishers, and merchants, and for activities like collecting, trading, legislating, and entertaining."[3] These new frameworks for writing scientific history in Latin America—narratives that must, importantly, include the stories of intermediary figures, as Carlos Quintero Toro also observes—will ultimately allow for a fuller understanding of how indigenous, imperial, and *criollo* actors converged across multiple sites and throughout periods that witnessed different waves of transformation in the relations between metropolitan centers and the colonies.[4]

Concurrent with these hugely important revisions to how we write history, we should nevertheless continue to interrogate how the Iberian Peninsula and its imperial nexus have been represented in histories of world science. Following historian Juan Pimentel Igea, we should question why science continues to be the ghost or specter (*fantasma*) of Hispanic culture or, put differently, why science figures as an indeterminate entity within peninsular and transatlantic collective memory and as marginal in global histories of science.[5] In the past three decades in particular, historians have increasingly sought to redress the virtual absence of Spain and Portugal from these accounts. This absence is surprising, given the cartographic advancements that led subjects from the peninsula to accomplish ground-breaking feats in navigation and exploration. This would place Spain and Portugal, one would assume, at the center of the history of modernity, which is also a history of the spread of capitalism.[6]

Nonetheless, a study of recent surveys in world science offers a rather sobering picture of how Spain and Portugal fared in the historical progress toward the Age of Enlightenment and, indeed, the scientific revolution. According to such historical outlines, scientific work appears to have stopped completely in the Iberian Peninsula during the Enlightenment. In volume 3 of his multivolume *Science and Technology in World History*, to name just one example, David Deming highlights the move of Europe's "intellectual center" to the north after the persecution of Galileo. Alongside this displacement in the seventeenth century, Deming notes, "Spain labored under a totalitarian government that waged war relentlessly against all culture, science, and learning. Even the mere possession of books made the owner an object of suspicion."[7] Global historians' treatments of science from this period have overwhelmingly focused on certain disciplines such as mathematics and physics, namely areas that saw great breakthroughs in Anglo-American and European countries.

The sciences that advanced within the Iberian world, such as cartography, have not until recently been given the same importance. What is more, such surveys obscure what Antonio Sánchez and Henrique Leitão call the paradigm of "Iberian science." This "new way of doing science," the authors contend, involved "the emergence of new ways of relating nature to its cognition"; "new ways to collect, accumulate, and circulate knowledge," as well as the perception of "novelty in the objects of nature and also in their descriptions (verbal, textual, graphical)."[8] In view of the transformations that took place within science in the Iberian world, what an account like Deming's leaves out is the kind of science that was flourishing as a result of the Iberians' navigation to the New World, their extensive experience within new ecosystems, and their contact with other worldviews.

Deming does put his finger on the delicate issue of the circulation of science writing. The problem of the printed word in a society so often stifled by religious persecution was compounded by an archival problem, one that historians like Richard Kagan, María M. Portuondo, and Jorge Cañizares-Esguerra have variously described as what Cañizares-Esguerra calls a "culture of *arcana imperii*, a tendency to keep details of the empire (maps, natural history reports, etc.) unpublished" in Habsburg Spain.[9] This estrangement of the historian from an archive, noted by Alexander von Humboldt in 1830, has long affected the chronicling of Iberian science in the peninsula and the Americas, as Cañizares-Esguerra has remarked.[10] The late colonial period, which was followed by emancipation and the establishment of new Latin American nations, coincides (as we see in Section III in this volume) with the Age of Enlightenment and, with it, an intensification of naturalist thinking across the Western world. During this period in Latin America, scientific knowledge served political aims; to understand new national spaces from the inside was a pathway out of colonization and into emancipation. What is important for contemporary readers of Latin America's history of science is to consider the continuities between the colonial period (and the relations forged within it) and the postindependence landscape in order to gain a fuller comprehension of the many ways in which science emerged out of the New World experience, evolving from the first scenes of transatlantic contact to the present.

The chapters in this section illustrate these continuities while also offering a corrective to repeated omissions of the Latin American colonial archive in histories of science. Analyzing a range of texts from the colonial period to the twentieth century, Heidi Scott, Yarí Pérez Marín, and Edward Chauca introduce a range of figures who discovered and disseminated forms of knowledge resulting from their experiences of living in regions exceptional in geography and culture. Scott's chapter on colonial mining in the Andes presents a nuanced reading of how politics, religion, and intercultural relations contribute to a specialized understanding of Andean nature; her work is founded on a new attention to archives that reveal the importance of everyday practice and experience in the formation of theories of nature that underpinned the logics of extraction.

Moving from studies of landscape to discourses on bodies, Pérez Marín's reading of Agustín Farfán's *Tractado breue de anothomia y chirvgia* (1579) demonstrates how medical writing in this period reflects the cultural heterogeneity of a community, displaying rich and complex attitudes toward compassionate health treatment. Chauca's chapter explores medical writing from the twentieth century, focusing on the figure of the Peruvian Hermilio Valdizán (1885–1929), a physician and psychiatrist who spent his career researching local and indigenous science in the country (himself delving into an extensive archive) in order to mount a direct challenge to assumptions about the mental capacities of indigenous Peruvians that had historically been made by European commentators.

The contributions by Pérez Marín and Chauca to this volume speak directly to what Marcos Cueto and Steven Palmer call the "medical pluralism" that has characterized Latin American science across centuries, the at times troubled yet gradual integration of "an interconnected complex" that is "both conflictual and complementary."[11] Read together, the three chapters in this second section offer a fuller and fascinating historical and cultural sense of how scientific observation has functioned within the Latin American region. They describe diverse ways in which knowledge has been produced through everyday intercultural interactions in colonial and postcolonial settings, stressing the importance of reading Latin American scientific practice as polycentric and relational.

Notes

1. de las Casas, "Capítulo CCXLV," 535–536.
2. Lozoya, "Natural History and Herbal Medicine," 30.
3. Achim, "Science in Translation," 107.
4. Quintero Toro, "¿En qué anda la historia de la ciencia?," 156–157.
5. Pimentel Igea, "El edificio de Villanueva."
6. See Dussel, "Europa, modernidad y eurocentrismo."
7. Deming, *Black Death*, 203.
8. Sánchez and Leitão, "Introduction," 110.
9. Cañizares-Esguerra, *Nature, Empire, and Nation*, 23.
10. Cañizares-Esguerra writes, "[Humboldt's] *Examen critique de l'histoire de la géographie du nouveau continent (Critical Examination of the History of Geography of the New Continent; 1836–39)* was a massive, five-volume history of the origins of the early modern Iberian expansion to the New World. His reconstruction of the history of late medieval and early modern geography was possible, he remarked, only because new archival sources had recently been made available by the Spanish scholar Martín Fernández de Navarrete. The sources Fernández de Navarrete published between 1825 and 1837 had in fact been collected some fifty years earlier by the Valencian Juan Bautista Muñoz.

Like Humboldt, Muñoz thought that the answer to a well-entrenched Enlightenment tradition that dismissed Spanish contributions to early modern knowledge lay in the archives" (Cañizares-Esguerra, Introduction, 2).

11. Cueto and Palmer, *Medicine and Public Health in Latin America*, 6.

References

Achim, Miruna. "Science in Translation: The Commerce of Facts and Artifacts in the Transatlantic Spanish World," *Journal of Spanish Cultural Studies* 8, no. 2 (July 2007): 107–115.

Cañizares-Esguerra, Jorge. Introduction to *Science in the Spanish and Portuguese Empires, 1500–1800*, edited by Daniela Bleichmar, Paula De Vos, Kristin Huffine, and Kevin Sheehan, 1–5. Stanford, CA: Stanford University Press, 2009.

———. *Nature, Empire, and Nation: Explorations of the History of Science in the Iberian World*. Stanford, CA: Stanford University Press, 2006.

Cueto, Marcus, and Steven Palmer. *Medicine and Public Health in Latin America: A History*. Cambridge, England: Cambridge University Press, 2015.

de las Casas, Bartolomé. "Capítulo CCXLV." In *La Historia de las Indias*, 535–536. Madrid: M. Ginesta, 1875–1876.

Deming, David. *The Black Death, the Renaissance, the Reformation, and the Scientific Revolution*. Vol. 3 of Deming, *Science and Technology in World History*. Jefferson, NC: McFarland, 2012.

Dussel, Enrique. "Europa, modernidad y eurocentrismo." In *La colonialidad del saber: Eurocentrismo y ciencias sociales. Perspectivas latinoamericanas*, edited by Edgardo Lander, 41–53. Buenos Aires: CLACSO (Consejo Latinoamericano de Ciencias Sociales), 2002.

Lozoya, Xavier. "Natural History and Herbal Medicine in Sixteenth-Century America." In *Science in Latin America: A History*, edited by Juan José Saldaña, 29–50. Austin: University of Texas Press, 2006.

Pimentel Igea, Juan. "El edificio de Villanueva y el fantasma de la ciencia española." In *El Museo del Prado y la política borbónica de instituciones culturales*, edited by Joaquín Álvarez Barrientos. Madrid: Museo del Prado, 2019.

Quintero Toro, Camilo. "¿En qué anda la historia de la ciencia y el imperialismo? Saberes locales, dinámicas colonials y el papel de los Estados Unidos en la ciencia en el siglo XX." *Historia Crítica* 31 (January–June 2006): 151–172.

Sánchez, Antonio, and Henrique Leitão. "Introduction: Revisiting Early Modern Iberian Science, from the Fifteenth to the Seventeenth Centuries," *Early Science and Medicine* 21, nos. 2–3 (June 2016): 107–112.

4

Empathy, Patients' Needs, and Therapeutic Innovation in the Medical Literature of Early Viceregal Mexico

YARÍ PÉREZ MARÍN

The *Tractado breue de Anothomia y Chirvgia* (1579) may be an unusual starting point for a discussion of innovation in the context of the history of science and medicine in early modern Latin America. Following Francisco Bravo's *Opera Medicinalia* (1570) and Alonso López de Hinojosos's *Svmma y Recopilacion de Chirvgia* (1578), Agustín Farfán's text misses out on watershed status in the history books, as neither the first medical book proper to be printed in the New World nor the first one to appear in a European vernacular language. Initially sparse in its discussion of indigenous medicine, the text has not generated as much interest among modern scholars as the work of contemporaries such as Francisco Hernández and Juan de Cárdenas, whose *Primera parte de los problemas y secretos marauillosos de las Indias* appeared a decade later, in 1591. And yet, when considered in the context of the first century of New World imprints, Farfán's project stands out as one of the few texts on secular subject matter from that group to have gone into multiple editions.[1]

While only a handful of copies survive today, the *Tractado* seems to have been a bestseller, published no fewer than three times in Mexico City between 1579 and 1610 and continuing to be a source referenced well into the seventeenth century.[2] If it is true that medicine constitutes not just "a varied form of cultural practice and production" in the early modern Hispanic world but also "a significant matrix for the intersection of a wide range of cultural phenomena (political, literary, religious, or otherwise)," as has recently been argued,[3] then the comparative analysis of the *Tractado*'s transformation from one in-demand version to another that was at least as successful, given the need for a third printing, makes a compelling case for treating it as a gateway to the sensibilities of the period readers with whom it resonated.[4]

Around forty-six years old when the first edition was published, Farfán announced in the first sentence of the preface to his *Tractado breue de Anothomia y Chirvgia, y de algvnas enfermedades, que mas comunmente suelen hauer en esta Nueua España* (1579) that he had been practicing "the science of medicine and surgery" for twenty-seven years already while also proudly declaring himself "a graduate of this illustrious University of Mexico,"[5] an institution that had just begun to offer medicine as a subject. Based on a comparison of archival documents, Germán Somolinos d'Ardois concludes that Farfán's decision to move to New Spain with his wife and daughter in 1557 was probably the result of having relatives who had already settled there. He would go on to have two more children once in Mexico but was a widower by 1568, when he joined the Augustinians, taking his vows a year later and changing his name from Pedro García to Agustín.[6]

Although written in Spanish and, according to its author, offering a "modo de curar claro, e inteligible. Para que todos los que quisieren y tuvieren necesidad se aprovechen de ella [cirugía]" (a clear and accessible method. So that anyone who wishes or has need [of surgical knowledge] can benefit from it),[7] the first edition of the *Tractado*, nonetheless, is not exactly a book for the everyday person. Viceroy Martín Enríquez's approving statement explains that the material had been vetted by "personas graves y doctas en la facultad" (important and serious authorities on the subject), as is borne out from the endorsing signatures. Bravo calls it a work of "much erudition and study."[8] The sentiment is echoed by Juan de la Fuente, the first professor to teach medicine at the University of Mexico and another of the authorities called upon to review the book; Fuente deems it "very erudite." "It has nothing that would go against Catholic doctrine," declares Fray Alonso de la Vera Cruz, the first professor to teach philosophy at the university. The elegant sonnet written in praise of the author by Francisco de Solís, son of one of the original *conquistadores* of Mexico, provides further evidence of Farfán being rather well positioned among New Spain's elite.[9] The first version of the *Tractado* did have features that enhanced its accessibility beyond its choice of language; a well-organized table of contents runs the length of two full folios,[10] and an extremely detailed alphabetized introductory table, at nine and a half folios, is searchable by illness, remedy, or body part. Despite these tools, a minimal level of expertise would have been required to make sense of the information and put it to good use.

Farfán's discussion, particularly in the first half of the book, is strongly invested in a nuanced critique of academic sources. Findings by Galen, Ibn Sīnā, Al-Zahrawi, Hippocrates, Rhazes, Giovanni da Vigo, Guy de Chauliac, Arnau de Vilanova, and others are meticulously explained or contested and continuously compared with the author's own experience as a practitioner. In addition, the *Tractado* includes three medical illustrations, a feature that thus far has not been remarked upon by historians of medicine or scholars of early modern visual culture. A small character of a cross is used twice in this manner, first on folio 22r to illustrate the placement of overlapping abdominal muscles and again in folio 151v, where it is shown alongside an uppercase tau, this time to explain the proper shape of an incision to expose the brain. The third instance is that of a tiny image of two juxtaposed half moons facing in opposite directions embedded within a section in which the author shows surgeons how nerves linking the brain and the eyes connect to one another (figure 4.1).[11] This level of detail on incisions and intraocular structures assumes a degree of medical knowledge and surgical dexterity that is unlikely to have been of much use to a reader who was not an expert medical practitioner.

There are signs already in the first edition that Farfán was concerned the specialized nature of his writing would not achieve his often-stated goal of making the book useful to a wider readership. He apologetically writes,

No es posible menos sino que en algunos capítulos habemos de ser algo largos, por requerirlo así la materia de lo que tratamos, y porque no se deje de decir en ellos lo que forzosamente conviene. Por esta causa lo he sido en los pasados, aunque mirándolo bien, todo ha sido menester, porque si abreviásemos, podría parecer oscura la materia.[12]

(It is unavoidable that in some chapters we have been somewhat lengthy because the matter being discussed thus requires it, and so as not to leave unsaid what must be mentioned. It is for this reason that I have been thus [prolix], although in retrospect, all has been necessary, because if we were to shorten it, the matter could remain unclear.)

But ultimately Farfán was unconvinced by his own justification for including exhaustive explanations and opted for a radical restructuring of the text in its second iteration. Whereas the 1579 version had been a quarto edition of 274

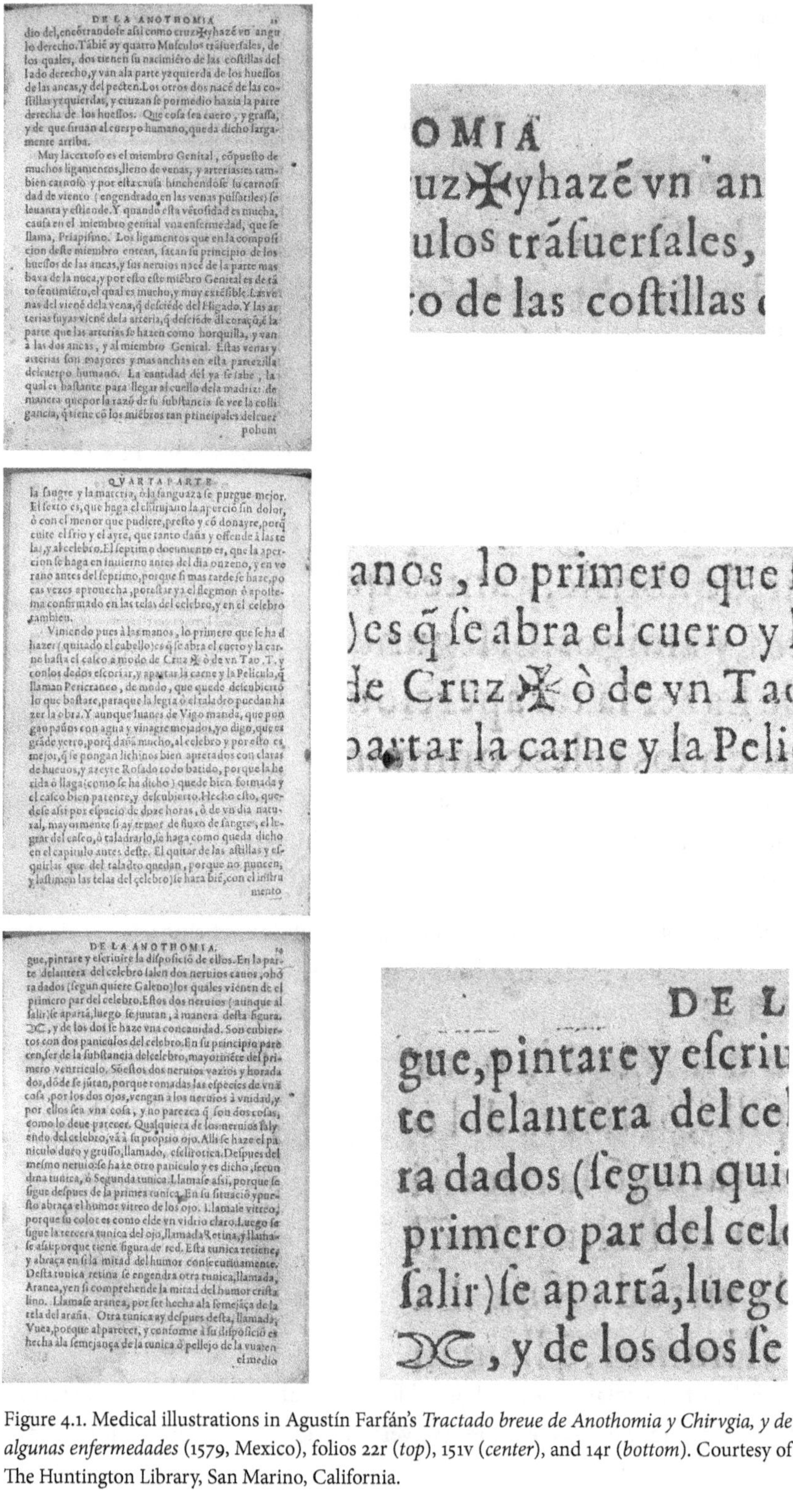

Figure 4.1. Medical illustrations in Agustín Farfán's *Tractado breue de Anothomia y Chirvgia, y de algunas enfermedades* (1579, Mexico), folios 22r (*top*), 151v (*center*), and 14r (*bottom*). Courtesy of The Huntington Library, San Marino, California.

folios, the 1592 version, also a quarto edition, was roughly 20 per cent longer, at 353 folios. The section on basic human anatomy required for surgical practice, which had gone first in the original sequence, was placed last, with Farfán choosing to begin instead with a treatise on digestive problems based on the "common illnesses" section of the first edition. The more informal tone and self-referential style that had progressively emerged in the 1579 text became a core feature of Farfán's prose in the 1592 version:

> De estos males de estómago no se quejan los viejos solos, y los no muy sanos, sino los mozos y los que parecen más robustos. Las causas de esta enfermedad son muchas, y la más común y mayor es el mal orden y mala regla que tenemos en el comer y beber, y si no me creyeren, díganme (por amor de Dios) ¿qué hombre hay, y yo el primero, que deje de comer lo que le sabe bien, aunque tenga experiencia que le hace mal? Cierto que son muy pocos. . . . ¿De qué nos quejamos? Pues, nos mata la gula.[13]

> (These stomach ailments afflict not just old and unhealthy men but also young ones and those who seem more robust. There are many causes for this sickness, and the most common is our disorder and lack of discipline in what we eat and drink, and if you do not believe me, tell me (for the love of God) what man, and I first among them, will refuse to eat what he fancies, despite having experienced its ill effects? Truly, very few. . . . Why do we complain? It is our own gluttony that kills us.)

The structural overhaul reorganized the material into five treatises: on stomach pain, menstrual irregularities, and problems such as pox, rashes, and colds (12 chapters); on *bubas*, skin afflictions, headaches, different kinds of pain, alcoholism, and paralysis (15 chapters); on fevers (8 chapters); on wounds (fresh and old appearing together this time) in various parts of the body including the face (12 chapters); and lastly, on anatomy (13 chapters). As Farfán himself notes, "Aunque otra vez impresa [la obra], sale la segunda [edición] reformada y añadida que es casi de nuevo en todo" (Despite being another printing [of the work], the second edition is refashioned and expanded so as to be almost completely new).[14]

Indeed, the rewriting is more extensive than first appears, not just adding material but also removing and condensing information to make space for new content. Medical authorities like Galen and Ibn Sīnā are still adduced, but the critique of finer points linked to their ideas is reduced or eliminated, as is the allusion to Farfán's younger days in Spain studying medicine.[15] Gone also are the two nonalphabetic cross symbols as well as the half moons, replaced with reversed parentheses and periods.[16] In the text, rebranded the *Tractado brebe de Medicina, y de todas las enfermedades*, attention turns to a plethora of indigenous remedies. The increase in scale alone is startling, jumping from eleven mentions of Nahuatl medical products in the first edition to fifty in the second.[17] Some sections are entirely new, such as the passages on the *cocolmeca* plant (used to treat a range of problems, from humoral obstructions to genital abscesses), ground armadillo tails (useful to treat earaches), and iguana bezoars known as *quacuetzpalintechutli* (used to treat pain and constipation).[18] In other instances, an allusion to a particularly effective indigenous remedy replaces what in the first edition is a lengthy discussion on a less reliable cure of European origin, as in the treatment of hemorrhoids.

On the surface, the ease with which Farfán refers to Francisco Hernández's work, leaving no doubt that he was familiar with his colleague's research, could explain the new interest in indigenous medicine as merely the next chapter in an ongoing, closed conversation between European medical voices. But Farfán's ability to supplement Hernández, clarifying more than once that the plants his colleague was familiar with also went by other Nahuatl names and including terms not used by Hernández, suggests that Farfán's knowledge of the subject was not limited to that of a reader. It would be plausible to explain the shift as the result of increased opportunities for direct contact between Farfán and indigenous practitioners unacknowledged in the text and from whom he would have learned about the medicinal properties of local products. These unnamed sources would stand in contrast to Hernández, who is given special attention and whom he credits with "having found many medicinal plants in this New Spain."[19] Yet, the frequency with which the commercial element enters into the second edition suggests a more commonplace mode of engagement by which access to indigenous medical knowledge could have occurred through multiple channels.

Despite the silence of earlier Mexican imprints on local remedies, by 1592 the absence of a favorable valuation of their efficacy in print perhaps had more to do with the social stratification of early colonial Mexico during the last third of the sixteenth century and the reluctance on the part of Spanish settlers to engage with indigenous spaces more broadly than with an unfavorable assessment of indigenous medicine on scientific grounds. It is telling that the inclusion of local products referred to in Nahuatl in the 1592 *Tractado* is often immediately followed by a mention of the indigenous marketplace: the "xicamilla que venden en los tiangues" (the croton weed that is sold in the *tianguis*), the *matlatin* flower, "que venden en los tiangues" (which is sold in the *tianguis*), the *etzpatli* (Jesuits' Tea) "y la hay en los tiangues" (and they have it in the *tianguis*), or in the case of remedies that called for both fresh and dried versions of an ingredient, like the cure for severe headaches that required both *picietl* (tobacco leaves) and "polvos de estornudar" (snuff), Farfán reassuringly tells his readers, "todo se vende en los tiangues" (everything is for sale in the *tianguis*).[20]

In many ways, Farfán's voice is similar to that of other period medical authors, and his claims are in keeping with emerging attitudes that stressed the importance of personal success with a given remedy in one's practice over information drawn from academic sources. However, the way he positions himself in relation to his subject matter in the second edition importantly pushes past the by-then familiar trope of a discovery claim demanding reward and recognition for the judicious European endorsement of indigenous praxis. *Tianguis*, local market events, had been a fixture of Aztec societies in pre-Hispanic times and were fundamental to the survival of the community. In his lifetime, Farfán would have had access to three main such spaces: the Tianguis of Mexico, located south of the city; the Tianguis of Santiago Tlatelolco, which "lay adjacent to that *altepetl's tecpan*, or palace of the indigenous government, and the monastery"; and the Tianguis of San Hipólito, which "sat at the western edge of the city, north of today's Alameda park."[21] The Tianguis of Mexico, "enormous, packed with vendors and buyers," as Barbara Mundy describes it, "was the commercial hub . . . [and] the mainstay of economic life of the city's indigenous people."[22] While earlier projects on Aztec medicine such as the Badianus Codex and Hernández's manuscripts include information about where to harvest plants or how to cultivate them, Farfan's context is decidedly urban, reflecting a thriving local market economy where medicinal simples were not only routinely bought and sold but already marketed as medicines.[23]

From an epistemological standpoint, the *Tractado* stands in contrast to narratives that continued to insist on framing an engagement with non-Western knowledge as interventionist and revelatory. "Cierto se debe a v.m. mucho, porque nos descubrió tan gran tesoro como éste" (Truly we owe your lordship a great deal, having revealed to us a treasure as great as this one), purportedly declared an admirer from Peru to Nicolás Monardes in a letter sent to him in Seville thanking the doctor for at last revealing the hidden usefulness of Andean bezoars to the people living in the region.[24] Monardes, whose *Historia medicinal* (1565) became one of the most widely read texts on American *materia medica* in Europe in its day, enjoying twenty-five printings in multiple languages in the sixteenth century alone, explicitly compares his contribution to that of a fellow Genoese, Columbus, in the very first sentence of that work. He goes on to claim that as "incredible" and profitable as was the unearthing of New World precious metals and stones, it paled in comparison to the true treasure trove of medicinal products to be had that "exceden mucho en valor & precio a todo lo susodicho" (far exceed in value and price everything else), as could be gleaned from his careful review of informants' accounts and the experiments he had performed with imported ingredients "en muchas & diversas personas" (on many and diverse persons) while in Spain.[25]

Meanwhile, Farfán's insistence on the relevance of *tianguis* seems more invested in coaxing the Spaniards of Mexico, as they were known to refer to themselves at the time, to enter into indigenous spaces where highly effective medical treatments were readily known and for sale. In fact, the mention of alternate modes of treatment in the text is often justified with the argument that some of his readers lived far from metropolitan centers like Mexico City, privileged locations because of their access to European-trained practitioners like Farfán and their close proximity to large-scale and well-stocked indigenous marketplaces that brought together an offer of natural products not circumscribed to any one area.

The critical eye dictating the choice of what to include or replace in the second version of the *Tractado*, be it about indigenous medicines or European forms of treatment, endeavored to reflect readers' needs, ceding far more textual space than before to the interaction with patients at the moment of diagnosis. Significant attention is devoted to follow-up care, an element that was not a defining trait in medieval surgical exempla in which the quick confirmation of a desired outcome was usually the end of the discussion. Rather than individual cases, the chapters are often presented as first-person reflections on how patients reacted to different medical treatments. The text partakes of what Michael Solomon calls the "contingent utility" of the vernacular recipe collection insofar as it "captured the imagination of the sickly reader by detailing how common and seemingly inert matter could acquire astonishing medical potency."[26] But it also provided a different kind of reassurance by opening a space of intimacy and complicity between the university-endorsed, experienced author (who could write in Latin but chose not to) and the savvy reader who employed a variety of strategies to remain healthy and could be just as critical as Farfán of local medical practitioners.[27]

The *Tractado* is set in opposition to both academic medicine on the one hand, through its choice of mundane subject matter and selective corrections of learned authorities, and quackery on the other, in its strong and repeated denunciations of untrained doctors passing for physicians in New Spain who profited from others' misery for monetary gain, men whom the author refers to as "matasanos" (slayers of the healthy) and "carniceros" (butchers).[28] Under the protective umbrella of Farfán's Christian charity and humility in his role as an Augustinian friar professing to seek only the common good, the book is clear in his willingness to tackle problems for which patients may have been reluctant to publicly seek out medical care despite being commonplace afflictions. Farfán blurs the line between doctor and patient, discussing the importance of empathy to determine a sufferer's level of discomfort and making himself a test subject for unremarkable ailments such as toothaches or constipation: "hombres fidedignos" (trustworthy men) had profited from the benefits of a product like the *quacuetzpalintechutli*, he writes, "y yo también" (and I too).[29]

The kind of medicine that most interests Farfán, especially in the second edition of the *Tractado*, does not involve complicated medical cases related to anatomy or physiology. Unlike many of his contemporaries, including those with a lower degree of academic training like López de Hinojosos, there is no noticeable steering of the discussion toward issues that would have enhanced his professional prestige or brought him fame. Instead, the text makes space for miscellaneous and ordinary problems faced by people in his community. Unseemly facial scars, chronic pain, hemorrhoids, hair loss, foul breath, erectile dysfunction, that is, the private physical struggles that in genres like the picaresque novel had been fodder for entertainment, are taken seriously by Farfán, resulting in a form of medical writing that is as much social commentary as it is science.

Sixteenth-century readers of a work like *Lazarillo de Tormes* (1554), a literary landmark in Spain's early modern canon, may have momentarily taken pity on its fictional *pícaro* protagonist when he recounts how his face had been "smashed in so hard with a clay jar that its pieces had gone into [his] face, tearing it in many parts, and breaking [his] teeth, without which [he is] still today,"[30] but the narrative still coaxed the reader into laughing alongside the frugal blind man who had cunningly figured out he was being cheated out of wine by his young guide and had punished him so severely. A text like the *Tractado* charts the movement from this Renaissance mindset into what increasingly comes into focus as the new set of sensibilities of the Hispanic Golden Age. It would ask readers to consider the social implications such physical damage would have had for a young man living thereafter with a face full of scars and ever-aching, toothless gums, at even greater risk than before of being ostracized in a society increasingly invested in physical appearance and physiognomic readings.[31]

Farfán's integrated approach to medical care speaks of a broader concern for health understood as well-being rather than merely bodily functions, offering a glimpse of the personal and intimate spheres then about to be explored with the arrival of the modern novel. "Dejando pues cuestiones de medicina, porque no escribo (como he dicho otras veces) para médicos" (Leaving behind, thus, questions of medicine, given that, as I have expressed before, I do not write for doctors),[32] the *Tractado* stands out in the context of early modern New World writing as a work that compassionately takes responsibility for the "griefs and groans" of others,[33] insisting on the importance of empathy for their suffering and of organizing information for the common good as central considerations in the advancement of scientific knowledge.

Notes

1. López de Hinojosos's surgical treatise is another example of this phenomenon, with his *Svmma* appearing in a second, expanded edition in 1595.

2. Each of the editions of the *Tractado* currently accounted for was handled by a different printer: Antonio Ricardo in 1579, Pedro Ocharte in 1592, and Jerónimo Balli in 1610, six years after the author's death. There are mentions dating as far back as the nineteenth century of yet another edition of the text from 1604, but it is possible this is an instance of an unchecked, repeated error, as there have been no confirmed sightings of such an edition on the part of modern scholars. Farfán is also mentioned in later medical texts such as Francisco Ximénez's *Qvatro Libros* (1615), which was presented as a translation of Francisco Hernández's work on the flora and fauna of New Spain but drew on earlier Mexican imprints such as the *Tractado* as well, even if Ximénez did not always acknowledge that he did so.

3. Slater, Pardo-Tomás, and Maríaluz López-Terrada, Introduction, 2.

4. Pardo-Tomás also notes as odd that the *Tractado* has not garnered more attention in the context of early modern cultural studies, given the work's editorial success at the time. He also highlights an important element to consider when assessing its popularity, which is the role played by the Augustinians in promoting the text's dissemination even after the author's death, given their support for the 1610 edition. See Pardo-Tomás, "Pluralismo médico," 764, 766–767.

5. Farfán, *Tractado breue de Anothomia y Chirvgia*, 1r, title page. This quotation is from an unnumbered folio that precedes the first numbered folio. Citing a document showing his involvement alongside Juan de la Fuente in the examination of a medical student at the University of Seville in the 1550s, Somolinos d'Ardois thought Farfán was a graduate of the University of Seville who had then spent time at the University of Alcalá ("Relación alfabética," 220). Farfán alludes in passing to his years in Alcalá (162v, 1579),

but he does not explicitly state that he obtained a degree there. Juan Comas addresses the inconsistencies in the author's biography (including the possibility that it was a different Farfán who arrived in New Spain in 1557), among them his assertion of being a graduate of the University of Mexico. Comas notes that medicine would not be taught at that institution officially until 1579, the year of the *Tractado*'s publication (Comas, "La influencia indígena," 108). This would support the view that Farfán was assessed by way of an exam since he was awarded the degree in 1567 and briefly served as *protomédico* in 1568 (Somolinos d'Ardois, "Relación alfabética," 221–222).

6. Somolinos d'Ardois, "Relación alfabética," 220–222.

7. Farfán, *Tractado breue de anothomia y chirvrgia*, 1r. Quotations from period sources have been adapted for modern Spanish grammar and syntax.

8. The endorsements quoted here from Bravo and others are from unnumbered folios that precede the first numbered folio in Farfán, *Tractado breue de Anothomia y Chirvgia* (1579).

9. Fray Alonso de la Vera Cruz was a leading intellectual figure in New Spain and the author of some of the earliest books on philosophy printed in the Americas, including the *Dialectica resolutio cum textu Aristotelis* (1554). It is almost certain that the author of the sonnet was one of Francisco de Solís's fourteen sons. Solís senior had been a companion of Cortés and held several major *encomiendas* south of the city. The parentage is supported by the poet's need to clarify that he is "Solís Mexicano," seeing as Francisco was the name of his father and his grandfather (Himmerich y Valencia, *Encomenderos of New Spain*, 244).

10. The book is divided into six sections: human anatomy (25 chapters); abscesses, inflammation, and tumors (20 chapters); tumors occurring specifically in the head, nose, mouth, testicles, breasts, and joints (12 chapters); fresh wounds (24 chapters), ulcers, and "bubas" (taken to be syphilis by some scholars today) (22 chapters); and illnesses "common in the land," the majority of which entail gastrointestinal problems (13 chapters).

11. Bravo's *Opera Medicinalia* (1570) features an image of the veins in the thorax copied from Vesalius's *Venesection Letter* (1539), which incorrectly reverses the direction of the vena cava, and also a botanical illustration of sarsaparilla. The second edition of López de Hinojosos's *Svmma y Recopilacion* (1595) would include an anatomical illustration of the intestines and the kidneys.

12. Farfán, *Tractado breue de Anothomia y Chirvgia*, 67r-67v.

13. Farfán, *Tractado brebe de Medicina*, 2r.

14. Farfán, *Tractado brebe de Medicina*. This quotation is from the unnumbered folio that precedes the first numbered folio.

15. Farfán mentions participating in demonstrations of surgical techniques in Alcalá de Henares in the first edition of the *Tractado*.

16. The structure is represented as ".)(." in the 1592 edition and changed yet again into a capital "X" in the 1610 edition.

17. In his "Nombres indígenas de plantas americanas," Manuel Galeote makes this numerical observation in his linguistic analysis of the use of Hispanicized Nahuatl terminology related to plants in the *Tractado*. Galeote's research findings helpfully chart

where these mentions occur in both texts and provides contextual information clarifying meaning in relation to similar words found in period sources.

18. For a fuller discussion of Farfán's use of indigenous medicine in the second edition of the *Tractado* specifically, see Comas, "La influencia indígena," 109–124.

19. Farfán, *Tractado brebe de Medicina*, 206r. For an in-depth analysis of the reach of Hernández's ideas in early modern botanical and medical texts more broadly, see López Piñero and Pardo Tomás, *La influencia de Francisco Hernández*.

20. Farfán, *Tractado brebe de Medicina*, 210v, 211r, 211r, 219r.

21. Mundy, *Death of Aztec Tenochtitlan*, 85.

22. Mundy, *Death of Aztec Tenochtitlan*, 88.

23. For a discussion that considers the significance of the Badianus Codex as a precursor to later medical sources, including print texts like Farfán's, see Viesca Treviño, "El Códice de la Cruz-Badiano."

24. In Monardes, *Herbolaria de Indias* [*Historia medicinal*], 197.

25. Monardes, *Dos libros* [*Historia medicinal*], 4r, 3v-4r.

26. Solomon, *Fictions of Well-Being*, 80.

27. Mauricio Sánchez-Menchero's research on sixteenth-century correspondence written by travelers and settlers in New Spain as they discuss illness and their attempts to seek medical care supports this characterization (Sánchez-Menchero, "From Where They Are Now").

28. Farfán, *Tractado brebe de Medicina*, 82r, 82v.

29. Farfán, *Tractado brebe de Medicina*, 238r.

30. When the blind man figures out that Lázaro has made a hole in the wine jug so that some of the liquid could pour into his mouth when sitting between the man's legs, his master retaliates by attacking the unsuspecting boy just as he prepared to drink. In Lázaro's words, "[Fue] el jarrazo tan grande, que los pedazos dél se me metieron por la cara, rompiéndomela por muchas partes, y me quebró los dientes, sin los cuales hasta hoy día me quedé" (*Lazarillo de Tormes*, 33). The tale of how Lázaro came to have a scarred face and its link to his desire for wine becomes a humorous recurring topic of conversation between the blind man and other characters in the novel: "Luego contaba cuántas veces me había descalabrado y arpado la cara. . . . Y reían mucho los que me lavaban, con esto, aunque yo renegaba" (He would later tell the story of how many times he had knocked me over the head and torn my face. . . . And [the innkeepers] who tended to me would laugh a great deal with this, though I fumed) (*Lazarillo de Tormes*, 43).

31. On the subject of how facial scars could be interpreted as evidence of criminal behavior, thus affecting legal proceedings, see Skinner, "Marking the Face," 199.

32. Farfán, *Tractado brebe de Medicina*, 33v.

33. The famous play on words in the opening lines of Miguel de Cervantes's *El ingenioso hidalgo Don Quixote de la Mancha*, with "duelos y quebrantos" being also the name of a modest dish made of eggs and remnants of animal parts.

References

Bravo, Francisco. *Opera Medicinalia, in quibus quam plurima extant scitu medico necessaria*. Mexico City: Pedro Ocharte, 1570.

Cárdenas, Juan de. *Primera parte de los problemas, y secretos marauillosos de las Indias*. Mexico City: Pedro Ocharte, 1591.

Cervantes, Miguel de. *Don Quijote de la Mancha*. Edited by Martín de Riquer. Barcelona: Planeta, 1994.

Comas, Juan. "La influencia indígena en la medicina hipocrática en la Nueva España del siglo XVI." In *El mestizaje cultural y la medicina novohispana del siglo XVI*, edited by José L. Fresquet Febrer and José M. López Piñero, 91–127. Valencia: Instituto de Estudios Documentales e Históricos sobre la Ciencia, Universitat de València, CSIC (Consejo Superior de Investigaciones Científicas), 1995.

Farfán, Agustín. *Tractado breue de Anothomia y Chirvgia, y de algvnas enfermedades, que mas comunmente suelen hauer en esta Nueua España*. Mexico City: Antonio Ricardo, 1579.

———. *Tractado brebe de Medicina, y de todas las enfermedades*. Mexico City: Pedro Ocharte, 1592.

Galeote, Manuel. "Nombres indígenas de plantas americanas en los tratados científicos de Fray Agustín Farfán." *Boletín de Filología de la Universidad de Chile* 36 (1997): 119–161.

Himmerich y Valencia, Robert. *The Encomenderos of New Spain, 1521–1555*. Austin: University of Texas Press, 1991.

Monardes, Nicolás. *Dos libros. El vno trata de todas las cosas que traen de nuestras Indias Occidentales, que siruen al vso de Medicina, y como se ha de vsar dela rayz del Mechoacan, purga excelentissima. El otro libro, trata de dos medicinas marauillosas que son contra todo Veneno, la piedra Bezaar, y la yerua Escuerçonera. Con la cura delos Venenados*. Seville: Sebastián Trujillo, 1565.

———. *Herbolaria de Indias*. Edited by Ernesto Denot and Nora Satanowsky. Mexico City: Turner, 1990.

Lazarillo de Tormes. Edited by Francisco Rico. Madrid: Cátedra, 2006.

López de Hinojosos, Alonso. *Svmma, y Recopilacion de Chirvgia, con vn Arte para sangrar muy vtil y prouechosa*. Mexico City: Antonio Ricardo, 1578.

López Piñero, José M., and José Pardo Tomás. *La influencia de Francisco Hernández (1515–1587) en la constitución de la botánica y la materia médica modernas*. Valencia: Instituto de Estudios Documentales e Históricos sobre la Ciencia, Universitat de València, CSIC (Consejo Superior de Investigaciones Científicas), 1996.

Mundy, Barbara E. *The Death of Aztec Tenochtitlan, the Life of Mexico City*. Austin: University of Texas Press, 2015.

Pardo-Tomás, José. "Pluralismo médico y medicina de la conversión: Fray Agustín Farfán y los agustinos en Nueva España, 1533-1610." *Hispania* 74, no. 248 (2014): 749–776.

Sánchez-Menchero, Mauricio. "'From Where They Are Now to Whence They Came From': News about Health and Disease in New Spain (1550–1615)." In *Medical Cultures of the Early Modern Spanish Empire*, edited by John Slater, Maríaluz López-Terrada, and José Pardo-Tomás, 91–105. Surrey, England: Ashgate, 2014.

Skinner, Patricia. "Marking the Face, Curing the Soul? Reading the Disfigurement of Women in the Later Middle Ages." In *Medicine, Religions, and Gender in Medieval Culture*, edited by Naoë Kukita Yoshikawa, 181–201. Suffolk, England: Boydell and Brewer, 2015.

Slater, John, José Pardo-Tomás, and Maríaluz López-Terrada. Introduction to *Medical Cultures of the Early Modern Spanish Empire*, edited by Slater, López-Terrada, and Pardo-Tomás, 1–17. Surrey, England: Ashgate, 2014.

Solomon, Michael. *Fictions of Well-Being: Sickly Readers and Vernacular Medical Writing in Late Medieval and Early Modern Spain*. Philadelphia: University of Pennsylvania Press, 2010.

Somolinos d'Ardois, Germán. "Relación alfabética de los profesionistas médicos, o en conexión con la medicina, que practicaron en territorio mexicano (1521–1618)." In *Capítulos de historia médica mexicana*. Vol. 3. Mexico City: Sociedad Mexicana de Historia y Filosofía de la Medicina, 1980.

Vera Cruz, Alonso de la. *Dialectica resolutio cum textu Aristotelis*. Mexico City: Juan Pablos, 1554.

Vesalius, Andreas. *Epistola, docens venam axillarem dextri cubiti in dolore laterali secandam: et melancholicum succum ex venae portae ramis ad sedem pertinentibus, purgari*. Venice: Comin da Trino, 1544.

Viesca Treviño, Carlos. "El Códice de la Cruz-Badiano, primer ejemplo de una medicina mestiza." In *Viejo y Nuevo Continente. La medicina en el encuentro de dos mundos*, edited by José M. López Piñero, 104–117. Madrid: SANED (Ministerio de Sanidad y Consumo), 1992.

Ximénez, Francisco. *Qvatro Libros. De la naturaleza, y virtudes de las plantas, y animales que estan receuidos en el vso de Medicina en la Nueua España*. Mexico City: Viuda de Diego López Dávalos, 1615.

5

Between Potosí and Nuevo Potosí

Mineral Riches and Observations of Nature
in the Colonial Andes, ca. 1590–1800

HEIDI V. SCOTT

In a tract composed in 1797 and addressed to the mining guild of Potosí, Pedro Vicente Cañete y Domínguez, a prominent Paraguayan-born government official at Potosí and, among other things, author of a detailed account of the city, its mines, and its province,[1] reflected on varying interpretations of the internal composition of the Cerro Rico, Potosí's silver-bearing mountain, and on the question of whether high-grade ore remained to be exploited inside the deepest reaches of the mountain.

For many years, Cañete explained, he, like the Jesuit José de Acosta in the sixteenth century and other, more recent commentators, among them Scottish historian William Robertson, adhered to the view that the lowest levels of the Cerro Rico would not yield significant riches. However, a painstaking survey of the mountain carried out by the Swedish mining expert Baron von Nordenflicht and his "Mineros científicos" (scientific miners) following their arrival in Potosí in 1789, had persuaded him to change his mind. Further, Cañete recalled that the Viceroy Francisco de Toledo, in the mining ordinances issued for Potosí in the 1570s, had predicted, "con parecer de los Mineros más aviles de ese tiempo" (with the views of the most capable miners of that time), that the Cerro Rico's silver veins would be found to converge and knit together at the base, there "formar en lo hondo la figura de Zepa; en cuyo caso devía aguardarse un manantial poderoso de riquesas" (to form in the depths the figure of a rootstock; in which case a powerful wellspring of riches should be anticipated).[2]

These late eighteenth-century deliberations over the geology of the Cerro Rico were closely enmeshed with debates among mine owners and officials about whether and where a new *socavón* (adit) should be opened to facilitate mine drainage and access to unworked silver deposits.[3] Indeed, Cañete's change of heart permitted him to favor the construction of an adit at the very base of the mountain, on the assumption that rich ore was still to be found at depth.[4] Amid late colonial endeavors to rehabilitate the mines of Potosí, the Cerro Rico became a focal point of intense scientific scrutiny, subject to repeated survey, mapping, and detailed description by a succession of engineers and officials.

Although these efforts to significantly revivify silver production at Potosí ultimately failed, the communications generated about them convey close observations of nature and reveal lively exchanges of ideas about the mountain and its mineral riches. As Cañete's invocation of Toledo's ordinances suggests, the intense study of the Cerro Rico in the late Bourbon era, far from being a novel Enlightenment phenomenon, drew on a trajectory of colonial-era knowledge production about the site and its silver deposits that had commenced centuries earlier. The theories and understandings of geology and metal formation that took shape at Potosí circulated outward to inform prospecting and mining operations elsewhere in the Andes, most famously through Álvaro Alonso Barba's 1640 mining treatise, *Arte de los metales* (Art of metals), but equally through the accumulated experiential knowledge of less prominent mine owners and mineworkers, Spanish and indigenous alike. Yet the very prominence of Potosí in shaping the social, environmental, and economic landscapes of the southern Andes, together with its potency as a symbol of Spanish imperialism, can easily overshadow the ways in which knowledge of nature also took shape in other colonial Andean mining centers.

A vast body of manuscript materials, preserved in Spanish and Latin American archives and generated during almost three hundred years of mining in the colonial Andes, holds records such as details of quotidian mining operations, official inspections and government correspondence, lawsuits, reports of mineral prospecting, proposals for reform, and petitions for assistance and reward. Historians of the Andes have predominantly drawn upon these rich archival sources to gain insights into technical change and innovation in mineral extraction and refining and labor practices as well as to illustrate the social, economic, and environmental reverberations of mining at a variety of geographic scales.[5]

For their part, historical interpreters most curious about colonial understandings of nature and in particular geology—that is, the qualities of minerals, their distribution and uses, the formation of precious metals, and the character-istics and causes of earthquakes and volcanoes—have mostly turned to formal published writings, especially natural histories, mining treatises, and scholarly journals.[6] Some of these writings—most notably, perhaps, José de Acosta's dis-cussion of mining and minerals in his 1590 *Historia natural y moral de las Indias* (*Natural and Moral History of the Indies*), and Barba's treatise—proved deeply influential, and their ideas reverberated throughout the colonial period in the Iberian world and beyond.

A narrow focus on prominent published texts, however, risks overshadowing the varied ways in which understandings of nature, as they related to the exploi-tation of precious metals, were shaped and deployed in wider quotidian con-texts by miners and mine owners, colonial officials, and others whose lives and livelihoods were deeply intertwined with mining. Historians of science as well as literary scholars, among them Antonio Barrera-Osorio and Allison Bigelow, are turning to the archives to flesh out networks of communication and col-laboration that contributed to knowledge within colonial mining communities and focus in particular on technical innovations such as the development and diffusion of new silver-refining techniques in the colonial Andes and beyond.[7] Although such technical processes were closely intertwined with wider under-standings of nature, the latter tend to remain hidden from view in this body of scholarship. While geological ideas and theories circulated widely, production of knowledge was both localized and highly relational, shaped by the particu-larities of place, and moreover, by competition and tension between established mines and new sites of mineral exploitation.

I take up a rich case in point, examining the efforts of a late sixteenth-century miner to persuade King Philip II (1556–1598) of the significance of silver de-posits he discovered in the central Andean region of Peru. More specifically, I demonstrate how, in the course of his account and petition, this miner placed his notions of mineral formation and distribution at the center of his persuasive tactics. The mines of Potosí loomed large in his account, both as a point of posi-tive comparison and as an obstacle to overcome.

Knowledge of Nature and Divine Providence at Nuevo Potosí, 1596

By the final decade of the sixteenth century Potosí's silver production, although still substantial, had begun a slow but steady decline due to the exhaustion of easily exploitable surface ores and the deepening of the mines, conditions that had caused work below ground to become increasingly difficult, dangerous, and costly.[8] In the following decades and indeed throughout the colonial era, the revival of production at Potosí, despite repeated predictions of the mines' demise and abandonment, remained a primary concern for the Spanish Crown, its representatives in America and, naturally enough, Potosí's mining community. Nevertheless, declining production at the Cerro Rico helped stimulate the search for substantial deposits of silver elsewhere in the Andean cordillera.

In 1596 a miner and mine owner named Juan Francisco de Hinestrosa informed the monarch that two years previously he had discovered important new silver deposits in the Cordillera de Pariacaca, a chain of mountains that marked the eastern boundary of the Peruvian province of Huarochirí.[9] Having promised to describe the new mining site "sin sajeraçion alguna mas que las razones de la verdad descalças" (without any form of exaggeration and [with] nothing more than the naked truth),[10] Hinestrosa wasted no time in warning Philip II that the end of Potosí's bonanza of silver was in sight and in offering a providential reading of his recent discovery. It was clear, he urged, that God willed these metallic riches to be revealed at this time,

> por q. segun las nuebas q. cada dia nos dan de la baja de los metales de potosi quiere que . . . aya otro nuebo potosi para de nuebo dar aliento a v. m.t y que se anime a ensalçar y acreçentar nra s.ta fee catolica.[11]

> (because according to the news we receive daily about the decline in metals at Potosí he [God] wishes there to be another new Potosí . . . to give hope once again to Your Majesty and encourage him to elevate and foster our holy Catholic faith.)

In producing a detailed *relación* (account) of his discovery Hinestrosa aimed to do more than simply inform. By 1596, a list of 129 individuals and institutions had already registered mines at Nuevo Potosí,[12] and Hinestrosa, who by this time had secured capital for his own mining operations in company with a wealthy man named Luis Rodríguez de la Serna,[13] proposed to the monarch a series of privileges and administrative arrangements for the nascent mining community. Prominent within the proposal was the allocation of *mitayos* (indigenous draft laborers) to ensure a supply of labor.[14] Hinestrosa already enjoyed the support of García Hurtado de Mendoza, the outgoing viceroy of Peru, who recommended that his successor assign between eight hundred and one thousand draft laborers to the mines.[15]

In particular, the miner hoped to convince the monarch that the four hundred *mitayos* then assigned to the mercury mines of Huancavelica from the jurisdictions of Hatun Jauja, Tarma, and Chinchaycocha should be allocated to Nuevo Potosí.[16] As Hinestrosa was well aware, however, he had to make a robust case for the viability of the new mining site in order to secure these forced workers along with other forms of support from the crown and the incumbent viceroy, Don Luis de Velasco. In light of continuing demographic decline in the Andes, the crown and colonial authorities were anxious to ensure that *mitayos* were allocated only to the most productive mines. In 1595, just months before Hinestrosa presented his *relación*, Hurtado de Mendoza informed the monarch that although Peru's mines were flourishing, it was impossible to apportion enough laborers to work them all.[17] Further, a succession of viceroys and other governors had developed a healthy sense of caution where reports of newly discovered silver deposits were concerned, for—as Hinestrosa himself observed—these discoveries frequently ended in disappointment.[18]

As we have already glimpsed, Hinestrosa placed his own discovery of precious metals within a providential framework, foretelling the decline and replacement of Potosí in order to construct a persuasive account for Philip II. At the same time, with the aim of persuading his royal recipient and councilors of the Indies, he sketched out the foundations of his knowledge of the formation of precious metals and their distribution in the Andes, enmeshing understandings of nature within his providential narrative. In doing so, Hinestrosa sought to convince the king that his own knowledge both challenged and offered a divinely inspired corrective to prevailing understandings of geology that guided the practices of miners and mineral prospectors in Peru.

By the time Hinestrosa first inspected the site of Nuevo Potosí, he was—or so he claimed—already a renowned mining expert. The claims appear to have been largely justified. At the silver mines of Urcococha and Choclococha, discovered in 1590, he had served as assayer and inspector by request of all the miners, leaving the mines in a flourishing state.[19] Four years later the viceroy dispatched him to the gold mines of Mataro in Huaylas province to offer his expertise. Coming into initial conflict with other miners over how to work the abandoned galleries, he eventually prevailed, "y los labre por la orden que yo di de que fue muy satisfecho vro visorrey" (and I worked them according to the orders which I myself gave, which greatly satisfied your viceroy).[20] Next, having received viceregal orders to inspect some abandoned silver mines, Hinestrosa succeeded where his predecessors had failed, bringing the mines back into a state of vigorous production.[21]

In 1594, aware of Hinestrosa's experience and reputation, Antonio Arias and Francisco Gómez sought him out, eager to show him "vna gran veteria" (a great mass of veins) they had found in the mountain ranges of Huarochirí at a site previously worked by the Incas.[22] Accompanying these men, Hinestrosa inspected the deposits and refined samples of ore, using "los benefiçios nuebos que yo sauia de la calçinaçion y otros que . . . fueron menester" (the new refining methods known as calcination, which I was familiar with, and others which . . . proved necessary), he determined that the indications of substantial subterranean wealth were extremely promising.[23] In Lima, however, the viceroy expressed reservations about the significance of the find and challenged the miner to explain. Why, he demanded to know, should he favor Hinestrosa's claims when other miners were bringing samples of much richer ore from sites across the kingdom? Rising to the challenge, Hinestrosa persuaded the viceroy with reasons formed on the basis of his personal experience, reasons that "son tan contrarias a todas las que [dan] los mineros . . . en este rreyno q no ay cosa mas contraria en el mundo" (so oppose all those given by the miners . . . in this kingdom that there is nothing in the world more antithetical).[24]

Other miners, he went on to explain, habitually sought out the richest metals they could find on the surface, ignoring the other surface ores that contained lower concentrations of silver. They did so, Hinestrosa tacitly acknowledged, partly because working those deposits was easier and less costly than penetrating the depths. However, he argued grandly, the rich ores—some of which were "mantos de metales" (mantles of metal) that the biblical flood had dislodged from high peaks in this mountain chain—frequently extended just two or three *estados* in depth before petering out.[25] Thus, the usual miners, paying attention to no more than what they saw before their eyes, obtained draft laborers and constructed refining mills only to lose all their investments and to be "tenidos por charlatanes" (considered charlatans).[26] What most miners failed to understand, Hinestrosa maintained, warming to his task, was that sites that contained rich metal on the surface were "minas de cabeça" (head mines), where the qualities of the earth allowed the formation of silver with the help of the influence of planets both large and small. Those planetary bodies, although they gave solid form to the metals and caused them to grow, "no pueden penetrar ni pasar a mas que hasta tres estados quando mas" (cannot penetrate nor pass more than three *estados* in depth at most).[27]

Hinestrosa subtly implied that those who insisted on seeking rich surface ores would do well to pay attention to the knowledge of native Andean miners. He had seen mines besides Nuevo Potosí that had been fruitfully worked by Indians and that began with much lead and low concentrations of silver but yielded great riches beneath.[28] Thus, he continued,

> tengo por regla çierta y aueriguada que las minas fijas an de ser pobres de ley encima de la tierra y si uan encapadas por debaxo de quemazones y de tierra son mejores aunque no ay rregla sin eçeçion que . . . ay vetas que desde su prinçipio llevan rricos metales como fueron los de potosi.[29]

> (I consider it a certain and proven rule that mines of substance must have low grade ore at the surface and if they are capped by *quemazones* [a type of ore] and earth,[30] they are better, although there is no rule without exception, such that . . . there are veins that from their very beginning contain rich metals, like those that used to be found at Potosí.)

In these "minas fijas" (mines of substance), Hinestrosa argued from experience and close observation, the silver matures and grows up toward the surface from what he called its "çepas" (rootstock), mixed with lead, tin, copper, and other minerals. Further down, the silver vein becomes wider and purer, and so the presence of those mixtures near the surface points to abundant silver below. If few mines of that kind were being worked, he observed, it was simply because they had been overlooked. Had it not been for the indigenous Andean miners who exploited those deposits, smelting the metals with their *huayras* (wind ovens), the abundance of Potosí's "minas de çepa" (rooted mines) would never have been discovered.[31] Hinestrosa took care to emphasize that Potosí's principal mines, which had once yielded astonishing wealth, were now so deep that they could no longer be expected to yield rich silver. Indeed, he exclaimed, the discovery of such metal would constitute a miracle "contra la orden de natu-raleza por sus honduras" (contrary to natural order as a result of their depth).[32]

At Nuevo Potosí the veins of ore he had inspected showed so many indications of abundant silver that, Hinestrosa waxed lyrical, "pareçe que se an me-dido con el deseo de mi ymaginaçion" (it is as if they have been measured by the desire of my imagination).[33] Across his account, Hinestrosa was eager to emphasize that he possessed privileged knowledge that allowed him to read and interpret these auspicious signs. It was vital, he underscored, to learn to identify the mixture of minerals within which the silver was formed and to follow the path they take underground.

Although other miners were dismayed by the abundance of sulfur, vitriol, iron pyrites, lead, and alcohol near the surface, Hinestrosa asserted, me da a mi mayor gusto y contento porque no querria yo ver en las minas metales muy linpios y acendrados sino suzios y encorporados con las dhas misturas porque es señal muy çierta que la fuerça de los metales y fuego dellos brotan y crian todo lo susdho.[34]

(that gives me the greatest pleasure and contentment because in the mines I did not wish to see very clean and pure metals but instead dirty ones that are incorporated with the said mixtures because it is a sure sign that the potency and fire of the metals [below] cause all of the said [substances above] to sprout and grow.)

Having set out the theory of metal formation upon which his claims were founded, Hinestrosa turned to describing the principal veins of silver and estimating the costs of bringing the mines into production. Later in this narrative he returns to his providential theme. God, he said, had been served in choosing him

como a ombre que aborreçia minas ricas en la haz de la tierra y apetesia vetas anchas y pobres para dar en la verdad de lo que todos mineros an aborreçido como lo an hecho en estas propias minas mucho numero de mineros.[35]

(as a man who detested rich mines on the surface of the earth and who desired wide and poor veins in order to uncover the truth that all miners have despised, as many miners have done in these very mines.)

In other words, God had opened Hinestrosa's eyes to the knowledge of true subterranean riches, permitting him to locate new mineral deposits that in their promise of substantial and sustained wealth resembled the Cerro Rico in its virgin state.

The miner Juan Francisco de Hinestrosa and the case he made to his sovereign rests within a broader context of experiential knowledge. Indeed, the archives offer tantalizing glimpses of ideas about mineral formation that circulated and appear to contradict Hinestrosa's theory or at least to qualify it in important ways. In 1608, for instance, the *corregidor* (regional magistrate) of Cuzco, Pedro de Córdoba Mexía, informed the viceroy that recently discovered gold deposits at a site known as Solimana near Arequipa in the province of Condesuyos had turned out to be far less promising than originally thought. Córdoba Mexía wrote that many hills in this region contained

gran cantidad de quemaçones y betas es la tierra tan seca que no ha llegado a dar birtud a los metales que tienen los çerros a que ayan criado oro ni otro metal mas que mucha fuerça de quemaçones con margajita cardenillo alcool y calichales que a los que no tienen esperiencia de minas an podido engañar que las señales referidas prometian betas de oro de mucha riqueza y fundamento.[36]

(a great quantity of *quemaçones* [*quemazones*] and veins, the earth is so dry that it has not managed to give virtue to the metals contained in the hills such that they generate gold or other [precious] metal and has produced nothing but a great quantity of *quemaçones* with iron pyrites, verdigris, alcohol, and saltpeter, and those who lack experience in mines have been fooled into thinking that the aforementioned signs promised very rich and substantial veins of gold.)

For Córdoba Mexía, then, the surface presence of the very minerals that Hinestrosa sought out by no means promised significant metallic wealth below. At Solimana, Córdoba Mexía insisted, the aridity of the earth was so pronounced that the formation of anything but *quemazones* mixed with a variety of minerals was impossible. Although he contended that *quemazones* and minerals were produced by precious metals, he maintained that those at Solimana had been generated below ground by a metallic vein where greater moisture was present and then forced to the surface by floods or earthquakes.[37] In other words, they were nothing more than relics of past meteorological or geological events.

It is doubtful that Hinestrosa's notions of how silver was formed and distributed were as antithetical as he suggested. García de Llanos's early seventeenth-century dictionary of mining terms that were in use at Potosí at that time provides a valuable point of comparison. Certain entries offer glimpses of generalized theories of mineral formation in the Andes that clearly resonated with Hinestrosa's belief in the greater abundance of precious metals at lower levels. The word *machacado*, Llanos explained, referred to the very richest silver ore, which tended to be found in deposits "de poca duración y fundamento, solamente criaderos en la haz de la tierra" (that are quickly exhausted and without foundation [and are] merely incubators on the surface of the earth).[38] His discussion of another term, "tomar humedad las minas," which described the presence of water in the depths of the mines suggested, like the *corregidor*'s report about Solimana, that a steady supply of moisture was essential for the formation of high concentrations of silver and hence that it was at lower levels where the greatest wealth was typically to be found.[39] Although Hinestrosa made no reference to the role of moisture in the formation of precious metals, both he and Llanos made similar observations about the distribution of rich ore beneath the surface.

Recognition of the striking correspondences between Hinestrosa's own claims and ideas that appear to have been circulating among miners in the early colonial Andes does not render his *relación* less interesting. On the contrary, with an unusual attention to detail, Hinestrosa's account is suggestive of how miners deployed and also challenged theories of the earth's mineral riches in pursuing their profit-focused enterprises. In Hinestrosa's case, knowledge and experience of nature was not solely an asset to be mobilized in prospecting for viable mineral deposits and in profitably mining and refining the ore. His understandings and observations also served as tools of persuasion in his efforts to secure support and privileges for Nuevo Potosí—support that in an era of ongoing demographic decline and continued privileging of Potosí was by no means guaranteed. After all, the very notion that the Cerro Rico's rich silver deposits were largely exhausted was strongly contested until the end of the colonial period. Hinestrosa's *relación* conveyed to his royal recipient that he had not only discovered a "new Potosí"; rather, thanks to his extensive experience of diverse mining sites, he possessed an overarching understanding of geology that held the key to improving prospecting and mineral extraction throughout Peru, in light of the decline of Potosí.

By the eighteenth century, the city and mines of Potosí, demographically and economically, had become a shadow of their former selves, yet they were considered too significant to be declared defunct. The official Pedro Vicente de Cañete y Domínguez exclaimed in 1797 that while the Cerro Rico's silver deposits formed "el tronco de donde germinan todas las vetas del Reyno del Perú" (the trunk from which all the [silver] veins of the Kingdom of Peru germinate), the Potosí mining guild had always been considered "el centro de la ciencia metálica" (the center of mining science). Hence, even if Potosí's mines were to yield no more substantial riches, he contended that they should nevertheless be maintained, "para decoro, para enseñanza, y para estímulo de los demás Minerales del Reyno" (for the purposes of decorum and instruction, and as an encouragement for other mining operations in the kingdom).[40] It would be senseless, Cañete continued, to go excavating holes in Peru's hillsides in search of other deposits of comparable wealth, as he doubted that they existed, and to abandon "lo cierto por lo dudoso" (the certain for the doubtful).[41] Cañete's conflicted sentiments reverberated widely across the late eighteenth-century

Andes. Just as crown and vice-regal policies in the era strongly encouraged precisely the systematic search for new mineral deposits throughout the Andes and beyond, so too the crown and many of its representatives refused to relinquish the idea of a Potosí renaissance.

In the 1780s, two hundred years after Hinestrosa walked a tightrope of comparing his own mines with those of Potosí while declaring the latter in terminal decline, the Cerro Rico continued to be a powerful presence in the ways knowledge of new mining sites was presented and circulated. Discussing the silver mine of Uspallata in his widely read *Compendio de la historia geográfica, natural y civil del Reyno de Chile* (Compendium of the geographical, natural, and civil history of the kingdom of Chile), the exiled former Jesuit and natural historian Juan Ignacio Molina maintained that the mine in the highlands of Chile held the promise of becoming one of the most celebrated mines anywhere in the Americas. The veins of Uspallata, he went on, were known to run northward for thirty leagues without any sign of diminishment in their wealth. Indeed, he declared, there were even those who believed that the veins ran all the way to Potosí and that the mine was nothing less than "una ramificación de aquella célebre mina del Perú" (a branch of that celebrated Peruvian mine).[42]

Conclusion

Debates over nature emerge in unexpected mining-related sources that go beyond formal natural histories and technical treatises such as the famous text by Álvaro Alonso Barba. Deeper exploration of varied archival sources—and of myriad mining-related petitions, reports, and administrative communications in particular—makes visible much wider and richer networks of experience and participation as understandings of Andean nature were forged. Hinestrosa's *relación* was generated by the pursuit of particular material and self-serving objectives. Yet the very embeddedness of Hinestrosa's natural historical reflections within such objectives exemplifies how in the everyday contexts of the mining milieu, rather than devaluing them, theories of nature were aired and shaped.

Such archival endeavors are never straightforwardly triumphal exercises, as in celebration of how ordinary members of colonial-era mining communities participated in the creation and circulation of natural historical knowledge. Orlando Bentancor demonstrates that understandings of nature were inseparable from the development of imperial logics that justified the large-scale exploitation of minerals in Spanish America regardless of the human and environmental costs.[43] These connections to social and economic realities are apparent in the sources discussed here no less than in the philosophical, juridical, and political writings that determined the legal and moral underpinnings of Spain's empire and its appropriation of New World resources. If in the late sixteenth century Acosta interpreted Peru's abundance of silver as a God-given dowry intended to attract the Spanish to the Andes, his near-contemporary Hinestrosa's empirical observations of nature in the Cordillera de Pariacaca, also set within a providential framework, were inseparable from his claims to forced labor.

The politics and realities of knowledge-making around mineral wealth were not lost on a broad range of people in the Andes throughout the colonial era. Francisco de Serra Canals, in his mining treatise completed in 1799, reflected on a trajectory of more than two hundred years of colonial mineral extraction in the Andes and elsewhere in Spanish America, lands in which the hills had been thoroughly "traqueadas" (furrowed), he observed, by so much mining activity. The author and mine owner contended that the inhabitants of the New World possessed much greater knowledge of minerals and how to identify and exploit them than was the case in peninsular Spain. This difference, he maintained, had nothing to do with natural variation, for Spain too was richly endowed with minerals. It rested, rather, with the workings of an imperial economy. Spanish indifference toward such knowledge "proviene de que allá no se necesita de esta penosa aplicación para vivir, por tener el arbitrio de dedicarse a otras muchas que les aseguran su subsistencia" (results from the fact that, over there, people do not need this penurious activity in order to live, for they have the choice to dedicate themselves to many others that ensure their subsistence).[44] In Spain, manufacturing offered the population "mines" of such quality that, unlike the mines of the New World, their riches were never exhausted.[45] The inhabitants of the Americas, Serra Canals insisted, could not afford the luxury of such distance from and ignorance about the earth beneath their feet.

Notes

1. Cañete y Domínguez, *Guía histórica.*

2. Doctor Pedro Vicente de Cañete a los Señores Diputados del Ilustre Gremio de Azogueros de Potosí. La Paz, 8 February 1797. Charcas 700, Archivo General de las Indias, Seville (henceforth AGI).

3. For an account of repeated efforts to construct a new adit in the second half of the eighteenth century, see Buechler, *Gobierno, minería y sociedad,* 65–73.

4. Cañete a los Señores Diputados, AGI.

5. It is impossible to offer a comprehensive overview of the literature. For Potosí, key studies include Bakewell, *Miners of the Red Mountain*; Buechler, *Gobierno, minería y sociedad*. For other Andes regions see Fisher, *Silver Mines*; Gavira Martínez, *Minería en Chayanta*; Povea Moreno, *Minería y reformismo*.

6. Bentancor, "Matter, Form, and the Generation of Metals"; Salazar-Soler, "Obras más que de gigantes."

7. Barrera-Osorio discusses mining briefly, focusing particularly on innovations in refining techniques (*Experiencing Nature,* 65–72). See also Bigelow, "La técnica de la colaboración"; Salazar-Soler, "Los 'expertos' de la corona."

8. Bakewell, *Miners of the Red Mountain*, 26–32. Production peaked in 1592 at 201 metric tons of silver, and after that, despite occasional surges, "the trend was consistently down" (26).

9. Hinestrosa, *Relaçion*, March 19, 1596, fol. 1, AGI. Hinestrosa's birthplace and background, along with other biographical details beyond those directly related to his mining activities, are not mentioned.

10. Hinestrosa, *Relaçion*, fol. 2.

11. *Relaçion*, fols. 2–2v.

12. Twenty-one women, one élite Andean named Don Lorenzo Chuquitapa, and two lay brotherhoods (El Santo Crucifixo de Burgos and Nuestra Señora de Copacabana) in addition to an Inquisition official were among those who had registered mines at Nuevo Potosí (Hinestrosa, *Relaçion*, fols. 22v-24).

13. Hinestrosa, *Relaçion*, fols. 18–19v. Rodríguez de la Serna was a prominent city official in Lima. According to Hinestrosa, Rodríguez had already provided 3,000 pesos to finance the initial phase of prospecting and assaying.

14. Hinestrosa, *Relaçion*, fols. 36–38v. Little appears to have been written about the history of Nuevo Potosí. It is clear, however, that the mines were still being actively worked in the 1750s (Spalding, *Huarochirí*, 191–193).

15. Hinestrosa, *Relaçion*, fol. 39.

16. Hinestrosa, *Relaçion*, fol. 37v.

17. Viceroy Mendoza to Philip II, Lima, November 8, 1595, Lima 33, no. 30, libro IV, fol. 134, AGI. This message was frequently reiterated. Three years later, within the context of inquiries into whether *mitayos* could be assigned to newly discovered mines, the Royal Audience of Lima recommended to the viceroy that it would be "buen gobierno" (good government) to send draft laborers only to the most productive and useful mines (Lima, April 9, 1598, Lima 33, no. 36, libro III, fol. 73, AGI.

18. Viceroy Velasco commented that many mine owners exaggerated the productivity of their mines with the sole purpose of securing *mitayos*, many of whom they illicitly put to work in the fields rather than in the mines (Viceroy Velasco to Philip III, Lima, December 4, 1602, Lima 34, no. 40, libro V, AGI).

19. Hinestrosa, *Relaçion*, fol. 3. In 1593 the viceroy informed the monarch that the Urcocha mines, situated close to the mercury mines of Huancavelica, were experiencing a significant upswing in production (Viceroy García de Mendoza to Philip II, Lima, 18 May 1593, Lima 33, no. 7, libro III, fol. 45, AGI).

20. Hinestrosa, *Relaçion*, fols. 3v-4.

21. Hinestrosa, *Relaçion*, fols. 5v-6. Hinestrosa does not indicate where these mines were located, only that they had been registered by two brothers named Hernán and Lucas Ramírez. The mines in question were first exploited in the 1560s and then quickly abandoned.

22. Hinestrosa, *Relaçion*, fols. 6v–7.

23. Hinestrosa, *Relaçion*, fols. 7v–8v. Hinestrosa's use of the term "calçinaçion" refers to a refining technique that was used to extract silver from a particular kind of silver ore known as *metales negrillos*. The technique involved toasting or burning ground before it was mixed with mercury. This was a variant of the patio process, introduced to the Potosí mines in the 1570s. For a colonial-era description of this process see Serra Canals, *El perito incógnito*, 102–105. Hinestrosa mentions having spent time in Potosí, and it is likely that he learned the technique there.

24. Hinestrosa, *Relaçion*, fols. 12–12v.

25. Hinestrosa, *Relaçion*, fols. 12v–13. An *estado* is an antiquated Spanish measure that corresponds to the height of an average man.

26. Hinestrosa, *Relaçion*, fols. 13–13v.

27. Hinestrosa, *Relaçion*, fols. 13v–14.

28. Hinestrosa, *Relaçion*, fol. 14v. He also noted that compared to German miners he had met in Potosí, Andean miners had much greater success in extracting silver from the local ore using their *huayras* (wind ovens). He was careful to add that the new refining technique known as calcination, which he had mastered, yielded twice as much silver as the *huayras*.

29. Hinestrosa, *Relaçion*, fol. 14v.

30. García de Llanos, the author of a dictionary of mining terms in use at Potosí in the early seventeenth century, identified *quemazones* as a type of ore found alongside almost all veins of metal. He explained that they resemble "escoria de fragua" (furnace slag), are metals that appear to be "pasado del punto y abrasado" (past their prime and burned), and usually are found on the surface. Some *quemazones*, he continued, are found at depth and are known by the native Andeans as *rupasca*, a Quechua term which means "burned" (Llanos, *Diccionario*, 89).

31. Hinestrosa, *Relaçion*, fols. 14v–15.

32. Hinestrosa, *Relaçion*, fol. 2v.

33. Hinestrosa, *Relaçion*, fol. 16.

34. Hinestrosa, *Relaçion*, fols. 16–16v.

35. Hinestrosa, *Relaçion*, fols. 29–29v.

36. Don Pedro Córdoba Mexía, corregidor del Cuzco, al virrey, sobre el suceso del cerro de Solimana, Cuzco (?), January 3, 1608, Lima 35, n.p., AGI.

37. Parecer de Don Pedro de Córdoba Mexía, corregidor de Cuzco, sobre las minas del cerro de Solimana, Solimana (?), November 29, 1607, Lima 35, n.p., AGI. Hinestrosa's views on the role of moisture in the formation of precious metals is not clear.

38. Llanos, *Diccionario*, 84. Like Hinestrosa, Llanos cites the mines of Vilcabamba as an example of these superficial silver deposits.

39. Llanos, *Diccionario*, 122.

40. Cañete a los Señores Diputados, AGI.

41. Cañete a los Señores Diputados, AGI.

42. Molina, *Compendio de la historia geográfica*, 105. The study was first published in Italian in 1782.

43. Bentancor, *Matter of Empire*.

44. Serra Canals, *El perito incógnito*, 146.

45. Serra Canals, *El perito incógnito*, 144–146. Serra Canals, who was born in Barcelona, began working silver mines in Uspallata (today in northwestern Argentina) in the 1770s. For biographical details see Edberto Oscar Acevedo's introductory study in Serra Canals, *El perito incógnito*, 10–16.

References

Acevedo, Edberto Oscar. "Estudio preliminar." In Francisco de Serra Canals, *El perito incógnito y el curioso aprovechado: Tratado inédito del Virreinato del Río de la Plata*, 9–58. Frankfurt: Vervuert/Iberoamericana, 1999.

Acosta, José de. *Historia natural y moral de las Indias*. Seville, 1590.

Bakewell, Peter. *Miners of the Red Mountain: Indian Labor in Potosí, 1545–1650*. Albuquerque: University of New Mexico Press, 1984.

Barba, Álvaro Alonso. *Arte de los metales en que enseña el verdadero beneficio del oro: y plata por açogue. El modo de fundirlos todos, y como se han de refinar, y apartar unos de otros*. Madrid, 1640.

Barrera-Osorio, Antonio. *Experiencing Nature: The Spanish American Empire and the Scientific Revolution*. Austin: University of Texas Press, 2006.

Bentancor, Orlando. "Matter, Form, and the Generation of Metals in Alvaro Alonso Barba's *Arte de los Metales*." *Journal of Spanish Cultural Studies* 8, no. 2 (2007): 117–133.

———. *The Matter of Empire: Metaphysics and Mining in Colonial Peru*. Pittsburgh, PA: Pittsburgh University Press, 2017.

Bigelow, Allison. "La técnica de la colaboración: Redes científicas e intercambios culturales de la minería y metalurgia colonial altoperuana." *Anuario de Estudios Bolivianos, Archivísticos y Bibliográficos* 18 (2012): 53–77.

Buechler, Rose Marie. *Gobierno, minería y sociedad: Potosí y el "Renacimiento" borbónico 1776–1810*. La Paz: Biblioteca Minera Boliviana, 1989.

Cañete y Domínguez, Pedro Vicente. *Guía histórica, geográfica, física, política, civil y legal del Gobierno e Intendencia de la Provincia de Potosí*. Potosí, Bolivia: Potosí, 1952 [1787].

Fisher, John. *Silver Mines and Silver Miners in Colonial Peru, 1776–1824*. Liverpool, England: University of Liverpool, 1977.

Gavira Martínez, María Concepción. *Minería en Chayanta: La sublevación indígena y el auge minero 1775–1792*. La Paz: Plural, 2013.

Hinestrosa, Juan Francisco de. *Relaçion del descubrimiento del nuebo Potossi*. Nuevo Potosí, 1596. Charcas 134, Archivo General de Indias.

Llanos, García de. *Diccionario y maneras de hablar que se usan en las minas y sus labores en los ingenios y beneficios de los metales*. Madrid: Consejo Superior de Colegios de Ingenieros de Minas, 2009 [Lima, 1611].

Molina, Juan Ignacio. *Compendio de la historia geográfica, natural y civil del Reyno de Chile. Primera parte*. Madrid: Don Antonio de Sancha, vol. 1, 1788.

Povea Moreno, Isabel M. *Minería y reformismo borbónico en el Perú: Estado, empresa y trabajadores en Huancavelica, 1784–1814.* Lima: Banco Central de Reserva del Perú, 2014.

Salazar-Soler, Carmen. "Los 'expertos' de la corona: Poder colonial y saber local en el Alto Perú de los siglos XVI y XVII." *De Re Metallica* 13 (2009): 83–94.

———. "'Obras más que de gigantes.' Los jesuitas y las ciencias de la tierra en el Virreinato del Perú (siglos XVI y XVII)." In *El saber de los jesuitas, historias naturales y el Nuevo Mundo,* edited by Luis Millones Figueroa and Domingo Ledezma, 147–172. Frankfurt: Vervuert/Iberoamericana, 2005.

Serra Canals, Francisco de. *El perito incógnito y el curioso aprovechado: Tratado inédito del Virreinato del Río de la Plata.* Frankfurt: Vervuert/Iberoamericana, 1999.

Spalding, Karen. *Huarochirí: An Andean Society under Inca and Spanish Rule.* Stanford, CA: Stanford University Press, 1984.

6

Indigenous Medicine and Nation-Building

Hermilio Valdizán's Medical Project

EDWARD CHAUCA

The physician Hermilio Valdizán (1885–1929) is considered one of the founding fathers of modern psychiatry in Peru. He devoted most of his research efforts to building a national archive of medical literature, with a particular focus on mental health. He sought to demonstrate a capacity for the production of medical knowledge in Peru that was not derivative of European science. His main interest was to find continuity between the medicine of the past and the official medicine of his present.[1] In order to do so, he traced the foundations of Peruvian medicine in traditional indigenous healing practices, in medical practices from colonial times, and also in contemporary urban folk medicine. He paid special attention to illnesses, afflictions, treatments, and toxins that were not studied by his European counterparts, hoping to open up new paths of investigation for national researchers as well as to teach urban physicians about public health and culture in Peru's rural areas.

My purpose is to study the role of Andean culture in Valdizán's project of creating and disseminating a national medical history. Following an extensive examination of his publications, I will focus here on a series of articles and two books he wrote about this topic, *La alienación mental entre los primitivos peruanos* (Mental illness among primitive Peruvians, 1915) and *La medicina popular peruana* (Peruvian folk medicine, 1922); the latter was co-written with the pharmacologist Ángel Maldonado and published in three volumes. Valdizán's interest in indigenous medicine and its healing treatments emerged as a critique of certain European intellectuals and physicians who suggested that people in the Americas were intrinsically inferior and unhealthy. Through the use of medical literature, *crónicas de Indias* (chronicles of the Indies), literary fiction, newspapers, dictionaries, and precolonial pottery, Valdizán defended indigenous peoples' intellectual capability, emphasizing how they categorized mental illnesses and their treatments. Although his method of analysis swung between description and speculation, praise and disdain, his groundbreaking research was the first attempt to insert traditional Andean medicine into the national history of medicine and mental health.[2]

Valdizán studied at the Universidad Mayor de San Marcos, where he received his bachelor's degree in medicine in 1909. From 1911 to 1914, he studied psychiatry and neurology in Bologna, Italy, and made some visits to hospitals in France and Switzerland. He returned to Peru at the beginning of World War I and received his doctoral degree in 1915. He was appointed director of the recently founded Asilo Colonia de Magdalena (later called Hospital Víctor Larco Herrera) in 1921. It was considered one of the most modern and prestigious mental health hospitals in Latin America at that time. Valdizán was part of the first generation of specialized professionals in Peru who abandoned the idea of pursuing a broad education in favor of more specialized knowledge in a specific field.[3]

However, to study a specific field did not imply following only one kind of methodology. Like many of his peers, Valdizán's medical approach was influenced by various branches of medical science, such as the hygienic movement, miasma theory, eugenics, phrenology, Pinel's moral treatment, and Charcot's anatomo-clinical method. The historians Marcos Cueto and Steven Palmer state, "The invocation of hygiene in Latin America was complex, mixed up with influences taken from Social Darwinism, evolutionism, and positivism, and for many of its supporters it was seen as complementary to medical treatment."[4] This approach was due to a number of factors. Some scientific advances that came from Europe were slow to spread in the Americas, and the delays produced overlaps between tendencies that were successive in Europe. New hygienic treatments appeared but did not replace previous ones.[5] Other factors include the long coexistence and cultural negotiation between European medicine and indigenous and African healing treatments in the region as well as the colonial Baroque's legacy of producing knowledge using a multidisciplinary approach. Indeed, the historian Jorge Cañizares-Esguerra has explained that in the eighteenth century most western European scholars questioned the veracity of the knowledge produced in Spanish about

the Americas written during the sixteenth and seventeenth centuries for imperial and theological reasons. In their search for new disciplines to produce
knowledge, Enlightenment scholars promoted the rise of specialized, uncon-
taminated sciences. Meanwhile, when scholars in Spain and its colonies faced
the discrediting of the knowledge produced for the previous two centuries
in their territories, they opted to pursue interdisciplinary approaches as the
best way to study nature and human affairs in the Americas.[6] Thus, it was not
unusual that despite Valdizán's decision to specialize in the history of medicine, he chose to combine different disciplines and approaches to carry out his
research; his doing so mirrors precisely how knowledge had been produced in
the region since the eighteenth century.

La alienación mental entre los primitivos peruanos was Valdizán's doctoral
dissertation. Some sections of it were published earlier in medical journals
while he was still a student in Europe and Peru. His work was a response to
"aquellos numerosos escritores que hicieron de los americanos objeto de las
más burdas invectivas y de las más extravagantes hipótesis que hayan sido
formuladas respecto a tópico alguno" (those many writers who made people
from the Americas the object of their grossest invectives and of the most extravagant hypotheses that have ever been developed about any kind of topic).[7]
Specifically, he referenced the eighteenth-century Dutch philosopher Cornelius de Pauw, who was considered one of the most influential experts on the
Americas. De Pauw's *Recherches philosophiques sur les Américains* (Philosophical inquiries about people from the Americas), published in 1771, influenced
histories and travel compilations written by Adam Smith, William Robertson,
and Alexander von Humboldt, among others.[8] Valdizán received de Pauw's
ideas from two secondhand sources, *Historia Antigua de Méjico* (Ancient history of Mexico) by the Novohispano Jesuit historian Francisco Javier Clavig-
ero and *Obras científicas y literarias* (Complete scientific and literary works)
by the Peruvian physician Hipolito Unánue. Valdizán states,

> Para de Pauw, los Americanos hallábanse caracterizados por la memoria
> débil, por el ingenio obtuso, la voluntad fría, el ánimo apocado y el en-
> tendimiento indolente, características psicológicas cuya exageración es
> apenas comparable a la de los sujetos que ponían en tela de juicio si el
> hombre americano era o no un racional.[9]

(For de Pauw, Americans' main characteristics were weak memory, ob-
tuse wit, lack of will, quiet mood, and indolent understanding: psycho-

logical characteristics whose exaggeration is almost comparable to that of those individuals who questioned whether Americans were or were not rational beings.)

The idea that the people and "cities in the tropical and semitropical atmospheres of the Americas were intrinsically unhealthy" and inferior was a widespread belief among scientists at the time.[10] De Pauw, in particular, was "convinced that the American climate had wiped out the will, foresight, and memory of natives," obstructing the development of their mental faculties.[11] The purpose of Valdizán's dissertation was to demonstrate that de Pauw's ideas were wrong by showing that indigenous peoples had developed mental abilities before the arrival of Europeans. To prove it, he argued, it would be necessary to demonstrate that they suffered, categorized, and treated mental illnesses.

Valdizán delved into *crónicas de Indias*, Quechua dictionaries, folklore, and medical publications to trace Quechua words that indicated different emotions and mental illnesses. Some of the examples he found include *llaqui* (suffering), *llaquiscca* (depressive), *llaquipayani* (having empathy for the suffering of somebody else), *muspay* (bewilderment), *muspani rimayta* (speech impairment), *muspani upiyaspa* (loss of judgment caused by alcoholism), and *utik* (insane).[12] The list contains specific descriptions and brief analyses of these afflictions. Besides his own good understanding of Quechua, the research on precolonial archaeology and indigenous healing practices carried out by other physicians such as Julio C. Tello, Ruiz Huidobro, Daniel Lavoreria, Antonio Lorena, and Edmundo Escomel proved crucial for enriching and organizing Valdizán's ideas.[13] I want to emphasize this because his work was not a rare case at that time. He was part of a whole generation of intellectuals who were immersed in what historian Alberto Flores Galindo calls "el horizonte utópico" (the utopic horizon),[14] the years in which *indigenismo* emerged as a cultural and political project in Peru, 1910–1930. What distinguishes Valdizán's work from his contemporaries' is the bridge he established between indigenous culture and mental health that essentially founded the field of medical anthropology in Peru.[15]

However, as Cueto indicates, Valdizán did not attempt to understand indigenous culture on its own terms, but in contemporary scientific terms.[16] As a consequence, his reader can detect a tension between his description of an old indigenous civilization as capable of producing knowledge and of present indigenous people as backward. Even in *La medicina popular peruana*, in which he and Maldonado value some contemporary indigenous healers' practices for their medical potential, the authors characterize indigenous people as "savages."[17] In Valdizán's 1918 article "La chicha, bebida de los primitivos peruanos" (Chicha, a beverage of aboriginal Peruvians), he refers to the indigenous race as "degenerated," condemning this beverage—which was brewed in pre-Republican times with a high percentage of alcohol—as being one of many factors that caused that condition. In his conclusion, however, he ambiguously exculpates contemporary chicha drinkers because by then the beverage was brewed with a very low percentage of alcohol and thus had no perceptible effect in the cases of alcohol poisoning and hallucinations registered in the Asilo Colonia de Magdalena.[18] Efraín Kristal has explained that this tension was not problematic at the time for many Peruvian intellectuals who idealized the Inca civilization and attributed the supposed degeneration of contemporary indigenous people to colonialism and the legacy of its social structures.[19]

Here is one example of Valdizán's style of interpretation: he describes how El Inca Garcilaso and Pedro Pizarro documented in their respective chronicles a delirium-causing fever that affected the Inca rulers Atahualpa and Huayna Capac. But to his dismay, Valdizán does not find in their accounts either categorizations or treatments for the different types of delirium caused by fever, as they were classified by his contemporary European psychiatrist Richard von Krafft-Ebbing, for whom the kind of delirium suffered depended on the type of illness or infection causing the fever. In *La alienación mental*, Valdizán states, "Es de creerse que los antiguos peruanos . . . conocieron del delirio febril pero que no lo relacionaran a determinada enfermedad y le aplicaron tratamiento común a todas las manifestaciones de *rupa* que de ellos fueron conocidas" (It is possible that ancient Peruvians . . . knew about delirium caused by fever but did not associate them with specific illnesses and applied the same treatment to all the cases).[20]

For Valdizán, Western science was the norm for establishing the taxonomy of mental illnesses. Examples such as the previous one as well as the Quechua words listed above were intended to demonstrate that ancient Peruvians were aware of mental illnesses but also that they were greatly confused regarding their categorization. It is not clear, he argues, whether they were able to differentiate between

> formas congénitas de déficit psíquico de aquellas adquiridas; tal vez sí
> llegaron a establecer diferencia entre los trastornos del carácter y aquellos
> de la inteligencia; pero no es posible aventurar que los conocimientos de
> los primitivos peruanos en materia de Siquiatría fueran más lejos.[21]

(inherited and acquired mental diseases; perhaps they were able to establish a difference between personality disorders and intellectual disabilities; but it is not possible to maintain that ancient Peruvians' knowledge of psychiatry went any further.)

Nevertheless, he also maintains that if the Andean cultures lacked more "refinement" in their mental afflictions—that refinement so common in Romantic poetry, in his words—it is because they were not an old and decayed culture like the European; they were somehow more natural and more authentic, which he viewed favorably. Even though Valdizán's attempt to validate indigenous knowledge did not see beyond European scientific standards of the day, it is important to recognize that he broke with those standards to a degree when he included indigenous knowledge as part of the medical debate. These contradictions can be understood as indicative of his two-way strategy to pave the way for a national history of medical sciences.

The subfield of sexual pathologies illustrates some of the complexities of Valdizán's project. In the nineteenth and early twentieth centuries, debates among medical professionals about so-called sexual perversions, which were considered mental health disorders, flooded the Western medical literature. Along these lines, and influenced by his friend the Peruvian physician and archaeologist Julio C. Tello, Valdizán used precolonial pottery, in particular the so-called *huacos eróticos* (erotic pottery), to study the reach of indigenous expertise on mental illness. For Valdizán, the knowledge of eroticism and sexual perversions would prove that ancient indigenous cultures had some sort of understanding regarding mental illnesses, which in turn would prove that they were capable of conceptual elaborations similar to those developed by contemporary European sexologists. Valdizán explains in *La alienación mental,*

> Las fotografías de huacos pornográficos que ofrecemos como ilustración de este trabajo, manifiestan bien claramente que en este campo de las aberraciones del espíritu . . . los primitivos peruanos no tenían nada que envidiar a los pueblos más corrompidos de la tierra. (84)

> (The pictures of pornographic pottery that illustrate this work clearly show that regarding perversions of the spirit . . . primitive Peruvians did not have any reason to be jealous of the most corrupt societies on earth.)

In his analysis of a pottery artifact that shows three men masturbating, he reproduces some of the moral biases of his time:

Este . . . huaco ofrece la particularidad de que los tres sujetos parecen rep-
resentaciones de muerte; son tres esqueletos o tres sujetos esqueléticos.
Diríase que el artista ha querido representar los efectos del abuso del vicio
sobre el organismo. (84)

(The distinctive feature of this artifact is that the three men look like representations of corpses. They are either three skeletons or three skeletal
men. It is possible to suggest that the artist wanted to represent the effects
of this vice on the body.)

He considers that the items of erotic pottery were "pruebas indiscutibles de
la existencia, entre los antiguos peruanos, de las perversiones sexuales, aun de
aquellas que vulgarmente vienen consideradas como verdaderos refinamien-
tos morbosos de la civilización actual" (97) (undeniable proof of the existence
among ancient Peruvians of sexual perversions, including those that are commonly considered true morbid refinements of current civilization). Valdizán's
method of analyzing pottery is completely speculative and guided by what he
believed ancient Andean people knew or did not know as compared to contemporary Western knowledge. His elaborations are grounded in the analysis of
strange gestures and disproportions in the faces and bodies represented in the
pottery. Although he briefly notes the possibility that some of the artifacts were
made with the main purpose of either entertainment or religiosity, he argues
that these possible interpretations do not change the fact that ancient Peruvians
had an irrefutable knowledge of sexual perversions.

But despite Valdizán's speculative approach, as Arthur Kleinman states, "a
psychiatric diagnosis implies a tacit categorization of some forms of human
misery as medical problems."[22] To carry out psychiatric diagnoses based on
historical characters and ceramic figures implies recognition of the suffer-
ings that affected ancient indigenous people as medical problems. Beyond
the conceptual limitations of his perspective on sexual perversions, Valdizán
maintained his attempt to underscore the relevance of indigenous people and
their cultural knowledge for the national history and practice of medicine.
As an example, he acknowledges the importance of the healing practices of
brujos and *curanderos*:

Antiguos peruanos adoptaron, en el tratamiento de la alienación mental,
algunos recursos terapeúticos empleados en el tratamiento de otras enfermedades: la *balneación*, la *sangría*, la *trepanación*. Pero es evidente que,
siempre, la mayor energía terapéutica fue confiada a la acción . . . de los he-

chiceros. Y que éstos ... recurrieron a las virtudes terapéuticas de algunos vegetales cuya acción sobre el sistema nervioso no les era desconocida.[23]

(Ancient Peruvians adopted in the treatment of mental derangement some therapeutic techniques employed in the treatment of other illnesses: *balneotherapy, blood-letting, trepanation.* But it is evident that they mostly trusted the work of *healers* for treatments. And that these [healers] ... resorted to the therapeutic virtues of some plants whose effect on the nervous system was not unknown to them.)

He emphasizes indigenous people's capacity to produce a knowledge that was not a by-product of European science. Seven years later, that idea is reinforced in *La medicina popular peruana*: "estos curanderos conocen algunas de las propiedades de ciertos vegetales y tal conocimiento les permite aprovecharlas en beneficio de los enfermos, orientados siempre en el sentido de la curación sintomática" (these healers know the beneficial properties of certain plants and how to use this knowledge to help sick people, always with the aim of alleviating symptoms).[24] This favorable medical perspective on indigenous healers was not new. As Cueto and Palmer have shown, it came from a long history of encounters between European medicine and indigenous and African healing practices from colonial times on. In seventeenth-century Lima, for example, "the majority of professors [of San Marcos University] opined that a new chair in medicine was unnecessary because of the efficiency of indigenous healers."[25] It is important to mention that in many cases the medical field eventually accepted the popular knowledge about the healing properties of native plants. The use of cinchona bark for the treatment of malaria is without a doubt the most celebrated of those cases.[26]

Valdizán did not explore how indigenous healing practices influenced European medicine. Rather, he mentioned the influence of American illnesses such as syphilis in Europe and the new knowledge and medical practices that developed around them. An important part of his study was to establish the geographical origin of certain illnesses in order to clarify which knowledge and healing practices were produced in the respective continents and which ones were assimilated after cultural contact. For him, the study of these native illnesses became a national priority. Cueto finds this to be a common characteristic of the production of scientific knowledge in the periphery:

The cases in which science has been fully integrated into developing countries suggest that it created a realm of its own within the local culture by relating research activities to particular issues concerning the country's interests. . . . Nationalism manifested itself, although not exclusively, not only through the thematic selection of research topics—such as the study of native diseases—but also in the content of science itself.[27]

To fulfill the national interests, Valdizán considered it important to circulate his ideas beyond the medical field. He advocated that the path to success for the mental health sciences was to become popular with the public so that people would eventually implement physicians' ideas in their private lives. It was in this spirit that Valdizán would publish a series of medical cases as stories in the newspaper *La Prensa* in 1915, gathered later in his book entitled *Historias de enfermos* (Stories of patients, 1923). In its introduction, he states that he had one and only purpose for publishing these medical cases, which he believed he had accomplished: "formar, en el público de Lima, una conciencia psiquiátrica . . . obligar al público de Lima a aceptar la especialización médica en Psiquiatría, de manera análoga a como la aceptaba . . . en Pediatría, en Ginecología, etc." (to form a psychiatric consciousness in the public of Lima . . . to force the public to accept the medical specialization of psychiatry in the same way that they accepted pediatrics, gynecology, etc.).[28] Cueto has indicated that the efforts by Valdizán and other scholars to disseminate scientific ideas within popular culture was crucial for generating social enthusiasm for science in Peru in the early twentieth century. This trend would change from the 1940s onward, when scholars prioritized isolated research and publication in specialized journals as the path to scientific excellence.[29]

Valdizán's interest in connecting science with local culture had two more objectives. One was to help doctors and medical students understand the particularities of their patients' worldview and how those could affect their diagnoses and treatments. In his 1918 article "La psicoterapia extrapsiquiátrica," Valdizán asks doctors to reflect on "el compromiso psíquico incuestionable que toda enfermedad representa" (the unquestionable psychic dimension that every disease implies) and the need to "atender a dicho compromiso, con idéntica solicitud e idéntico afecto con que atiende al órgano enfermo" (address this dimension, with the same solicitude with which [they] take care of a sick organ).[30] He adds,

Tendrá ventajas indiscutibles para ustedes conocer . . . nuestra Medicina Popular, para poderse dar cuenta exacta de muchas consultas que les serán dirigidas por el enfermo o por la familia del enfermo, a propósito de ciertos agentes terapéuticos que obran verdaderos milagros psicoterápicos en manos de los curiosos de la Medicina Popular. . . . [Los estudios del folklore médico], aparte de su importancia desde un punto de vista netamente antropológico, tienen la no pequeña importancia de ponernos a los médicos en relación estrecha con el criterio médico de nuestros enfermos. Y la ventaja no es despreciable, como queda dicho: Ustedes hallarán muchos enfermos agotados que implorarán de ustedes la licencia necesaria para tomar unos caldos de determinada especie de pescado. . . . Ellos tienen una leyenda de milagrosas curaciones y de maravillosas derrotas de la enfermedad y de la muerte. . . . Ustedes no harán mal en tomar en serio las virtudes freno-tónicas de estos caldos; . . . no harán mal en permitir, si no hay indicación formal en contrario, el empleo de estos caldos; . . . no harán mal tampoco en participar de los entusiasmos del enfermo al considerarse beneficiado por la popular preparación.[31]

(Discovering our traditional medicine will bring undisputed advantages to help you respond accurately to the many queries that will be directed to you by the patient or by the patient's family about certain therapeutic agents that perform true psychotherapeutic miracles in the hands of people curious about our traditional medicine. . . . [Studying our folk medicine], quite apart from its importance from an anthropological perspective, is very important in bringing doctors closer to understanding their patients' own medical opinions. And this advantage is not irrelevant, as already explained: you will find many exhausted patients who will implore you for permission to take some broth of a certain species of fish . . . about which there are legends of miraculous cures and wonderful victories over illness and death. . . . You will do well to take seriously the virtues of these broths; . . . you will not do wrong in allowing, if there is no formal contraindication, the use of these broths; . . . neither will you do wrong in echoing your patient's enthusiasm when he considers that he has improved thanks to the layman's broth cure.)

The study of folk medicine would allow doctors to grasp patients' approach to medicine and thus to implement medical treatments that their patients would not refuse.

The other main objective of Valdizán's publications on local culture and traditional medicine was to put medical students in contact with "problemas netamente nuestros" (problems that are specifically our own) that would benefit from zealous investigation on the part of young people.[32] In his 1919 book *Locos de la colonia* (Lunatics from the colony), he acknowledges the strong nationalist sentiment that propelled his writing, dedicating his work to youth,

> para que ellos, huyendo de las tentaciones de la investigación extranjera, aborden el estudio de los problemas nacionales, de aquellos que, de continuar en nuestra apatía y en nuestra indolencia, nos vendrán estudiados algún día, de fuera, como lección y como reproche.[33]

> (so that they, refusing the temptations of research abroad, should study national problems; those problems that, if we continue in our apathy and indolence, will one day be studied from abroad, as a lesson to be learned.)

Valdizán's words point in two directions: first, praising those brilliant students who decided to stay and do research in Peru (probably a reference to his then colleague Honorio Delgado), and second, urging students to prioritize the study of illnesses that had a national character. He expected future generations to continue the same path of scientific excellence as those doctors who had studied *la verruga peruana* (Peruvian wart, also called Carrion's disease), an illness that received national relevance in scientific and social circles. Therefore, in *La medicina popular peruana*, he and Maldonado advise medical students to focus their research efforts on national illnesses such *el susto*, *el aire*, cocainism, *la nevada*, chicha intoxication, and *chamico* (datura plant) intoxication.

It is important to note that Valdizán and Maldonado discuss not only rural but also urban folk medicine in their book, dedicating dozens of pages to descriptions of how the upper and middle classes in coastal cities like Lima, Trujillo, and Piura made sense of and treated some of their traditional afflictions. The case study of *el aire* typifies the authors' approach:

> En el elemento blanco de la población y en el mestizo existe también la creencia muy generalizada de que el aire puede dar lugar a toda una serie de enfermedades . . . afecciones óseas, articulares, musculares y nerviosas. . . . La gran mayoría de los dolores que no hallan explicación satisfactoria en los conocimientos del vulgo, son atribuidos al aire, a un golpe de aire o a un viento colado y son tratados mediante fricciones de yerbas maceradas en alcohol o mediante la aplicación de azufre en flor.[34]

> (Within the white and mestizo population there is the very popular belief that the air can cause a whole series of diseases . . . bone, joint, muscle, and nerve complaints. . . . Most pains that do not have a satisfactory ex-

planation in people's mind are attributed to air, a blast of air, or a draft, and are treated by massages with herbs soaked in alcohol or by applying sulfur powder.)

To clarify, the authors did not consider a disease such as *el aire* to be scientifically accurate per se but acknowledge that behind its different manifestations there were real physical and mental health issues as well as remedies, treatments, and placebo effects that deserved the serious attention of the medical field. The authors also explored how religion affected people's medical criteria, underscoring the role of Catholicism in traditional Peruvian medicine throughout social classes and ethnic groups. They did not consider one group to be more modern than or innately superior to the others. They even compared some of Peru's traditional medicine with Europe's, finding similarities and differences. In their view, the role of Peruvian scholars was to reappropriate those differences as scientific concerns in order to build a more solid national history and practice of medicine.

To conclude, in his writings Valdizán faced an ongoing dilemma within the medical sphere: the separation between mind sciences and clinical medicine.[35] He advocated for the urgency of taking into consideration patients' mental health when physicians carried out any kind of clinical treatment. This approach allowed him to explore factors and practices beyond the limits of modern medicine. Within a cultural framework that emphasized the construction of a national character, he underscored the relevance of indigenous knowledge in building a national medical science. For him, indigenous healing practices and other forms of traditional medicine had an inherent scientific value that was crucial to an understanding of the nation's peculiarities. Peru's contribution to global medicine, as Valdizán observed, lay in the study of these peculiarities and not in blindly copying European medicine. His work on indigenous medicine marked the beginning of medical anthropology in Peru.

While few were to follow in his path in the twentieth century, there has been a significant rise of interest in this field and in traditional medicine in more recent years.[36] The work of Kimberly Theidon and the studies undertaken for Peru's Truth and Reconciliation Commission, for example, have highlighted the importance of medical anthropology in a broader understanding of political violence, memory, transitional justice, and reconciliation in contemporary Peru. The Peruvian government has added two institutes for the research of traditional medicine within its health care system, the Instituto de Medicina

Tradicional (Institute of Traditional Medicine) in 1991 and the Centro de Investigación Clínica de Medicina Complementaria (Clinical Research Center of Complementary Medicine) in 2016. Almost a century after his work, these state institutes are finally and officially carrying out Valdizán's medical project.

Notes

1. See Cueto, *Excelencia científica en la periferia,* 22.
2. Mariátegui, *Hermilio Valdizán,* 85–93.
3. Cueto, *Excelencia científica en la periferia,* 81–90.
4. Cueto and Palmer, *Medicine and Public Health,* 68.
5. Latour, *Pasteurization of France,* 19–22.
6. Cañizares-Esguerra, *How to Write the History.*
7. Valdizán, *La alienación mental,* 9.
8. Cañizares-Esguerra, *How to Write the History,* 13.
9. Valdizán, *La alienación mental,* 9.
10. Cueto and Palmer, *Medicine and Public Health,* 87.
11. Cañizares-Esguerra, *How to Write the History,* 118–119.
12. Valdizán, *La alienación mental,* 10–20.
13. Edmundo Escomel's work did not influence Valdizán's dissertation in 1915, but it was extremely influential in his and Ángel Maldonado's 1922 book, *La medicina popular peruana.*
14. Flores Galindo, *Buscando un Inca,* 237–284.
15. Even though Valdizán's advances in this area would be affected by his early death, other physicians such as Edmundo Escomel, Humberto Rotondo, Carlos Seguín, and Oscar Valdivia would continue exploring and produce sporadic publications throughout the twentieth century. In the mid-twentieth century, most of the mental health debates about indigenous people were focused on how internal migration affected them.
16. Cueto, *Excelencia científica en la periferia,* 24.
17. Valdizán and Maldonado, *La medicina popular peruana,* 9.
18. Valdizán, "La chicha."
19. Kristal, *Una visión urbana,* 30–35.
20. Valdizán, *La alienación mental,* 57.
21. Valdizán, *La alienación mental,* 20.
22. Kleinman, *Rethinking Psychiatry,* 8.
23. Valdizán, *La alienación mental,* 97. Emphasis in the original.
24. Valdizán and Maldonado, *La medicina popular peruana,* vi.
25. Cueto and Palmer, *Medicine and Public Health,* 14.
26. See Crawford, *Andean Wonder Drug.*
27. Cueto, "Excellence in Twentieth-Century Medical Science," 235.
28. Valdizán, *Historias de enfermos,* vii.
29. Cueto, *Excelencia científica,* 112–113.
30. Valdizán, "La psicoterapia extrapsiquiátrica," 250.

31. Valdizán, "La psicoterapia extrapsiquiátrica," 265.

32. Valdizán and Maldonado, *La medicina popular peruana*, 396.

33. Valdizán, *Locos de la colonia*, 8.

34. Valdizán and Maldonado, *La medicina popular peruana*, 94.

35. Rhodri Hayward states, "In clinical work, psychological phenomena are used to define the limits of medical knowledge" ("Medicine and the Mind," 524).

36. In their article "Medicina tradicional y medicina moderna en México y el Perú: Valorización y explotación," Douglas Sharon and Rainer Bussmann trace some of the major advancements and struggles in the institutionalization of traditional medicine in Peru since the late 1970s.

References

Cañizares-Esguerra, Jorge. *How to Write the History of the New World: Histories, Epistemologies, and Identities in the Eighteenth-century Atlantic World*. Stanford, CA: Stanford University Press, 2001.

Crawford, Matthew James. *The Andean Wonder Drug: Cinchona Bark and Imperial Science in the Spanish Atlantic, 1630–1800*. Pittsburgh, PA: University of Pittsburgh Press, 2016.

Cueto, Marcos. *Excelencia científica en la periferia: Actividades científicas e investigación biomédica en el Perú, 1890–1950*. Lima: CONCYTEC (Consejo Nacional de Ciencia, Tecnología e Innovación Tecnológica), 1989.

———. "Excellence in Twentieth-Century Biomedical Science." In *Science in Latin America: A History*, edited by Juan José Saldaña, 231–240. Austin: University of Texas Press, 2006.

Cueto, Marcos, and Steven Palmer. *Medicine and Public Health in Latin America: A History*. New York: Cambridge University Press, 2015.

Flores Galindo, Alberto. *Buscando un Inca: Identidad y utopía en los Andes*. Lima: Horizonte, 1988.

Hayward, Rhodri. "Medicine and the Mind." In *The Oxford Handbook of the History of Medicine*, edited by Mark Jackson, 524–542. Oxford: Oxford University Press, 2011.

Kleinman, Arthur. *Rethinking Psychiatry: From Cultural Category to Personal Experience*. New York: Free Press, 1988.

Kristal, Efraín. *Una visión urbana de los Andes: Génesis y desarrollo del indigenismo en el Perú, 1848–1930*. Lima: Instituto de Apoyo Agrario, 1991.

Latour, Bruno. *The Pasteurization of France*. Cambridge, MA: Harvard University Press, 1993.

Mariátegui, Javier. *Hermilio Valdizán: El proyecto de una psiquiatría peruana*. Lima: Minerva, 1981.

Pauw, Cornelius de. *Recherches philosophiques sur les Américains, ou Mémoires intéressants pour servir à l'Histoire de l'Espèce Humaine. Avec une Dissertation sur l'Amérique & les Américains*. London, 1771.

Sharon, Douglas, and Rainer Bussmann. "Medicina tradicional y medicina moderna en México y el Perú: Valorización y explotación." In *Por la mano del hombre: Prácticas y creencias sobre chamanismo y curandería en México y Perú*, edited by Silvia Limón

Olvera, Luis Millones, and Claudia Rosas Lauro, 425–458. Lima: Fondo Editorial de la Asamblea Nacional de Rectores, 2014.

Valdizán, Hermilio. *La alienación mental entre los primitivos peruanos*. Lima: Universidad Mayor de San Marcos, 1915.

———. "La chicha, bebida de los primitivos peruanos." *Revista de psiquiatría y disciplinas conexas*, no. 2 (1918): 62–77.

———. *Historias de enfermos*. Lima: Talleres Gráficos del Asilo Víctor Larco Herrera, 1923.

———. *Locos de la colonia*. Lima: Instituto Nacional de Cultura, 1988 [1919].

———. "La psicoterapia extrapsiquiátrica (lección inaugural del curso de 1918)." *Anales de la Facultad de Medicina* 1 (1918): 250–271.

Valdizán, Hermilio, and Ángel Maldonado. *La medicina popular peruana*. Lima: Consejo Indio de Sudamérica, 1985 [1922].

III

Science and the Modern Nation

Introduction to Section III

MARÍA DEL PILAR BLANCO AND JOANNA PAGE

As outlined in the introduction to this volume, one of the characteristic qualities of Latin American science is its close and complex imbrication with politics in the region. Among other spaces, eighteenth-century *criollo* critiques of metropolitan political power in the Americas were also played out in the theater of science. This is evident, as Antonio Lafuente has demonstrated, in such examples as the botanical study carried out in Mexico between 1801 and 1804 by Mariano Mociño and Luis Montaña to test the validity of each plant's medicinal qualities; this set of experiments became a platform on which local scientists contested the imposition of European methods and theories by the metropolis.[1] As Lafuente explains, such experiments reveal to the historian of science that elite *criollos* believed there was "not just one scientific center, but many."[2] This perspective held by eighteenth-century Latin Americans casts productive doubt on the universality of science while signaling the importance of rationalizing scientific discovery and observation as culturally, socially, and locationally diverse.

The interconnection between science and politics, while constant across centuries, takes on a number of different shapes as Latin American nations gained independence in the nineteenth century. For many thinkers across that postcolonial period, scientific progress went hand in hand with notions of national identity and prosperity. Francisco Miró Quesada describes this phenomenon as a "mitoide" (mythoid), which allowed the state to impose an "*ethos* colectivo" (collective ethos) that prioritized the absolute importance of science to the organization of the modern nation.[3] The chapters in this section explore this relationship, underlining the epistemological problems that lie at the heart of science as "mythoid" in post-independence Latin America.

It would be difficult to overlook the significant role played by the Prussian naturalist Alexander von Humboldt (discussed briefly in the introduction to Section I and in Carlos Fonseca's chapter) in the construction of scientific imaginaries in and of Latin America. In his travels across the so-called torrid zone at the end of the nineteenth century, at that point a nearly impossible feat for a non-Spanish subject, Humboldt alerted the Spanish American intelligentsia to the singularity of the natural and cultural wealth of their region, as well as the many possibilities the region offered in terms of scientific discovery. Humboldt was a model of the Enlightened man of science; open to the wonders of the natural universe and equipped with scientific instruments, he was meticulous in his empirical observations and painstaking in his taxonomies. Among Humboldt's admirers were many of the figures who would later play key roles in the wars of independence that broke out across South and Central America in 1809, including Andrés Bello, one of the intellectual architects of Spanish American independence. When Humboldt visited Caracas in 1799–1800, he invited Bello, then finishing his university studies, to join him and Aimé Bonpland on an expedition to ascend Mount Ávila in the Caracas Valley.[4] Bello's admiration for Humboldt and his contributions are clear in the copious writings he produced throughout his career. His famous "Silva a la zona tórrida" (1826), in which Bello implores the "jóvenes naciones" (young nations) of Latin America to close the wounds of war and come close to nature, is a well-known example of the powerful links he forged between natural observation, on the one hand, and intellectual and political emancipation, on the other.

However, the interest in Humboldt among nineteenth-century Latin Americans also points to a persistent Eurocentrism in their own approach to scientific and national advancement. This is evident in the profound legacies of scientific philosophies such as Comtean positivism, which was adopted to varying degrees by a number of burgeoning Latin American states as a way of imposing order in these societies. It is also manifest in the solidification among many Latin American thinkers in the late nineteenth century of a theory of underdevelopment to account for Latin America's relations with Europe and the United States. This persistent look northward to the United States and across the Atlantic to western Europe rather than to the new nations themselves for scientific inventiveness and development often posed a problem for historians of Latin American science in the twentieth century, as Juan José Saldaña observed in the

1990s. All too often, accounts of the region's scientific culture focused on the underdevelopment of scientific infrastructures, leading to the representation of a *desire* for science rather than a working science. Furthermore, Saldaña argues, the Latin American historian of science continuously faces an "embarrassing situation": "on the one hand, the assertion of the universal and positive character of scientific knowledge; on the other, the generally recognized contextual nature of scientific activity."[5]

The chapters brought together in this section recognize the conundrum that Saldaña poses, but they also turn it on its head, advancing as they do a broader, more historical awareness of Latin American scientific endeavor that entertains science's hegemonic power and simultaneously interrogates its local and popular relevance. For one, the authors draw attention to varying forms of activity and interest in science in different Latin American locations at different points in the nineteenth and twentieth centuries. Honing in on specific national contexts, they also explore how scientific thinking gave shape to political thinking and, in particular, to the development of the "jóvenes naciones" of Latin America. While pointing to a received perception of universality of science in the nineteenth century, the authors of these essays also give necessary historical nuance to the different, original, and often surprising uses and forms that this combination of science and political thinking take, from the days of emancipation to the first days of the Cold War. Miguel de Asúa's essay outlines the shape taken by scientific discourse in the poetics of post-independence Argentina. Studying anthems and poems written by men of science in this period, de Asúa offers a granular view of how scientific discourse combines with the literary at a moment of deep political hopes. Taking us to the early days of post-independence Colombia, Lina del Castillo draws attention to the figure of José María Samper (1828–1888) and his research into forms of early republican science, including agrimensure, that attempted to undo the legacies of colonialism and its pernicious institutions such as slavery. Del Castillo's analysis of a science that is local but with universal designs offers a more accurate version of postcolonial Latin America's relation to the sciences as one that perceives local science as a way of overcoming the missteps and crises brought on by Spanish colonialism.

While these first two chapters trace scientific and intellectual activity in the decades that followed the wars of independence, Hernán Comastri's opens an archive from Juan Domingo Perón's first presidency in Argentina, which coincided with the first days of the Cold War. This archive, containing letters and plans sent to the government describing new scientific and technical inventions proposed by nonspecialist Argentine citizens, narrates a different moment in the state-science configuration, one that has shifted from the *letrados* or intelligentsias of the nineteenth century to the popular and nonspecialized classes of the mid-twentieth century, showing how an interest in science and technology becomes heterogeneous and, as Comastri argues, nonvertical.

In plotting these different—and to a great extent unknown—histories of Latin America's relations with science, the authors provide a fuller sense not only of the region's long-standing interest in developing a scientific culture but also of the many ways in which this culture has evolved or been implemented over time. If science has indeed been a mythoid, it is one that, instead of purely serving a hegemonic purpose, has had the capacity to encourage and develop creativity among Latin American citizens and to underpin imaginaries that are emancipatory, popular, and democratic.

Notes

1. See Lafuente, "Enlightenment in an Imperial Context," 161–166.
2. Lafuente, "Enlightenment in an Imperial Context," 171.
3. Miró Quesada, "Ciencia y técnica," 74–75.
4. Jáksic, *Andrés Bello*, 36.
5. Saldaña, "Introduction," 4.

References

Jáksic, Iván. *Andrés Bello: la pasión por el orden*. Santiago, Chile: Universitaria, 2001.

Lafuente, Antonio. "Enlightenment in an Imperial Context: Local Science in the Late-Eighteenth-Century Hispanic World," *Osiris* 15 (2000): 155–173.

Miró Quesada, Francisco. "Ciencia y técnica: Ideas o mitoides." In *América Latina en sus ideas*, edited by Leopoldo Zea, 72–94. Mexico City: Siglo XXI, 2006.

Saldaña, Juan José. "Introduction: The Latin American Scientific Theatre." In *Science in Latin America: A History*, edited by Saldaña, 1–28. Austin: University of Texas Press, 2006.

7

Postcolonial Social Sciences of Nineteenth-Century Spanish America

Land Surveys, Comparative Political Sociology, and the Malleability of Race

LINA del CASTILLO

Everyone he knew in Europe deemed the republics of the "Hispano-Colombian" continent either a mystery at best or a monster with fifteen deformed, discordant heads at worst.[1] New Granada's native son José María Samper attempted to demystify "Colombia" for Europeans in 1861.[2] He did this despite the staggering number of works that were circulating in Europe by Spanish American publicists, historians, geographers, novelists, economists, and jurists from cities like Caracas, Bogotá, Santiago de Chile, Buenos Aires, Mexico City, Quito, and Lima. These men had already elucidated the historical, social, political, economic, and ethnographic conditions of what Samper termed the "Colombian" republics.[3] They did so as part of an effort to undo what, according to these nineteenth-century Spanish American figures, were the destructive legacies left by three hundred years of Spanish colonialism.[4] With so many innovative studies by Spanish Americans available, Samper rhetorically wondered why Europe remained deaf to the sociopolitical complexity and innovation of Colombian republics. He knew the unfortunate answer: the confusing and confounding noise produced by the political storms that crashed through Colombia during the first half of the nineteenth century.

Samper wrote his treatise at a time when narratives about chaos, violence, and caudillos drowned out any discussion about how the "Colombian republics" were experimenting with democracy, sovereignty, republican equity, self-determination, and the expansion of suffrage. These experiments allowed Spanish Americans to see themselves at the vanguard of political modernity.[5] Perhaps inadvertently, one of the most resilient inventions that came out of these experiments was rhetorical—the portrayal of the obscurantist, tyrannical, and corrupt legacies of Spanish colonialism. Such representations helped Spanish Americans legitimize the intellectual and political work in which they engaged to root out these colonial legacies. During the first half of the nineteenth century, Spanish American intellectuals believed science could diagnose, treat, and excise those legacies. They pursued a radical new form of political modernity that would allow them to take a systematic approach to understanding and changing their society, their economic structures, and their political processes. Spanish Americans crafted their original contributions to what we now might call modern social sciences at a time when the idea of modern science was gradually emerging as universal yet remained primarily Western.[6] Men like Samper developed unique postcolonial approaches to a variety of *ciencias*, and they did not see themselves as doing so from the margins.[7] Spanish Americans saw their innovations in political, economic, and social structures as quintessentially modern; these were, at their core, scientific innovations.

But these nineteenth-century postcolonial social sciences were not those of the late twentieth century. Early republican Spanish American innovators in the social sciences did not seek to deconstruct power-laden orientalizing discourses in the manner pursued by Edward Said in *Orientalism*, for example.[8] Instead, their aim was to pinpoint supposed lingering colonial legacies that needed to be rooted out such as slavery, corporate land tenure, and limited political access. They saw their postcolonial work as transformative and offered their societies and experiences as examples for the world to emulate. Rather than arguing that the West constructed a colonizable other, nineteenth-century postcolonial intellectuals in Spanish America argued that their societies were further on the path toward political modernity than those anywhere else on the planet. That was precisely because of the work of decolonization performed by the sciences in these new republics. For these intellectuals, *ciencias* encompassed a variety of practices, including journalism, historical writing, cartography, land surveys, jurisprudence, constitutional law, government administration, novelistic social commentary, and popular education.[9] In short, during the early to mid-nineteenth century, Spanish American intellectuals developed nuanced, pluralistic, and original contributions to global social sciences. They did so to undergird their localized republican experiments with an eye toward engendering change throughout the world.

And yet, the innovative nature of Spanish American postcolonial contributions to nineteenth-century social sciences was as difficult to recognize and understand then as it is now. As Samper suggested, the conflict and chaos unleashed by republican experimentation drowned out any favorable evaluations of Spanish American efforts to excise the colonial legacy. Generations of scholars have unwittingly appropriated an array of Spanish colonial legacies invented by Spanish Americans like Samper, only to frame nineteenth-century inventions as the objectively enduring colonial legacies of the region.[10] In the wake of World War II, layer upon layer of narratives reshaped Samper's Colombian continent into a Latin America that belongs in the developing Third World, a place on the periphery of centers of calculation and scientific excellence.[11] Postcolonial approaches to history, political science, and literary criticism further rendered contributions by nineteenth-century Spanish Americans to global social sciences largely invisible.[12] Even recent studies on intellectuals of the period underscore how Samper's writings primarily evince racist, *comprador-*bourgeoisie efforts to elaborate rigid, neocolonial nation-building binaries that demarcated a (European) civilization distant—and that distanced itself—from a frozen-in-time (Colombian) barbarism.[13] This is not to say that Samper or other nineteenth-century writers like him did not formulate rigid binaries, including those pitting civilization against barbarism.

However, I suggest that current postcolonial approaches have so deeply shaped recent scholarship that we have yet to fully grasp the nature of nineteenth-century Spanish American intellectual postcoloniality. Mid-nineteenth-century postcolonial intellectuals working in, from, and on Spanish America sought to root out supposed colonial legacies through an array of technologies, from print culture to land surveying. They challenged suppositions and stereotypes of backwardness and disorder. As noted above, Samper rightly identified scores of intellectuals working in this line. Examining these works as efforts to contribute to a Spanish American nineteenth-century postcoloniality rooted in republicanism allows for alternative historiographical trajectories and interpretations. I offer two examples toward this end.

The first emerges from an 1830s land-surveying catechism developed by the noted New Granadan educator and publicist Lorenzo María Lleras.[14] Lleras was interested in contributing to the array of legal, technological, and sociological tools Spanish Americans developed to carry out the privatization of indigenous communal lands, called *resguardos* in New Granada. Lleras's catechism focused on producing equity in *resguardo* land surveys. His booklet also contributed to the nineteenth-century invention of the *indígena* as a malleable being out of the colonial-era *indio*. After benefiting from a trained land surveyor's detailed calculus of equity, the *indígena* would be transformed into a republican citizen like any other.

The essay then turns to Samper's mid-century invention of comparative political sociology to consider how Spanish Americans offered up their innovative experiments as beneficial to humanity. Samper and Lleras, when taken together, exemplify how early nineteenth-century Spanish Americans not only developed new approaches to emerging social sciences but did so in ways that challenged prevailing European and Anglo-American ideas about race well before the late nineteenth-century adoption and adaptation of eugenics.[15] Rather than argue that human racial characteristics were dependent on genealogically inherited characteristics, Spanish American intellectuals like Lleras and Samper contended that the deployment of sciences in society would transform the institutions that produced human racial groups. The result would be the emergence of a new race of democratic republicans.

Catecismo de Agrimensura

In the midst of the dissolution of Spain's Atlantic monarchy, as the liberal Cortes de Cádiz sought legitimacy and support from Spanish kingdoms in the Americas, they abolished indigenous tribute and granted citizenship to indigenous subjects.[16] In their own bid for legitimacy in the wake of independence, Spanish Americans redeployed the liberal principles of the Cádiz constitution for their own ends. Most notably, they characterized tribute and *resguardos* as two prime colonial-era legacies that needed to be fully excised to make all citizens of the republic equal. Independence transformed colonial-era *indios* into protected *indígenas* who would gain full rights to citizenship once their communal lands were measured, divided, parceled out, and privatized. This transition, together with the abolition of slavery, would allow republicanism to be fully realized.[17] Spanish Americans thereby transformed a liberal measure first implemented by the Cortes de Cádiz to maintain monarchy into a quintessential marker of republicanism. Concrete, scientifically informed practices of land surveying would bring these republican laws to equitable fruition.

Like other new republics in the hemisphere including the United States, New Granada needed to find reliable ways to measure and parcel out land. Land surveys facilitated the identification and sale of government-owned lands in exchange for relief of debt incurred during the wars of independence.[18] Public lands, once surveyed, would attract settlers and colonists from within and from abroad. Borderlands, through surveys and settlement, could become bordered lands.[19]

Lleras was cognizant of international trends in land surveying, yet he was no distanced liberal intellectual seeking to impose foreign models on New Granada. New Granadan land surveyors needed to go beyond abstract trigonometry and geometry to include historical, ecological, economic, political, legal, and social forces, that is, if their land-surveying work was to be effective. Lleras's catechism reveals how he adapted existing international land-surveying methods to New Granada realities.

The full title gives away the central argument; this catechism on land surveying was written specifically for "granadinos," citizens of New Granada.[20] A few key aspects of global land-surveying approaches nevertheless remained. For instance, Lleras painstakingly translated European units of length into the emerging New Granada republic's system of measurement.[21] Divergent regional and local units of measure could be reconciled by using the French decimal metric system. In this sense, Lleras's text was part of a larger effort to facilitate homogeneity and circulation among New Granada's internal markets. A shared system of measurement and a single official currency would allow exchange among different regions within New Granada. Lleras's generation came to be obsessed with breaking down what they saw as colonial-era obstacles to cross-regional circulation. Lleras's proposed solutions joined efforts to build a networked system of roads, steamboat navigation along the Magdalena, and ports that pointed toward internal circulation.[22]

Lleras's text nevertheless went beyond an effort to standardize measurements. His catechism launched an innovative methodology intended to calculate republican equity. Within New Granada, trained surveyors were responsible for transforming lands held in common by indigenous groups in the highlands into individual, privately owned plots. New Granada's future as a republic depended on avoiding what many of Lleras's generation diagnosed as another colonial-era legacy: the perpetuation of a parallel and separate republic of Indians within the polity.[23] Lleras's manual on land surveying taught readers about early nineteenth-century laws that transformed colonial-era *indios* into New Granada's republican-era *indígenas*. Once indigenous communal lands were measured, parceled out, and privatized, then nineteenth-century *indígenas* would be converted into New Granada citizens. Land surveying sought to create individual, private landowners out of the Indians who supposedly had been oppressed by colonial-era lords, corporate bodies, and laws that had forced indigenous populations into ethnic and economic endogamy.

Such conversion would avoid reproducing all the alleged pathologies of Spanish colonialism. Those supposed pathologies were widely accepted as sociological fact among several Spanish American intellectuals.[24] Communal, indivisible indigenous landholding had purportedly disincentivized Indians from working in anything other than agriculture, yet the benefits derived from agriculture were negligible because no tangible individual benefit came from working communal land. Most deplorably, they argued, colonial-era laws on *resguardos* isolated indigenous peoples from wider society and established inheritance rights to *resguardo* land along maternal lines. The effect of these laws perpetuated what Samper later called "autogenésia de la raza," a legal, institutional situation that forced indigenous peoples to avoid mixing with people of a different "raza o casta" (race or caste).[25]

Like Samper, Lleras believed that the equitable distribution of *resguardo* land among *indígenas* was critical for republicanism. Such equity could be scientifically calculated, avoiding grave errors and injustices. Equitable land parceling depended on determining the size of the *resguardo* in its entirety. Different values would then be assigned to lands along the mountainsides that had distinct levels of fertility within the *resguardo*. A census of indigenous people would then facilitate the calculation of how much land was allotted to indigenous families. According to Lleras, however, surveyors needed to be careful interpreters of New Granadan laws concerning census-taking in *resguardos*. Lleras explains, "La lei solo dice, que se atienda al número de individuos, pero, si hubiese querido que la distribución se hiciese entre ellos por partes iguales, lo habría dicho sin dar lugar a dudas" (The law only states that the number of individuals in a family must be accounted for, but if the law intended that the distribution were to be made among each of them equally, then it would have said so clearly to avoid any doubt).[26] The problem for Lleras was that by assigning equal amounts of land to each individual *indígena* within a family, a surveyor would fail to take into account how different people had different rights and obligations in society. Equity, for Lleras, did not mean equality among all *resguardo* inhabitants. The base unit for an individual family member needed to be the single male with no children. All other family members would be counted as fractions of that base unit. The final tally would supposedly yield a proper amount of land for each specific family.

Lleras's *Catecismo de agrimensura* taught land surveyors how to transform sets of ecological, historical, communal, and family variables into tables of numbers and fractions. Land surveying became a complex calculus of republican equity, if not equality. Lleras taught future surveyors the skills they needed to transform a New Granadan race of *indígenas* into Colombian republican citizens. Despite Lleras's best intentions, however, the generation of surveyors trained on his catechism did not settle disputes over the division of indigenous communal lands; disagreements over the process continued for generations. Yet, problems with the division and parceling out of *resguardo* lands surprisingly did not explode in civil war, at least in the highlands surrounding New Granada's capital city, Bogotá. Instead, thousands upon thousands of people—indigenous or not—with a stake in *resguardo* lands appealed to a variety of levers of the state, from provincial governors to legislatures, courts, and appeals, to have their cases heard.[27] Day-to-day transactions seeking to privatize communal land, together with concomitant efforts by some to resist privatization through the emerging New Granada republican state, had indeed forged New Granadan *indígenas* out of colonial-era *indios*. These were *indígenas* who were cognizant of their rights as republican citizens and who understood how republican governance functioned.

Lleras's calculus of equity did not manage to break up corporate Indian communities by the middle of the nineteenth century. Instead, and somewhat ironically, republican national state-makers like Lleras inadvertently created the legal and practical mechanisms that uncannily resembled colonial-era traditions that pitted multiple centers of justice against each other. Lleras's catechism, from his own point of view, may have produced equity by counting single tribute-paying males as whole persons and all others as fractions. After being implemented on the ground by land surveyors, governors' offices were overwhelmed by indigenous petitions claiming that the surveyors had miscounted the number of *indígenas* in each family. Furthermore, while some indigenous community members may have favored *resguardo* privatization, others blocked the process of *resguardo* measurement, census-taking, parcelization, and privatization at every turn. The supposedly ethnically closed corporate Indians did not magically mutate into Lleras's individual property holders engaged in the construction of internal markets and vibrant new republics of racially mixed equals.

Lleras's new mathematics of land surveying nevertheless did contemplate *indígenas* and landownership in radically different ways than did another republic to the north, the American cotton kingdom, which was expanding its territory.[28] The United States offered mathematical calculus of equity for some—settlers moving westward—through land surveying. Western lands became empty vessels, or rather, vessels emptied of their indigenous inhabitants. For Spanish American postcolonial intellectuals like Lleras, though, the mathematics of social equity meant that land could not be assumed to be empty or disembedded from complex local histories of indigenous appropriation, improvements, and rents. Abstract rectangles of equal size like those adopted by republicans in the United States would fail to account for *indígenas* as future republican citizens. Common lands needed to be divided according to variegated spatial patterns that took into account local microclimates, family structures, previous histories of land claims, and the built environment. Lleras, along with fellow Spanish American postcolonial intellectuals, understood that in order for his republic to be just or to even merely assume the definition of republic, it needed to be very different from that of the United States of America.

Postcolonial Political Sociology

While Lleras concentrated on the concrete, localized realities that affected *resguardo* privatization in New Granada, Samper focused on showing how Spanish America's experiments with republicanism could benefit the world. In the 1860s Samper highlighted the successes of Colombian republics where Europe, the United States, and Brazil had failed. Colombians had dismantled the legacies of the colonial period through the abolition of slavery, the advent of democratic republicanism, and the establishment of free trade. While the growing ability of the republican state to arbitrate offered peaceful paths for conflict resolution over *resguardo* privatization, several other republican measures did spark violent civil wars. However, for Samper, violence in Hispano-Colombia was a sign of historical success rather than failure. He invented comparative political sociology to explain the transformative nature of violence in Hispano-Colombia.

The history of science tends to emphasize the European and Anglo-American origins of foundational disciplines. Sociology emerges as a European science, invented in France in the nineteenth century by the likes of Emmanuel-Joseph Sieyés, Auguste Comte, Adolphe Quetelet, Alexis de Tocqueville, and Emile Durkheim.[29] Few in the Western academic world have heard of Samper.

Those who have would be reluctant to categorize him as an original thinker capable of developing a new discipline of comparative sociology. But that is what Samper did. His *Ensayo sobre las revoluciones políticas* focuses on societies, not individuals, as the proper unit of analysis.[30] Samper ably deployed the category of "hecho social" (social fact) as a tool of analysis well before Durkheim.[31] He underscored how political climate and institutions played an overwhelming role in shaping individual potential and actions. To illustrate his point, Samper distinguishes the region he calls "Hispano-Colombia" from the "America" appropriated by the United States.[32] This distinction mattered if European readers wished to understand the nature of the political revolutions that transformed the Colombian republics.

Samper succinctly summarized several egregiously mistaken assumptions about early nineteenth-century Colombian revolutions that many Europeans had taken as fact.[33] Although Samper did not directly cite the Europeans against whom he arrayed his scathing arguments, several nineteenth-century conservative French proponents of the concept of the "Latin Race" did argue that "Latin" nations were not suited for democratic rule, requiring authoritarian rule instead.[34] Samper argued that politically motivated fear and baseless disdain lay behind the European argument that the Spanish American revolutions were evidence of incompetence. In his *Ensayo*, Samper describes how Europeans feared that the consolidation and progress of Colombian democracy "podia tarde ó temprano hacer irrupcion en Europa y destruir, ó por lo ménos socavar profundamente los tronos y las aristocracias é instituciones europeas" (could, sooner or later, emerge in Europe and destroy, or at least thoroughly undermine, European thrones, aristocracies, and institutions).[35] Mired in the oligarchical and aristocratic values and social pathologies of the *ancien regime*, Europe lagged far behind the former Spanish American colonies. According to Samper, Europe's fear of Hispano-Colombian republicanism and its misguided interpretations harmed not only Hispano-Colombia, which might otherwise have shone as a universal beacon of modernity, but also Europe, and the general progress of civilization (12). Europe was not the only place to mischaracterize the nature of Hispano-Colombia's political transformations.

Samper explains in his *Ensayo* how the southern states of America unanimously—and erroneously—believed that Spanish America's abolition of slavery was the root cause of Colombia's problems (211). The widest divergence between the political system of the United States of America and that of Hispano-Colombia by the early 1860s related precisely to the question of slavery. Samper's explanation vigorously rejected any effort to equate the Colombian republics with the constitutional framework employed in the United States. The United States offered only a weak influence over the region, Samper argued. That was because the Colombian republics understood the racist roots of the American revolution:

> Desde luego, la revolucion de los Americanos habia comenzado por una cuestion de derechos sobre el té y otros artículos, y terminando por el reconocimiento de la esclavitud. . . . Como no fué heróica ni generosa, no ejerció fascinacion ninguna sobre los hispano-colombianos. (139)

> (After all, the revolution of the Americans had begun with conflict over taxes on tea and other goods, and ended by accepting slavery. . . . Since it was neither heroic nor generous, this revolution did not have any fascination for Hispano-Colombians.)

The French Revolution, on the other hand, which promised citizenship rights and solidarity with all oppressed peoples, captured Spanish American imaginations.

By the mid-nineteenth century, the Hispano-Colombian revolutions had gone much further than either the French or the American Revolutions. The abolition of slavery, universal male suffrage, racial equality, and *mestizaje* all pointed the Hispano-Colombian republics toward political modernity. These transformations were profound, and conflict was inevitable. Samper argues in the *Ensayo* that passing through these tempests was critical for Hispano-Colombia to resolve problems rooted in the colonial era. Violence was destabilizing, but he contends it would eventually deliver the fully fledged arrival of political republicanism, racial democracy, and free trade in the world (217).

By comparison, he asserts, the United States of America and Brazil, despite their progress in material and intellectual arenas, suffered from the continued existence of those colonial legacies that the Hispano-Colombian republics were in the process of extricating from their midst (218–219). For the United States, the civil war that had exploded over slavery would not easily erase the fact that the "antagonismo entre la raza blanca y los hombres de color es un obstáculo inmenso para el progreso Cristiano y la verdadera democracia, que producirá, lo tememos mucho, las mas terribles calamidades no muy tarde" (218) (antagonism of the white race against men of color is an immense obstacle for Christian progress and true democracy, an antagonism that we fear will produce before long the most terrible calamities). Without a fundamental transformation, the

United States was structurally incapable of incorporating significant sectors of the polity into the promise of democracy.

Brazil's case was similar. Samper drew parallel arguments between the outbreak of civil war in the United States and the potential for war in Brazil: "¿No se deberá temer que un día la Guerra política y social estalle en el Brasil, motivada por la esclavitud, las diferencias de razas y las aspiraciones republicanas?" (219) (Should one not fear that one day a political and social war will explode in Brazil, motivated by slavery, racial difference, and republican aspirations?). For Samper, Brazil was not part of the vanguard of republicanism in the world. It shared with Europe the pathologies of a monarchical regime. With the United States, it shared the social illnesses that came with slavery. As for the rest of the hemisphere that was still under colonial rule, Samper had an overarching prediction: the historical arc of the New World pointed toward self-determination. The struggle would be long, but independence, self-determination, and popular suffrage would come.

Samper's writings are especially interesting precisely because they reject the notion that human races had developed unchanging, inherited characteristics. Samper, writing the same year Franz Boas was born, produced an uncannily similar argument in 1858 to the one that the German-Jewish newborn would later pen in 1911 as an anthropologist.[36] Like Boas, Samper used the terminology of race, yet Samper argued that racial difference did not depend on unchanging inherited characteristics but rather on differences in a society's institutions.[37] In "La cuestión de las razas," Samper emphasizes racial malleability to question the very existence of human races:

> La raza latina no existe en Colombia, como no existe la indígena, ni la africana, llevada con la esclavitud en mala hora. Lo que se encuentra en la gran mayoria de Colombia, es la raza moderna de los demócratas republicanos.[38]
>
> (The Latin race does not exist in Colombia. Neither does the indigenous race, nor the African race, which was brought under the misfortune of slavery. Instead, within the great majority of Colombia, we find the modern race of democratic republicanism).

Similarly, any claim that an Anglo-Saxon "race" existed in the United States worked to "olvidar completamente la historia de la colonizacion y las inmigraciones en aquel pais" (2) (completely forget the history of colonization and immigration of that country).

What, then, explained the differences between the United States, Hispano-Colombia, and Europe? Samper's answer in "La cuestión de las razas" is as simple as it is eloquent: "Es que lo que une ó separa á los pueblos, no es su genealogía, sino su civilización, es decir, los principios que los impulsan y mejoran, ó los estancan y debilitan" (2) (That which unites or distinguishes societies is not due to genealogy but rather to civilization, that is, the principles that propel and improve societies, or that stagnate and weaken them). Institutions could make or break humans. For instance, the honorable, cultured, republican institutions of the northeastern United States, were, unfortunately, "en minoria, y el impulso de la nacion entera pertenece á los defensores de la esclavatura, del filibusterismo y del espíritu invasor del Yankee" (in the minority, and the impulse of the nation as a whole is in the hands of the defenders of slavery, filibusterism, and the invading spirit of the Yankee).[39] Bandits and human traffickers operated with impunity and pushed the United States toward a violent absorption of other societies in the New World. In the wake of the US-Mexican War and the filibustering endeavors of Narciso López and William Walker, Samper joined other Spanish Americans seeking to forge an alliance with Europe so as to "mantener en sus justos limites a esa potencia desbordada, cuyas hordas de bárbaros deshonran la democracia americana" (5) (keep in check this unruly power, whose hordes of barbarians dishonor American democracy).

Samper's sociology addressed twin dangers to democracy posed by a US republic—namely, slavery and imperialism—and did so in far more critical ways than did other European intellectuals like Alexis de Tocqueville, Karl Marx, and Friedrich Engels.[40] Tocqueville's *Democracy in America*, written after a nine-month journey through the United States in the 1830s, may have identified some of the strengths and threats to democracy, yet his treatise offers an apologetic explanation for slavery in the US South.[41] Already between 1858 and 1861, Samper understood that the United States had become a pathological polity, structurally incapable of republican equity because it was built on the twin foundations of industrialized slavery and imperialistic expansion.

Conclusion

The historiographical and geographic categories inherited from the mid- to late twentieth century have managed to render Samper, and the world he was calling into being, almost invisible. Samper's Hispano-Colombia has been replaced with Latin America, a regional category that assumes racial distinctions between "Latins" versus "Anglos."[42] This category hardened in the wake of World War II with the advent of US-based social sciences that refused to see modernity and democracy in a region that resisted intervention by the United States.[43] Postcolonial approaches of the later twentieth century offered incisive readings of colonizing, orientalizing discourses that constructed racialized others, but in doing so, they muted the breadth and complexity of contributions by nineteenth-century postcolonial intellectuals.

Samper's and Lleras's contributions to the social sciences and how they have come to be selectively read or ignored illustrate the geopolitical imbalances that characterize epistemology, particularly when it comes to establishing canons, narratives, and genealogies of knowledge and creativity. Neither Samper nor Lleras stood alone. There were dozens of others like them whose creativity and innovation produced wholly new postcolonial sciences in support of emerging forms of republicanism, yet whose contributions have been rendered largely invisible. We need the epistemological audacity of intellectuals like Samper and Lleras to understand the nuances, the borrowings, and the complexity of the postcolonial period of Spanish American history.

Notes

1. For Samper, Colombia was continental in scope, from Cape Horn through Mexico, and included the islands of the Caribbean. Samper also articulated distinct ethnographic Colombias: Spanish Colombia, Portuguese Colombia, French Colombia, British Colombia, Dutch Colombia, and so forth (Samper, *Ensayo*, xiv–xv).

2. Samper, *Ensayo*, 1.

3. For Venezuela: Rafael María Baralt, Ramón Diaz, Fermín Toro, Pedro José Rojas, José Heriberto García de Quevedo, and the geographer Agustín Codazzi. For New Granada: José María Vergara y Vergara, Florentino González, Cerbeleón Pinzon, José Manuel Restrepo, Joaquín Acosta, José Antonio Plaza, Manuel Ancízar, José Manuel Royo, and Eziquiel Uricoechea. For Ecuador: José Joaquín Olmedo and Manuel Villavicencio. For Chile, Peru, Buenos Aires, Mexico, and other republics: Andrés Bello, José Victorio Lastarria, Miguel Luis Amunategui, Benjamín Vicuña Mackenna, Domingo Faustino Sarmiento, Francisco Bilbao, José María La Fragua, and Alejandro Magariños-Cervantes (Samper, *Ensayo*, 8).

4. Del Castillo, *Crafting a Republic*, 1–26.

5. Sanders, *The Vanguard*, 5–23.

6. For the emergence of modern science as a Western idea during the nineteenth century, see Dear, "What Is the History of Science the History Of?"; Elshakry, "When Science Became Western." See also Raj, "Beyond Postcolonialism."

7. Samper, *Cuaderno*, 1.

8. Also see Varisco, *Reading Orientalism*.

9. Samper's definition of the science of constitutional writing offers an example (*Cuaderno*, 1): "Solución.—Una ciencia es un conjunto de principios puestos en mutual relacion i formando un sistema. . . . La ciencia constitucional es el conjunto de principios que determinan la organización política de los Estados" (Solution: A science is a body of principles that are put into a mutual relationship with each other thereby forming a system. . . . Constitutional science is a body of principles that determine the political organization of states).

10. Stein and Stein, *Colonial Heritage*; Viotti da Costa, *The Brazilian Empire*; Schwartz and Bulmer-Thomas, *Economic History*; Williams, *The Columbia Guide*; Wright, *Latin America*; Eakin, *History of Latin America*. The result has been a "poverty of progress" that implicitly suggests a "poverty of theory" among nineteenth-century historical actors (Burns, *Poverty of Progress*). More recent readings continue to take for granted the supposed resilience of colonial legacies, ignoring the role nineteenth-century actors played in their invention; among these are Adelman, *Colonial Legacies*; Quijano, "América Latina en la economía mundial"; Quijano "Coloniality of Power"; Larson, *Trials of Nation Making*.

11. This rhetorical operation may be seen, for example, in Gobat, "The Invention of Latin America"; Cardoso and Faletto, *Dependencia y desarrollo*; Grosfoguel, "Developmentalism, Modernity"; Basalla, "Spread of Western Science"; Cueto, *Excellence in the Periphery*; Gootenberg, "Forgotten Case."

12. Ashcroft, Griffiths, and Tiffin, *Empire Writes Back, 1–13*; Seth, "Colonial History," 63–85.

13. Martínez-Pinzón, *Una cultura de invernadero*, 13–20 and 51–86; Colmenares, *Partidos políticos*; Delpar, *Liberal Party*, 1–42; König, *En el camino de la nación*; Guillén Martínez, *El poder político en Colombia*; Viñas, *Literatura argentina*.

14. Davis, "Education in New Granada."

15. Leys Stepan, *Hour of Eugenics*.

16. Grandin, *Blood of Guatemala*, 71–72; Echeverri, "Race, Citizenship, and the Cádiz Constitution in Popayán (New Granada)."

17. Samper, *Ensayo*, 60–61.

18. Linklater, *Measuring America*.

19. Hubbard, *American Boundaries*; Linklater, *Fabric of America*; Rebert, *La Gran Línea*; Adelman and Aron, "From Borderlands to Borders"; Dym, "More Calculated."

20. Lleras, *Catecismo de agrimensura*, 1.

21. Lleras, *Catecismo de agrimensura*, 4.

22. Del Castillo, *Crafting a Republic for the World*, 76–121.

23. Thurner, *From Two Republics*, 151.

24. Larson, *Trials of Nation Making*, 71–102.

25. Samper, *Ensayo*, 63.

26. Lleras, *Catecismo de agrimensura*, 29–30.

27. Del Castillo, *Crafting a Republic for the World*, 122–158.

28. Only recently have US scholars tackled these questions. See Johnson, *River of Dark Dreams*; Torget, *Seeds of Empire*; Karp, *Vast Southern Empire*; Baptist, *Half Has Never Been Told*.

29. Heilbron, *French Sociology*.

30. Samper, *Ensayo*, xi.

31. Samper, *Ensayo*, 65, 141; Schmaus, *Durkheim's Philosophy of Science.*

32. Samper, *Ensayo*, xiv–xv.

33. Samper, *Ensayo*, 211.

34. The most famous of these were Michel Chevalier and Jules Michelet. See Chevalier, *Examen du systéme*, 9, 17–18, 61; Chevalier, *France, Mexico*; Gobat, "Invention of Latin America," 1356–1370; Michelet, *France before Europe*; Tenorio-Trillo, *Latin America*, 54.

35. Samper, *Ensayo*, 5.

36. Boas, "Instability of Human Types."

37. Gossett, *Race*, 419–430.

38. Samper, "La cuestión de las razas," 2.

39. Samper, "La cuestión de las razas," 3. This citation retains the original nineteenth-century spellings.

40. Nimtz, *Marx, Tocqueville, and Race in America.*

41. Tocqueville, *Democracy in America*, 354–378. See also Resh, "Alexis De Tocqueville and the Negro."

42. Gobat, "Invention of Latin America."

43. Grandin, "Your Americanism and Mine."

References

Adelman, Jeremy, ed. *Colonial Legacies: The Problem of Persistence in Latin American History.* New York: Routledge, 1999.

Adelman, Jeremy, and Stephen Aron. "From Borderlands to Borders: Empires, Nation-States, and the Peoples in Between in North American History." *American Historical Review* 104, no. 3 (1999): 814–841.

Ashcroft, Bill, Gareth Griffiths, and Helen Tiffin. *The Empire Writes Back: Theory and Practices in Post-Colonial Literatures.* 2nd edition. London: Routledge, 2002.

Baptist, Edward. *The Half Has Never Been Told: Slavery and the Making of American Capitalism.* New York: Basic, 2014.

Basalla, George. "The Spread of Western Science." *Science*, no. 156 (1967): 611–622.

Boas, Franz. "The Instability of Human Types." In *Papers on Interracial Problems Communicated to the First Universal Races Congress Held at the University of London, July 26–29, 1911*, edited by Gustav Spiller, 99–103. Boston: Ginn, 1912.

Burns, E. Bradford. *The Poverty of Progress: Latin America in the Nineteenth Century.* Berkeley: University of California Press, 1980.

Cardoso, Fernando Henrique, and Enzo Faletto. *Dependencia y desarrollo en América Latina.* Mexico City: Siglo XXI, 1969.

Chevalier, Michel. *Examen due systéme commercial connu sous le nom de systéme protecteur.* Paris: Librairie de Guillaumin, 1852.

——. *France, Mexico, and the Confederate States.* Translation by W. M. Henry Hurbut. New York: C. B. Richardson, 1863.

Colmenares, Germán. *Partidos políticos y clases sociales.* Bogotá: Uniandes, 1967.

Cueto, Marcos. *Excelencia Científica en la Periferia: Actividades Científicas e Investig-ación Biomédica en el Perú, 1890–1950*. Lima: GRADE and CONCYTEC, 1989.

Davis, Robert H. "Education in New Granada: Lorenzo María Lleras and the Colegio del Espíritu Santo, 1846–1853." *The Americas* 33, no. 3 (January 1977): 490–503.

De Tocqueville, Alexis. *Democracy in America*. Translation by Henry Reeve. 3rd American edition. New York: George Adlard, 1839.

Dear, Peter. "What Is the History of Science the History Of? Early Modern Roots of the Ideology of Modern Science." *Isis* 96, no. 3 (2005): 390–406.

Del Castillo, Lina. *Crafting a Republic for the World: Scientific, Geographic, and Historio-graphic Inventions of Colombia*. Lincoln: University of Nebraska Press, 2018.

Delpar, Helen. *The Liberal Party in Colombian Politics 1863–1899*. Tuscaloosa: University of Alabama Press, 1981.

Dym, Jordana. "'More Calculated to Mislead Than Inform': Travel Writers and the Mapping of Central America, 1821–1945." *Journal of Historical Geography* 30, no. 2 (2004): 340–363.

Eakin, Marshall. *The History of Latin America: Collision of Cultures*. New York: St. Martin's Griffin, 2007.

Echeverri, Marcela. "Race, Citizenship, and the Cádiz Constitution in Popayán (New Granada)." In *The Rise of Constitutional Government in the Iberian Atlantic World: The Impact of the Cádiz Constitution of 1812*, edited by Scott Eastman and Natalia Sobrevilla Perea, 91–110. Tuscaloosa: The University of Alabama Press, 2015.

Elshakry, Marwa. "When Science Became Western: Historiographical Reflections." *Isis* 101, no. 1 (2010): 98–109.

Gobat, Michel. "The Invention of Latin America: A Transnational History of Anti-Imperialism, Democracy, and Race." *American Historical Review* 118, no. 5 (2013): 1345–1375.

Gootenberg, Paul. "A Forgotten Case of 'Scientific Excellence on the Periphery': The National Cocaine Science of Alfredo Bignon, 1884–1887." *Comparative Studies in Society and History* 49, no. 1 (January 2007): 202–232.

Gossett, Thomas. *Race: The History of an Idea in America*. New York: Oxford University Press, 1997.

Grandin, Greg. *The Blood of Guatemala: A History of Race and Nation*. Durham, NC: Duke University Press, 2000.

———. "Your Americanism and Mine: Americanism and Anti-Americanism in the Americas." *American Historical Review* 111, no. 4 (2006): 1042–1066.

Grosfoguel, Ramón. "Developmentalism, Modernity, and Dependency Theory in Latin America." *Nepantla: Views from South* 1, no. 2 (2000): 347–374.

Guillén Martínez, Fernando. *El poder político en Colombia*. Bogotá: Planeta, 2008.

Heilbron, Johan. *French Sociology*. Ithaca, NY: Cornell University Press, 2015.

Hubbard, Bill. *American Boundaries: The Nation, the States, the Rectangular Survey*. Chicago: University of Chicago Press, 2008.

Johnson, Walter. *River of Dark Dreams: Slavery and Empire in the Cotton Kingdom*. Cambridge, MA: Harvard University Press, 2013.

Karp, Matthew. *The Vast Southern Empire: Slaveholders at the Helm of American Foreign Policy*. Cambridge, MA: Harvard University Press, 2016.

König, Hans-Joachim. *En el camino de la nación: Nacionalismo en el proceso de formación del Estado y de la nación de la Nueva Granada, 1760–1856*. Bogotá: Banco de la República, 1994.

Larson, Brooke. *Trials of Nation Making: Liberalism, Race, and Ethnicity in the Andes, 1810–1910*. New York: Cambridge University Press, 2004.

Leys-Stepan, Nancy. *The Hour of Eugenics: Race, Gender, and Nation in Latin America*. Ithaca, NY: Cornell University Press, 1991.

Linklater, Andro. *The Fabric of America: How Our Borders and Boundaries Shaped the Country and Forged Our National Identity*. New York: Walker, 2007.

———. *Measuring America: How an Untamed Wilderness Shaped the United States and Fulfilled the Promise of Democracy*. New York: Walker, 2002.

Lleras, Lorenzo María. *Catecismo de agrimensura, apropiado al uso de los granadinos*. Bogotá: Imprenta de la Universidad, 1834.

Martínez-Pinzón, Felipe. *Una cultura de invernadero: Trópico y civilización en Colombia (1808–1928)*. Madrid: Iberoamericana-Vervuert, 2016.

Michelet, Jules. *France before Europe*. Translated from the French. Boston: Roberts, Brothers, 1871.

Nimtz, August H. Jr. *Marx, Tocqueville, and Race in America: The "Absolute Democracy" or "Defiled Republic."* Lanham, MD: Lexington, 2003.

Quijano, Aníbal. "América Latina en la economía mundial." *Problemas del desarrollo* 24 (1993): 5–18.

———. "Coloniality of Power, Eurocentrism, and Latin America." *Nepantla: Views from South* 1, no. 3 (2000): 533–580.

Raj, Kapil. "Beyond Postcolonialism . . . and Postpositivism: Circulation and the Global History of Science." *Isis* 104, no. 2 (2013): 337–347.

Rebert, Paula. *La Gran Línea: Mapping the United States–Mexico Boundary, 1849–1857*. Austin: University of Texas Press, 2001.

Resh, Richard. "Alexis De Tocqueville and the Negro: Democracy in America Reconsidered." *Journal of Negro History* 48, no. 4 (October 1963): 251–259.

Said, Edward. *Orientalism*. New York: Vintage, 1976.

Samper, José María. *Cuaderno: Que contiene la esplicación de los principios cardinals de la Ciencia Constitucional*. Bogotá: Reimpreso en la Imprenta Imparcial, 1852. Biblioteca Nacional de Colombia, Fondo Pineda.

———. "La cuestión de las razas." In *La America: Crónica Hispano-Americana* 2, no. 17 (Madrid, November 8, 1858): 2.

———. *Ensayo sobre las revoluciones políticas y la condición social de las Repúblicas colombianas (hispano-americanas): Con un apéndice sobre la orografía y la población de la Confederación Granadina*. Paris: E. Thunot, 1861.

Sanders, James. *The Vanguard of the Atlantic World: Creating Modernity, Nation, and Democracy in Nineteenth-Century Latin America*. Durham, NC: Duke University Press, 2014.

Schmaus, Warren. *Durkheim's Philosophy of Science and the Sociology of Knowledge: Creating an Intellectual Niche*. Chicago: University of Chicago Press, 1994.

Schwartz, Roberto, and Victor Blumer-Thomas. *The Economic History of Latin America since Independence*. Cambridge, England: Cambridge University Press, 2014.

Seth, Suman. "Colonial History and Postcolonial Science Studies." *Radical History Review* 127 (January 2017): 63–85.

Stein, Stanley, and Barbara Stein. *The Colonial Heritage of Latin America: Essays on Economic Dependence in Perspective*. Oxford, England: Oxford University Press, 1970.

Tenorio-Trillo, Mauricio. *Latin America: The Allure and Power of an Idea*. Chicago: University of Chicago Press, 2017.

Thurner, Mark. *From Two Republics to One Divided: Contradictions of Postcolonial Nationmaking in Andean Peru*. Durham, NC: Duke University Press, 1997.

Torget, Andrew J. *Seeds of Empire: Cotton, Slavery, and the Transformation of the Texas Borderlands, 1800–1850*. Chapel Hill: University of North Carolina Press, 2015.

Varisco, Daniel Martin. *Reading Orientalism: Said and the Unsaid*. Seattle: University of Washington Press, 2007.

Viñas, David. *Literatura argentina y realidad política*. Buenos Aires: Jorge Álvarez, 1970.

Viotti da Costa, Emilia. *The Brazilian Empire: Myths and Histories*. Rev. ed. Chapel Hill: University of North Carolina Press, 2000.

Williams, Raymond. *The Columbia Guide to the Latin American Novel Since 1945*. New York: Columbia University Press, 2007.

Wright, Thomas C. *Latin America since Independence: Two Centuries of Continuity and Change*. Maryland: Rowman and Littlefield, 2017.

8

"Una nueva y gloriosa nación"

Patriotic Lyrics and Scientific Culture in the Forging
of Political Emancipation in Río de la Plata

MIGUEL de ASÚA

Science was a significant presence in the Latin American revolutions of independence, both as a social practice and a symbol of the values of emancipation dear to the Enlightenment.[1] In colonial and early independent Río de la
Plata, *ciencia* could be said in many ways. First, there was *scientia*, the kind of
natural philosophy taught in the University of Córdoba, Argentina, and the
schools in Buenos Aires, broadly based on the Aristotelian world picture with
borrowings to varying extents from seventeenth-century mechanical science.
There was also science in the sense of the experimental inquiry into nature,
as could be found in Joseph Redhead's *Memoria sobre la dilatación progresiva
del aire atmosférico* (1819), the first scientific work printed in Río de la Plata.
Lastly, one can identify *la science*, that characteristic product of the French
Enlightenment with its overtones of progress and rationality. These different
but not necessarily incompatible cultures of science were embodied in networks of sociability, institutions, and texts with an informative, political, and
aesthetic import.[2]

The idiom and imaginary of science emerged in the prose of cultural journalism, which left its mark on the period 1810–1830. In particular, three authors
of patriotic marches written in the aftermath of the first emancipatory political
movement in May 1810 were also embodiments of scientific life in Río de la
Plata: Esteban de Luca, Cayetano Rodríguez, and Vicente López y Planes. The

use of scientific diction in patriotic anthems and other transactions between the postrevolutionary literary corpus and the early attempts at generating a new culture of science are the subject of this essay.

Patriotic Marches

One of the earliest compositions celebrating the first cry of political freedom on May 25, 1810, was the "Marcha patriótica" by Esteban de Luca, published in the *Gaceta de Buenos-Ayres* on November 15, 1810.[3] The plain, rhythmic march was chanted to the accompaniment of music probably composed by the Catalan Blas Parera. Solar images, a token of the patriotic poetry of the period, open and close the poem. In the refrain, the "aurora feliz" (happy dawn) of a new "patria" (homeland) is shining, while in the last stanza the sun in its zenith presides at the gathering of those who hasten to gird the sword in defense of "la patria en cadenas" (the homeland in chains).[4]

On July 22, 1812, the First Triumvirate in charge of the executive power asked the *cabildo* of Buenos Aires to make arrangements for the composition of an anthem. The Franciscan friar Cayetano Rodríguez was commissioned with the task. He wrote the lyrics of a composition set to music by Parera that was performed on November 1 of that year by a children's choir. It seems that this song was not successful, for Vicente López y Planes was asked to write a new anthem by the Asamblea del Año XIII, a constitutional body that proclaimed the liberation of the children of slaves and the freedom of the press while abolishing the titles of nobility and the Inquisition. The anthem was finished on May 9, 1813, and approved two days later. Again, Parera composed the music. It was publicly performed for the first time on May 28 in the Coliseo Provisional, the only theater in Buenos Aires at that time. The first stanza of Vicente López's "Marcha patriótica" proclaims the birth of a new nation attended by a display of traditional symbols representing the three kingdoms of nature through the use of prosopopoeia and synecdoche: "la faz de la tierra" (the face of the earth), the classical laurel wreath, and the vanquished lion figuring the Spanish Empire.[5] The mold is neoclassical. The verse is not uninspired. The register is martial, at times brutal, at times alleviated with a dignified tone, always expressive of the heightened pace of the War of Independence. The elements of humoral medical theory contribute to the bodily pathos of the lyrics: blood, of course, but also the "pestífera hiel" (pestiferous gall) of the "fieros tiranos" (fierce tyrants) and the reviving of burning bones of the Incas. Images turning around bodily combat and the bloody encounter between a warrior and a fierce beast—the valiant "brazo argentino" (Argentine arm), the "fuertes pechos" (strong chests) of the braves—convey a dynamic of liberation.

A Republican Astronomy

Vicente López y Planes must have been equipped with a balanced and politically shrewd personality and endowed with a compromising disposition; serving in one or another capacity in almost all of the governments from 1810 until his death, he navigated safely over the tempestuous waters of the country's political life.[6] One of the more intriguing aspects of López's multifaceted personality was his youthful interest in the sciences, which coincided with his more productive years as a poet. The issue of *La Prensa Argentina* of June 25, 1816, communicated that Vicente Lopez, at that time thirty-one years old, together with Father Bartolomé Muñoz, had observed a total eclipse of the moon on June 9 with an achromatic Dollond refractor telescope. There is evidence that during 1816–1817 López did some serious private study of trigonometry, astronomy, and the use of instruments of observation.[7] For the prediction of the eclipse the observers had used the lunar calendar (*Lunario de un siglo*) calculated by the Jesuit astronomer Buenaventura Suárez, who spent all his working life in the Guaraní reductions.[8] The article, probably authored by Muñoz, underlines that Suárez was an "astrónomo Americano" and presents him as a model of emulation for the "patriotic youth."

Little more than three weeks after the eclipse, the representatives of the United Provinces gathered in a congress in Tucumán on July 9 to sign the Declaration of Independence. Two years later, in his poem "A la Patria en la victoria de Maypo," López indulged in an astronomical reference. In this chant of victory, the persona of the poet, driving the chariot of Phaeton across the heavenly orb, sings the victory of the South American army led by San Martín in the Battle of Maipú, Chile. To the verses "Yo entretanto ocupando / del Grande Tauro el hiperbóreo alcazar" (While I am dwelling / in the hyperborean fortress of the Great Taurus), López adds an instructive footnote explaining that at that time the sun was in the constellation of Taurus.[9]

Cayetano Rodríguez was born in the province of Buenos Aires and in 1777 joined the Franciscans. One year before he was ordained as a priest, he taught the *Physica generalis* course at the Colegio de Montserrat in Córdoba. He eventually moved to Buenos Aires, where he delivered a course of lectures on

Physica particularis that has been preserved.[10] Cayetano Rodríguez is mostly remembered for his political participation in the events following the May Revolution as one of the signers of the Declaration of Independence in 1816 as well as for his journalistic activity and vocal opposition to President Bernardino Rivadavia's secularizing measures in the 1820s.[11] Rodríguez's civic poetry, driven by an unceasing enthusiasm, sounds contrived; a somewhat jarring mechanics of verse is all too evident. Analogous limits apply to his teaching, in which his uncontrolled, eclectic embrace of traditional and modern natural philosophies carried him to inconsistencies. For instance, in the section on cosmology in his *Physica particularis* course, Rodríguez discussed the Ptolemaic, the Copernican, and the Tychonic systems of the world but in the end favored the proposal of an obscure Spanish dilettante who assumed that the Earth was at the center of the universe but oscillated in the direction of the polar axis.[12]

While in courses on natural philosophy the sun was either the center of the universe or an astral body circling around Earth, in the literary corpus of revolutionary Río de la Plata it was a symbol of liberty, of a clear beginning, of glory and enlightenment.[13] In one of his notebooks, Vicente López copied a series of poems from Espronceda, beginning with "Al Sol, himno," first published in 1834: "Para y óyeme ¡oh Sol! Yo te saludo" (Stop and hear me, oh Sun! I hail you).[14] The "Sol de Mayo" (May Sun), in its course from dawn to a triumphal zenith, is a core motif in the poetry of independence in Río de la Plata.[15] In Esteban de Luca's patriotic poem the culminating sun presides at the gathering of the sons of America ("el sol os preside / en su alto zenit").[16] In the poetry of Cayetano Rodríguez the solar topos acquires a particular force: "Saludo al astro bello, / que hoy fija con su luz nuestro destino. / ¡Ah! Su hermoso destello / es muda voz que dice 'Americanos, / no es este el día, no, de los tiranos'" (I hail the fair star / Whose light marks our destiny. / Oh! Its beautiful glittering / Is a mute voice that claims: "Americans, / This is not, oh no, the day of the tyrants").[17] The sun itself presides at the apotheosis of Carlos María de Alvear, victor in the siege of Montevideo, who in turn emulates the bright light of the star.[18] Cayetano's diction is also hospitable to terms pertaining to the vocabulary of natural philosophy, such as the four elements, the celestial orb, the sphere, the Austro and Aquilon.[19] The Franciscan friar never quite forgot his teaching on the system of the world.

Astronomical references in poetry did not pass unnoticed. In the *Argos de Buenos Aires* issue of June 2, 1821, Vicente López y Planes and Felipe Senillosa published their observations of a comet seen in April in the constellation of the Whale. Senillosa was a liberal Spanish military engineer forged in the Napoleonic armies who had acted as the director of an academy of mathematics for officers in the patriotic armies in Buenos Aires.[20] The authors proclaim that the main object of the article is to show that "Buenos-Ayres libre no sea un punto perdido para las ciencias" (the free city of Buenos Aires should not be a forsaken dwelling for the sciences).[21] In a paragraph favoring the survey of the skies from the south, they recall that the astronomer Jerôme Lalande had claimed that winter nights in the meridional countries are longer. The argument and its literary implications are contested in a brief note in the next issue of the periodical (June 9, 1821) by a certain "Britanniqus," who points out that when talking about "meridional" countries, Lalande did not mean the Southern Hemisphere but those regions of Europe near the equator. The same mistake, continues the (apparently British) anonymous critic, is evident "in the good and bad verses of your compatriots," in which they use "mediodía" to mean the south, "taking their figure not from nature as they should, but from the European poets."[22]

On the basis of their observations, Senillosa and López calculated the orbit of the comet and made it known in the *Argos* issue of June 23, 1821. In it, Senillosa (his inflated prose is easily recognizable) claims that Buenos Aires should inscribe its name "besides those of the cultured nations [of the world] in the catalogue of the 120 comets calculated until 1814."[23] He adds that the comet seen from Buenos Aires was different to all of these. At the same time a comet had been observed from Valparaíso from April 8 to May 3, 1821, by Captain Basil Hall of the British Royal Navy and its orbit calculated by J. Brinkley from Trinity College, Dublin; this comet had been previously seen in Europe. In an article that compiled astronomical observations made from Buenos Aires, the astronomer Baron von Zach, then at Genoa, bitterly ironizes about a comet seen at Buenos Aires while a different one was simultaneously observed from Chile: "C'est, depuis que notre planète existe, le premier example d'un tel événement" (This would be the first case of such an event since our planet has existed).[24] In a second paper on the subject, von Zach qualifies the data from Río de la Plata as "grossly false." The Hungarian noble von Zach, who spent much of his life at the service of German grandees, perceived the nexus that connected Senillosa's claim to novel discoveries and his republican rhetoric. The astronomical results of Senillosa and López were in his view just "trickery and boastfulness" to pull the wool over the eyes of "ces nouveaux républicaines" (these new republicans). If all the transactions of the new country were to be of this kind, claimed the cosmopolitan von Zach, "la nouvelle république ira bientôt prendre le chemin de la comète" (the new republic will soon take the path of the comet).[25]

Botanical Dialogues

The *Argos de Buenos Aires* for February 9, 1822, found the carnival taking place in the city to be a "barbarous" festival in which "the most distinguished persons" looked as if they had lost their reason and mixed with "la plebe más grosera" (the roughest rabble).[26] This could explain why Vicente López spent the following year's carnival secluded in his house, patiently annotating with Linnaean nomenclature the botanical illustrations of one of the volumes of José Quer's *Flora española* (Madrid, 1762–1764); Quer had used for his treatise the taxonomy of Joseph Pitton de Tournefort.[27] It seems that López enjoyed local fame as a botanist. In one of the reports on the expedition to the southern frontier of Buenos Aires undertaken by Juan Manuel de Rosas, Juan Lavalle, and the surveying engineer Felipe Senillosa, it is mentioned that samples of the grasses collected south of the Salado River and near Sierra de la Ventana had been given to Vicente López, "whose interest in botanical studies warrants that he has the expertise demanded for their classification."[28]

Bartolomé Muñoz, who joined Vicente López in his astronomical pursuits, was also his associate in botanical studies. Muñoz was a priest who early and forcefully took the side of the emancipatory movement and served as a chaplain during the siege of Montevideo. He was part of an informal but active circle of clerics who cultivated a knowledge of the natural history of Buenos Aires, at the center of which was the Uruguayan patriot Dámaso Larrañaga.[29] Both were among the earlier, if not the earliest, promoters of Linnaeus in Río de la Plata. Very much like his friend Vicente López and other contemporaries, Muñoz was a minor polymath: cartographer, poet, naturalist, editor of patriotic almanacs (1819–1829), amateur astronomer, meteorologist, and chronicler.[30]

Ricardo Rojas has denied that Muñoz's patriotic poems had any value and dismissed him as "subalterno en la historia de una literatura" (subaltern in the history of a literature).[31] Unable to share Rojas's adamantine critical certitude, I think that at least one of Muñoz's poems manifests a feeling for nature rarely heard in Río de la Plata.[32] In the spring of 1826, Vicente invited Muñoz to enjoy the pleasures of botanical conversation in a garden: "Ven, y a la Quinta de Balcanio iremos, / y antiguo culto a Flora tributando / allí también oiremos / al pajarillo" (Come, / And to Balcanio's country villa we shall go / Where celebrating the ancient worship of Flora / We shall hear the small bird).[33] Muñoz's poem responding that he was not able to attend the meeting suggests an appreciation of the natural world that has a ring of authenticity to it: "Antes de mucho tiempo yo ví a Flora. / A Flora ví, y vila que ostentaba / Su mayor esplendor, su mayor gala" (Before long I saw Flora. / Flora I saw, and saw her flaunting / Her greatest splendor and greatest pomp).[34]

Iron

There is an element of parallelism in the lives of Vicente López and Esteban de Luca.[35] They both studied at the Colegio de San Carlos and joined the Patricios regiment and saw combat during the British invasions of Río de la Plata. While López studied law in the University of Chuquisaca, de Luca entered the Academy of Mathematics of the Consulado created by Manuel Belgrano and later served as an apprentice to the Spanish Ángel Monasterio, a founder of cannons and munition whom Mitre famously dubbed "el Arquímedes de la Revolución" (the Archimedes of the Revolution).[36] De Luca, López, and Senillosa were active members of two societies that embodied the furor of sociability that spread in Buenos Aires in the era of Rivadavia, the Sociedad Literaria (1822–1824) and the Sociedad de Ciencias Físico-Matemáticas (1822–1823). The fields of action of both groups were fluid and the exchanges frequent; papers and communications were published in the monthly magazine *La abeja argentina* and *Argos*.[37]

From 1816, Esteban de Luca was in charge of the gun factory (Fábrica de Armas del Estado, also known as Fábrica de Fusiles) in Buenos Aires. One of the major problems he faced was the lack of iron. To compensate for this, he received a large fragment, around 780 kg, of an iron meteorite carried from Chaco. De Luca was able to fabricate two pistols from it that were presented as a diplomatic gift to the incumbent president of the United States, James Madison. At that time, the origin of the iron spread over the vast area of what is now known as Campo del Cielo was an unresolved problem that de Luca addresses in "Dissertation on the Iron from Tucumán," dated February 10, 1816. After lengthy speculations that end in the admission that the origin of the iron is "enshrouded in a veil of mystery," de Luca claims that he expects the new climate of freedom to favor the unveiling of the secrets of nature, in contrast to what happened in the old regime, when the Spanish government "pérfidamente ha privado a los Americanos del estudio de las ciencias naturales, tan útiles y recomendables para la prosperidad de todos los países"[38] (perfidiously deprived the Americans of the study of the natural sciences, so useful and recommendable for the prosperity of all countries). The notion that the Spanish authorities had prevented the education of the people of Río de la Plata with the aim of exploiting their natural resources and dominating them more efficiently was a standard formula in public discourse.

When on December 1, 1813, *La Gaceta de Buenos-Ayres* announced the start of courses to be taught at the Instituto Médico Militar, it proclaimed that "en el bárbaro plan de ignorancia sistemática adoptado por la política antigua, entraba, también, el designio de perpetuar en la América toda especie de enfermedades, impidiendo el progreso del arte de curarlas, como éste debía resultar del examen de las ciencias" (as part of the barbaric plan of systematic ignorance adopted by the old regime, it was intended that all kinds of disease should spread throughout America, by hindering the advance of the art of healing, which would result from the study of the sciences).[39]

De Luca eulogizes the victories of the patriots in lines illuminated by the radiance of the furnace and the density of metallic materials. In his "Canto" to San Martín, the hero of the Battle of Maipú, there is a tactile presence of silver, gold, iron, bronze, and steel.[40] His poem to the victory of Lord Cochrane in El Callao depicts "aterrantes rayos" (terrifying rays) of fiery volcanic matter flowing over the Spanish-held fortress of El Callao.[41] The generous breast of Lord Cochrane also becomes a volcano, inflamed by heroic and sublime fire ("inflamado / de un fuego tan heroico, tan sublime").[42] The theater of the battle is an "Etna mugidor en noche oscura" (roaring Etna in a dark night), vomiting a sea of ardent lava against a background of lightning and the clapping of thunder; the poem is a celebration of Cochrane's incendiary rockets.[43] Resounding thunders, "acero fulminante" (fulminating steel), and lightning bolts sent from the sky also abound in De Luca's more famous "Canto lírico a la libertad de Lima."[44] In an ode written in 1817, he sings of the glories of sword making: "Cíclopes son sus hijos [de Cochabamba] / que de Vulcano mejorando el arte/activos acicalan las espadas" (Cyclops are her [Cochabamba's] sons / Who improving Vulcan's art / Among hard and well accomplished works / They keenly burnish the swords).[45]

Manuel Moreno, who during his exile in North America had studied medicine at the University of Maryland, was named professor of the subject in the newly created Universidad de Buenos Aires.[46] In the October 15, 1822, issue of *La abeja argentina*, Moreno takes up the question of the origin of the iron from Tucumán.[47] He proposes the idea that the metallic fragments were "meteoric stones of different magnitudes" that had been formed in the upper regions of the atmosphere. This hypothesis had been already advanced in an 1820 article by Edward Howard in the *Philosophical Transactions of the Royal Society*.[48] Learned opinion was divided on this issue. Two Englishmen in Río de la Plata, Joseph Redhead, better known as the personal doctor of Manuel Belgrano, and Woodbine Parish, the British consul in Río de la Plata from 1825 to 1833 who devoted much of his time to natural history and the trading in fossils, were of the idea that the iron of Tucumán came from volcanic eruptions. The problem was that there were no volcanoes in the vicinity of Campo del Cielo.[49] Moreno had the right instinct, and his report shows that the capacity for scientific discernment among the Argentine-born elite was not negligible. In this regard, at the beginning of his report he points out that he has decided not to address a theoretical topic in chemistry, for the study of this thus-far neglected science "recién es que empieza a introducirse en este país" (had only recently begun to be introduced in this country).[50]

Epilogue

There is no need to proceed beyond Jean le Rond d'Alembert's prologue to the *Encyclopédie* to realize that science was the epistemological backbone of the Enlightenment. As practiced in Río de la Plata, enlightened science was a mild affair, far from the radical overtones that distinguished the vibrant scientific culture of early revolutionary France. But even so, it exhibited two features that made it suitable as an ingredient of literary patriotic discourse in the years surrounding the Declaration of Independence. As a source of intellectual authority, of certainty and reasonability, science chimed in well with the neoclassical register employed by the authors to legitimize the building of a new order through the invocation of the hallowed ideals of ancient republican grandeur.[51] This had a stabilizing function. There was another side to the notion of science, with its connotations of dynamism, progress, and liberation from tradition. In his "Canto al fuego," Bartolomé Muñoz salutes the "benigno fuego" (benign fire), for chemistry owes its progress to it ("La química te debe / sus adelantamientos"), and physics, aerostatics, and pharmaceutical art could have accomplished nothing were it not for the existence of fire.[52] This was a dynamic rhetorical function of science, one that evoked a circle of meanings associated with freedom and newness.

The United States' Declaration of Independence (1776) has been likened to a "scientific paper."[53] Certainly nothing like that could be claimed of the Declaration of Independence of the United Provinces of Río de la Plata (1816), written perhaps by the friar Cayetano Rodríguez. But the common scientific culture associated with the Ibero-American version of Enlightenment permeated the learned elite in Buenos Aires to such a point that it found expression in the nascent mythology of civic symbols expressive of the new state of things.

Juan Bautista Alberdi was the first to point out the lack of harmony between neoclassical forms associated with the colonial past and the revolutionary contents of the poetry of independence, a claim linked to his notion that Argentine literature began with his own generation, the Romantics.[54] This view and questions relating to it have been influential through to the present moment. On the other hand, Juan María Gutiérrez, another member of the Generación de 1837, looked benignly on the employment of neoclassical instruments for the expression of revolutionary ideas. Where others perceived a chiasm that called for explanation, Gutiérrez saw the fulfillment of André Chénier's call to write "versos antiguos sobre asuntos contemporáneos" (ancient verses on contemporary affairs).[55] Among all the critics who have studied this period, he has perhaps been the only one to perceive that there was a scientific dimension to the literature of independence. He contended that the two main preoccupations of Esteban de Luca were "la literatura y las ciencias" (literature and the sciences) and imagined Vicente López's muse as "siempre armada del telescopio de Galileo y penetra con él las constelaciones cuyos nombres griegos son tan armoniosos como su lira" (always armed with Galileo's telescope, piercing with it those constellations whose Greek names are as harmonious as his lyre).[56] It is only logical that Gutiérrez, a person of letters and leader of Department of Exact Sciences of the Universidad de Buenos Aires, the most advanced scientific educational institution of the country at the time of his writing, was able to understand that the use of neoclassical forms associated with the values of Enlightenment, of which science was a constitutive element, was not a totally inappropriate medium for conveying emancipatory meanings.

Notes

1. Glick, "Science and Independence."

2. Asúa, *La ciencia de mayo*.

3. [Esteban de Luca], "Marcha patriótica compuesta por un ciudadano de Buenos-Ayres," *Gaceta de Buenos-Ayres*, November 15, 1810, 382. The "Marcha" has been edited as "Canción" in *La Lira Argentina*. For a history of the national anthem, see Buch, *O juremos con gloria morir*, 13–72, and Fernández Calvo, "Breve reseña del nacimiento."

4. Luca, "Canción patriótica," 23–25, lines 4, 29.

5. López y Planes, "Marcha patriótica," 11–17.

6. Piccirilli, *Los López*, 9–21, 94–99.

7. Bartolomé Muñoz and Vicente López, "Observación astronómica del eclipse total de Luna del día 9 de junio de 1916, hecha en Buenos-Ayres," *La Prensa Argentina*, June 15, 1816, 6-9. López worked on geometry problems in January 1816 and copied a series of astronomical tables for 1817 (documents 3852 and 3855, room VII, box 2360, Archivo General de la Nación [hereafter AGN]).

8. On Suárez, see Asúa, *Science in the Vanished Arcadia*, 222–253.

9. López y Planes, "A la Patria," 216, lines 41–42.

10. Rodríguez, *Secunda phisicae pars*.

11. On Rodríguez's life, see Romero Carranza, "Fray Cayetano Rodríguez."

12. Asúa, "Los phisicos modernos."

13. Ricardo Rojas affirms that the solar symbol collapsed the Inti of the Incas with the Apollo of the Greeks, that is, America and Europe (Rojas, *La literatura* argentina, 944). Rojas was smuggling his own intellectual program into this comment.

14. Doc. 3919, ff. 94–96, room VII, box 2362, AGN.

15. Barcia, "Las poesías de 'La lira argentina,'" lxvi.

16. Luca, "Canción patriótica," 25, lines 39–40.

17. Rodríguez, "Oda al día augusto de la Patria," 97, lines 91–95.

18. Rodríguez, "Oda. Al Brigadier Don Carlos María de Alvear," 71, lines 49–51.

19. See, respectively, Rodríguez's "Cuento al caso," 87, lines 171–172 ("los cuatro elementos"); "Himno en las Fiestas Mayas," 131, line 5 ("Del celestial orbe"); and "Oda al día augusto de la patria," 96, lines 46, 63–64 ("en torno de la esfera," "recio aquilón, austro tan fuerte").

20. Asúa, *La ciencia de mayo*, 38–48.

21. López y Planes and Senillosa, "Ciencias," June 2, 1821.

22. "Britanniqus," "Artículo comunicado," *Argos de Buenos Aires*, June 9, 1821, 28–29.

23. López y Planes and Senillosa, "Ciencias," June 23, 1821.

24. Zach, "Lettre VI," 119. See Multhauf, "Zach, Franz Xaver von," 582–583.

25. Zach, "Comète de l'an 1821," 199–201.

26. "Carnaval," *Argos de Buenos Aires*, February 9, 1822, 4.

27. Furlong Cardiff, *Naturalistas argentinos*, 412.

28. Senillosa and Lavalle, "Memoria," 41.

29. Asúa, *La ciencia de mayo*, 117–123.

30. Furlong Cardiff [Beck, pseudonym], "Un benemérito de las ciencias," 53–80.

31. Rojas, *La literatura argentina*, 920–921.

32. It was the poet and historian Emilio A. Breda who first compiled, studied, and revalued Muñoz's poetry (Breda, "Bartolomé Muñoz," 297–330).

33. López y Planes, copied at the end of a letter addressed to him by Bartolomé Muñoz, October 23, 1826, document 3724, room VII, box 2358, AGN.

34. Muñoz, letter to López y Planes, October 23, 1826, document 3724, room VII, box 2358, AGN.

35. For de Luca, see Blasi Brambilla, "Esteban de Luca."

36. Mitre, *Historia de Belgrano*, 211.

37. González Bernaldo de Quirós, "Sociabilidad y opinión pública."

38. Luca, "Disertación sobre el hierro," 554.

39. "Instituto de Medicina," *Gaceta de Buenos-Ayres*, December 1, 1813.

40. Luca, "Al vencedor de Maypo."

41. Luca, "Al triunfo del Vicealmirante Lord Cochrane," 415, line 74.

42. Luca, "Al triunfo del Vicealmirante Lord Cochrane," 413, lines 19 and 8–9.

43. Luca, "Al triunfo del Vicealmirante Lord Cochrane," 416, lines 105–106.

44. Luca, "Canto lírico a la libertad de Lima," 500 and 504–505, lines 71, 178, 213–214.

45. Luca, "Al Superior Gobierno," 156.

46. Quiroga, *Manuel Moreno*, 91–94, 105–113.

47. Moreno, "Memoria sobre el fierro nativo."

48. Moreno, "Memoria sobre el fierro nativo," 278. Cf. Howard, "Experiments and observations."

49. Asúa, *La ciencia de mayo*, 147–153.

50. Moreno, "Memoria sobre el fierro nativo," 278.

51. For a different interpretation, see Poch, "Neoclasicismo y nación."

52. Muñoz, "Canto al fuego," in Breda, "Bartolomé Muñoz," 327–329.

53. Wills, *Inventing America*, 93–166.

54. See the citation in Giusti, "Las letras durante la Revolución," 276–277.

55. In Gutiérrez, "Estudios sobre las obras," 267. The original is "sur des pensers nouveaux faisons des vers antiques" (Chénier, "L'Invention," 186).

56. Gutiérrez, "Esteban de Luca," 81, 61.

References

Asúa, Miguel de. *La ciencia de mayo: La cultura científica en el Río de la Plata.* Buenos Aires: Fondo de Cultura Económica, 2010.

——. "'Los phisicos modernos quasi todos son copernicanos': Copernicanism and Its Discontents in Colonial Río de la Plata." *Journal for the History of Astronomy* 48 (2017): 160–179.

——. *Science in the Vanished Arcadia: Knowledge of Nature in the Jesuit Missions of Paraguay and Río de la Plata.* Leiden, Netherlands: Brill, 2014.

Barcia, Pedro L., ed. *La lira argentina.* Buenos Aires: Academia Argentina de Letras, 1982.

——. "Las poesías de 'La lira argentina' y sus motivos poéticos." In *La lira argentina,* edited by Barcia, lxii–lxxxiv.

Blasi Brambilla, Alberto. "Esteban de Luca: El poeta y la pólvora." *Todo Es Historia* no. 77 (October 1973): 54–65.

Breda, Emilio A. "Bartolomé Muñoz, el poeta de la independencia (17??-1831)." *Investigaciones y Ensayos* 16 (January–June 1974): 297–330.

Buch, Esteban. *O juremos con gloria morir. Historia de una Épica de Estado.* Buenos Aires: Sudamericana, 1994.

Chénier, André. "L'Invention." In *Poésies de André Chénier*, 181–192. Paris: Charpentier, 1840.

Fernández Calvo, Diana. "Breve reseña del nacimiento del Himno Nacional Argentino." *Boletín de la Academia Nacional de la Historia* 2, no. 6 (March 2013): 4–7.

Furlong Cardiff, Guillermo [Eugenio Beck, pseudonym]. "Un benemérito de las ciencias en el Río de la Plata. Bartolomé Doroteo Muñoz (1831–1931)." *Revista de la Sociedad Amigos de la Arqueología* 5 (1931): 53–80.

———. *Naturalistas argentinos durante la dominación hispánica.* Buenos Aires: Huarpes, 1948.

Giusti, Roberto F. "Las letras durante la Revolución y el período de la Independencia." In *Historia de la literatura argentina*, edited by Rafael A. Arrieta, 1:263–422. Buenos Aires: Peuser, 1958.

Glick, Thomas. "Science and Independence in Latin America (with Special Reference to New Granada)." *Hispanic American Historical Review* 71 (1991): 307–334.

González Bernaldo de Quirós, Pilar. "Sociabilidad y opinión pública en Buenos Aires (1821–1852)." *Historia Contemporánea* 27 (2003): 663–694.

Gutiérrez, Juan María. "Esteban de Luca. Noticias sobre su vida y escritos." In Juan María Gutiérrez, *Los poetas de la revolución*, 27–86. Buenos Aires: Academia Argentina de Letras, 1941.

———. "Estudios sobre las obras y la persona del literato y publicista argentino don Juan de la Cruz Varela." In Juan María Gutiérrez, *Los poetas de la revolución*, 131–511. Buenos Aires: Academia Argentina de Letras, 1941.

———. *Los poetas de la revolución.* Buenos Aires: Academia Argentina de Letras, 1941.

Howard, Edward. "Experiments and observations on certain stones and metalline substances, which at different times are said to have fallen on the Earth; also on various kinds of native iron." *Philosophical Transactions of the Royal Society* 92 (1802): 168–212.

Lamas, Andrés. "El aerolito del Chaco." *Revista del Río de la Plata* 1, no. 4 (1871): 533–555.

López y Planes, Vicente. "A la Patria en la Victoria de Maypo." In *La lira argentina*, edited by Barcia, 214–219.

Luca, Esteban de. "Al Superior Gobierno de estas Provincias en loor de los valientes cochabambinos." In *La revolución*, vol. 2 of *Antología de poetas argentinos*, Juan de la Cruz Puig, 154–157. Buenos Aires: Biedma, 1910.

———. "Al triunfo del Vicealmirante Lord Cochrane, sobre El Callao." In *La lira argentina*, edited by Pedro L. Barcia, 412–418. Buenos Aires: Academia Argentina de Letras, 1982.

———. "Al vencedor de Maypo." In *La lira argentina*, edited by Barcia, 220–229.

———. "Canción patriótica." In *La lira argentina*, edited by Barcia, 23–25.

———. "Canto lírico a la libertad de Lima." In *La lira argentina*, edited by Barcia, 496–515.

———. "Disertación sobre el hierro de Tucumán dirigida al Exmo. Director del Estado en 10 de Febrero de 1816." In Lamas, "El aerolito del Chaco," 549–554.

———. "Marcha patriótica." In *La lira argentina*, edited by Barcia, 11–17.

López y Planes, Vicente, and Felipe Senillosa. "Ciencias." *Argos de Buenos Aires*, June 2, 1821, 19–20.

———. "Ciencias." *Argos de Buenos Aires*, June 23, 1821, 45–46.

Mitre, Bartolomé. *Historia de Belgrano y de la Independencia Argentina*. Buenos Aires: Anaconda, 1950.

Muñoz, Bartolomé, and Vicente López. "Observación astronómica del eclipse total de Luna del día 9 de junio de 1916, hecha en Buenos-Ayres." *La Prensa Argentina*, June 15, 1816, 6–9.

Moreno, Manuel. "Memoria sobre el fierro nativo que se encuentra en los campos del Gran Chaco." In *La abeja argentina*, October 15, 1822, 278–287.

Multhauf, Lettie S. "Zach, Franz Xaver von." In *Dictionary of Scientific Biography*, edited by Charles V. Gillispie, 14:582–583. New York: Charles Scribner's Sons, 1981.

Piccirilli, Ricardo. *Los López: Una dinastía intellectual*. Buenos Aires: Eudeba, 1972.

Poch, Susana. "Neoclasicismo y nación (1806–1827)." In *Una patria literaria*, edited by Cristina Iglesia y Loreley El Jaber, 105–127. Vol. 1 of *Historia crítica de la literatura argentina*, edited by Noé Jitrik. Buenos Aires: Emecé, 2014.

Quiroga, Marcial. *Manuel Moreno*. Buenos Aires: Eudeba, 1972.

Rodríguez, Cayetano. "Cuento al caso." In *La lira argentina*, edited by Barcia, 81–88.

———. "Himno en las Fiestas Mayas." In *La lira argentina*, edited by Barcia, 131–134.

———. "Oda. Al Brigadier Don Carlos María de Alvear." In *La lira argentina*, edited by Barcia, 69–73.

———. "Oda al día augusto de la Patria." In *La lira argentina*, edited by Barcia, 94–98.

———. *Secunda phisicae pars, seu phisica particularis*. Manuscript, n.d. Fondo Antiguo de la Compañía de Jesús, Buenos Aires.

Rojas, Ricardo. *La literatura argentina*. Vols. 3–4: *Los coloniales*. 2nd edition. Buenos Aires: La Facultad, 1924.

Romero Carranza, Ambrosio. "Fray Cayetano Rodríguez." In *El Congreso de Tucumán*, edited by Seminario de Estudios de Historia Argentina, 241–290. Buenos Aires: Club de Lectores, 1966.

Senillosa, Felipe, and Juan Lavalle. "Memoria." In *El pensamiento político de Juan Manuel de Rosas*, edited by Andrés M. Carretero, 36–41. Buenos Aires: Platero, 1970.

Wills, Garry. *Inventing America. Jefferson's Declaration of Independence*. London: Athlone, 1980.

Zach, Franz Xaver von. "Comète de l'an 1821." *Correspondance astronomique . . . du Baron de Zach* 11, no. 2 (1824):199–201.

———. "Lettre VI [Observations astronomiques faites à Buenos Ayres]." *Correspondance astronomique . . . du Baron de Zach* 10, no. 11 (1824): 101–120.

9

Inventions and Discoveries in Letters to Perón

Dialogue and Autonomy in the Popular Technical Imagination in Argentina in the 1940s and 1950s

HERNÁN COMASTRI

The first governments of Juan Domingo Perón (1946–1955) marked a turning point in Argentine history, both because of the policies that Perón tested out across diverse fields and because of the experiences that he, voluntarily or involuntarily, facilitated among the popular sectors. This essay will focus on the dialogue between official policies and popular culture and specifically as this relates to the imaginaries associated with science and technology (henceforth ST). In particular, I will analyze the degree of autonomy wielded by the popular technical imagination in relation to public discourses espoused by the state and the press.[1] To do so, I will discuss an archive that offers an exceptional point of entrance into representations of a popular culture that has very rarely been translated into written records.

This archive is comprised of letters sent to Perón from citizens with inventions, ideas, advice, and strategies in response to an official call made by the president through the media in December 1951. Each of these initiatives was received by the Technical Secretariat of the Presidency (Secretaría Técnica de la Presidencia), turned over to an appropriate technical commission and, in the case of a favorable evaluation, included in the government's Second Five-Year Plan, which at that time was still being drafted. From a group of more than twenty thousand letters detailing a very wide range of requests, a subset of more than five hundred letters has been identified that focus specifically on ST. These letters provide an account of the particular forms in which the popular sectors that comprised a significant majority of those writing to Perón appropriated and resignified or even rejected the discourses and images that circulated in the public space of mid-twentieth-century Argentina.

Science in Argentina during the First Peronist Governments

Although the study of scientific-technical imaginaries is a relatively new area of investigation in Argentina, several existing works have reconstructed the forms of scientific imagination and practice of a heterogeneous group of actors that fell beyond the limits of what was deemed, in each historical moment, to be legitimate science. This includes the work of Sandra Gasparini and of Soledad Quereilhac, both of whom offer accounts of criticisms directed toward the discourse of progress espoused since the time a state policy on ST was formed, beginning in the last third of the nineteenth century.[2] The 1920s and 1930s are the subject of the pioneering work of Beatriz Sarlo on the popular-technical imagination in Buenos Aires.[3] In contrast to the works by Gasparini and Quereilhac, Sarlo's study highlights a fascination with practical know-how and a modern technical aesthetic, even if such interest remained in dialogue with the press and literature of its time.

Taking the three authors' works together, it is clear that the period from the 1870s to the 1930s does not lack studies of imaginaries relating to ST. However, studies stop precisely at the point at which the consolidation of an expanding mass market and the first Peronist regime's policies on ST, coupled with international changes in forms of understanding and practicing science, had the effect of multiplying the number of popular imaginaries that emerged from spaces outside what was considered legitimate science. Indeed, after 1945, the world of ST was undergoing a radical transformation, traversed by the tensions of the Cold War, which forcefully imposed an agenda on scientific and technological research. ST was also marked by the new model of Big Science; this led to a reconfiguration of the relations between states, scientific-academic institutions, and private industry.

The transformations that marked ST at a local level echoed in a popular technical imagination animated by questions related to the news of foreign advancements and discoveries as well as to Peronist projects. In the large majority of cases, these news items were accompanied by significant propaganda campaigns. Some of the most promoted ST policies of the first Peronist regime involved the nuclear sector, the aeronautical and automotive industries, and productivity.

The nuclear sector was an undisputed protagonist of Peronist propaganda about the advances of the new Argentina in the area of ST. It had a prominent presence in the mass media, ranging from periodicals and scientific magazines to cinematographic newsreels. Coverage featured authorities visiting the laboratories where Project Huemul was developed and the project director, Ronald Richter, receiving the Peronist medal from the president himself. After the discovery that this same man had committed fraud, media showed Perón launching new jobs in the National Atomic Energy Commission (Comisión Nacional de Energía Atómica) in Buenos Aires.[4]

Only one other aspect of ST competed for this level of attention in the media and in official propaganda: the activity of the aeronautical industry in Córdoba and, specifically, the design and testing of Pulqui II, the world's fifth fighter jet. This jet has been the object of numerous studies,[5] and it is only useful for me to add here that its development and use as a symbol of national power coincided with global images of a booming aeronautical industry.

Another major ST policy related to advances in transport technology concerned the exploration of Antarctic territory claimed by the state. The scientific nature of these expeditions (which did not deny but rather reinforced the claim of sovereignty) was consolidated in the creation of the Argentine Antarctic Institute in 1951 and in the planning of Antarctic campaigns as state policies. The events were accompanied by extensive coverage in graphic media, on the radio, and in cinematic newscasts that highlighted accounts of the international race for the "white continent."[6]

Finally, policies aimed at the problem of productivity were also of importance. Productivity was undoubtedly the major and most public challenge of official economic policy since 1950, and it gained particular visibility during the Productivity and Social Welfare Congress of 1955. One of the objectives of the congress was to increase the technological level of Argentine industry, and amateur inventors responded with designs of new and more efficient tools ready to be put into production.

Letters to Perón on Science and Technology

The designs of new smelting furnaces, electric motors, compressors, or packaging machines by those who sent their initiatives to the secretariat referred, usually explicitly, to personal experience of factory work and to the objective of increasing production in the context, from 1951 onward, of a growing trade deficit. The deficit made it impossible to access the foreign currency needed to continue importing new machines and tools for industry. The Peronist government attempted to address the problem caused by this external constraint through seeking innovations geared toward a rise in productivity rather than by devaluing the peso, which would have adversely affected living standards and consumption levels in the general population.

Another major objective of the policy to promote productivity was the technification of agricultural production as a mechanism for increasing exportable sales and alleviating the bottlenecks caused by the trade deficit. The initiatives pertaining to agricultural machinery are varied and came from different parts of the country and from cities as well as from small towns in the interior. They include plans for a "culti-sembradora" (culti-sower), a "cosechadora-desgranadora de maíz" (corn harvesting and threshing machine), a "máquina de agricultura" (agricultural machine), a "camión-cosechadora-trilladora" (truck-harvester-thresher), peanut and potato harvesters, a "matachispa" (spark arrestor) for tractors or harvesters, a sugarcane harvesting machine, sheds for storing cereal, new types of plows, systems of brands and signs for farms, and various forms of unspecified agricultural machinery.[7]

Some of the initiatives are mere ideas lacking any accompanying study, diagram, plan, or calculation. In one, for example, the author simply proposes the design of an Argentine tractor named "El Gaucho" and requests that the state contribute some "capital inizial" (sic; initial capital) to set up a cooperative. Others present prototypes and include detailed sketches, descriptions, photographs of the authors posing with their inventions, and even clippings from local papers celebrating the inventions.[8] Again, many of the proposed ideas stem from direct experience in fieldwork or local mechanical workshops in towns that depended on agricultural work.

Much less connected to direct experience, initiatives related to the exploration and transformation of national territory addressed governmental preoccupations concerning the claim to sovereignty over part of Antarctica. Without direct experience of the territory, those who wrote about this theme were captured by the mystery and conjecture evoked by the Antarctic as a space that was still uninhabited and largely unexplored. The letters contain different pseudoscientific theories about the formation of the white continent or policies the national government should follow with regard to it.[9] In other letters, the intent of the author is to be considered a participant in future expeditions. This is the case with a young Italian writer who arrived in Argentina in 1949 and who at that time was studying radiotelephony and radio assembly, skills he believed would be useful to the Navy in its exploration of Antarctica.[10] However, the resources poured into these explorations by the government also gave rise to other types of projects that were more ambitious and in which the lines dividing science and fiction are less clearly apparent.[11]

At times, initiatives on the topic of atomic energy acquired similar science-fiction undertones, as in the case of a proposed atomic gun capable of disintegrating bodies.[12] On other occasions, by contrast, what took precedence was the desire of those proposing the initiatives to join, in whatever capacity, the work carried out by the government on the island of Huemul or in the National Atomic Energy Commission. Writers even openly accepted the necessity on the part of the state of keeping such investigations secret. Indeed, in the specific context of the Cold War, conspiracy and espionage were incorporated into the popular technical imagination as external symbols of the value of invention. I will not dwell here on the topic of nuclear energy in the popular imaginary, as I have studied this topic in depth elsewhere.[13]

The aeronautical industry, the other great symbol of Peronist ST in official propaganda, also had an impact on correspondence addressed to Perón, although, as in the case of Antarctic exploration, the limited material resources of amateur inventors are more obvious in these letters. If austral explorations presented an exotic scenario in the popular technical imagination, in which national greatness, scientific curiosity, or the imagination were put in play and to which little could be contributed but the simple will to serve the projects of the state, something similar happened regarding aeronautical advances, which were so in vogue during the 1940s and 1950s.

Given that the amateur inventor approaches the topic of aeronautics with very limited tools and without practical experience, letters on the subject of this industry make up only seven out of the total of five hundred, and four of these seven are no more than mere expressions of ideas—an "avión longitudinal" (longitudinal plane), a "salvavidas para aviones" (life raft for planes), a new type of turbine, and a machine for "vuelo individual" (individual flight)— for which no descriptions, blueprints, or calculations are offered.[14] Of the other proposals, one initiative is accompanied by calculations worked out by the author for the design of a new type of "turbina a explosion" (combustive turbine) that technicians at the Instituto Aerotécnico de Córdoba (Aeronautical Institute of Córdoba) analyzed in detail and found interesting; the proposal was eventually rejected for not conforming to the institute's "línea actual de desarrollo" (current line of development).[15] In another case, a cooling system for aircraft nozzles was proposed that reviewers at the secretariat judged to be "inaplicable" (unworkable).[16] And finally, there was the case of a hand drawing of a "triciclo" (tricycle) powered by wind energy and capable of transforming itself into a plane.[17] This idea was submitted by a woman, which in itself is exceptional, as women's participation in the archive represents less than 2 percent of the total, and did not receive serious consideration by the secretariat.

Precisely the opposite occurred in relation to developments aimed toward the automobile industry, which were significantly more numerous and in which mechanics played the most important role.[18] Largely, initiatives related to the automotive industry arose from direct experience in mechanic shops; these repair shops were a necessary element of the expanding market but also a space of experimentation prompted by the flaws of the same market. With commercial circuits that had not yet overcome the dislocations caused by the war, problems of trade balance and a lack of foreign currency periodically caused complications when it came to accessing imported parts, and local industry could only partially supply the demand for parts. By contrast to initiatives focused on other objectives, here foreign technology neither concealed a mystery nor incited fantasy; from the moment auto parts arrived in the mechanic shop they could be studied, copied, and even altered and perfected. The inventiveness of workers with respect to automotive technology is not that of a spectator of technical marvels but rather the product of practical resourcefulness. These inventions were marked by an adaptation to a specific historical conjuncture and the appropriation of a form of technology that was within reach of local workshops.

In this widespread set of initiatives, practical experience with a technology that became increasingly common with the expansion of the automobile industry is combined with the strong media presence of the industry that went beyond mere advertisements. This period saw a huge increase in the special supplements dedicated to automobiles as well as magazines for hobbyists aimed at car maintenance and repair, while Perón's government enthusiastically promoted Puma motorcycles and two models of sports car, the Gran Sport Clásico and the Justicialista Sport, in addition to other products created by the Industrias Aeronáuticas y Mecánicas del Estado (IAME, State Aeronautical and Mechanical Industries).

Whether they talked about the automotive industry, developments in atomic energy, or other aspects of the technical imagination of the era, magazines for hobbyists, popular science magazines, and science fiction magazines exerted an important influence on this correspondence. The popular classes obtained ideas from these sources as well as technical argumentation that could be used to defend their projects and a discursive model that could be emulated when presenting themselves as inventors in letters to Perón. One particularly recurrent idea expressed in both the letters and the magazines is the trope of geniuses "tildados de locos" (labeled crazy) by the scientific community of their time, only to gain recognition later. Examples of such figures range from Christopher Columbus to Thomas Edison, but the lesson is always the same, namely, that discovery and invention are products of curiosity, creativity, and effort, and as such they can come from anywhere and anyone. Almost without exception, the letters reproduce this idea, claiming the legitimacy of the writers' contributions even as they recognize, often explicitly, the distance that separates popular inventors from academics and professional scientists. However, the relation between the popular technical imagination and these public discourses was far from one reduced to mere passive consumption and reproduction.

The Autonomy of the Popular Technical Imagination

Continuing on the subject of automotive designs, one letter arrived from the Buenos Aires province of Pergamino, signed by a thirty-six-year-old bricklayer who reports having built an "automóvil armado en su totalidad con materiales fuera de uso, injertados y modificados" (car assembled in its entirety with obsolete, grafted, and modified materials). The relation between amateur inventors and the press continued to hold a prominent place in this initiative, as shown in the inclusion of a clipping from an unidentified local newspaper that celebrates the invention of the car. However, the letter writer explains that he did not seek to have his design patented or to achieve commercial success, nor did he wish to collaborate with the plans of the government or national industry; rather, he was concerned with the simple pleasure of invention and the use value of the constructed car. He did not write to Perón to offer his design to IAME or to seek economic assistance to produce his design on a larger scale; he simply wrote to point to the bureaucratic impossibility of patenting his car, which prevented him from traveling in it outside of the city radius (figure 9.1). The note explains that the motor was from a 1938 Ford; the differential belonged to a 1937 Chevrolet pickup truck; the front axle was from a Plymouth car and the chassis from a Studebaker, while the bodywork was "diseñada y en gran parte construida" (designed and in large part constructed) by the owner. All of this he achieved through "ingenio, voluntad y paciencia" (ingenuity, will, and patience). The newspaper article explains,

> Se fue improvisando chapista y aprendió a soldar a medida que avanzaba en su trabajo, casi sin contar con herramientas. Su profesión—es constructor—es muy distinta, y sólo su entusiasmo, su capacidad de trabajo y ese ingenio tan criollo que resuelve a fuerza de intuición las dificultades, explican su éxito.[19]

> (He went on to improvise as a sheet metal worker and learned to weld as he worked, almost without tools. His profession—he is a builder—is very different, and it is only his enthusiasm, his capacity for work, and his *criollo* ingenuity that enabled him to resolve difficulties by virtue of intuition, that explain his success.)

Rather than being understood as a response to public discourses and images about automobile technology, this initiative should be understood as the result of the leisure-time productivity of an amateur who embodies what Sarlo describes as "la moral del artesano-*bricoleur*" (the morality of the artisan-*bricoleur*), in which repair, recycling, and the use of waste materials is not only a material necessity but also an aesthetic pursuit. The construction of the car from disused parts is not a mechanism for achieving economic success or for collaborating with the work of the government, but an end in itself, aimed at an individual satisfaction that has little to do with the discussions taking place in the press or in the sphere of public policy.

This capacity for autonomous action in the face of the public agenda of national and international ST is also verified by the absence in these letters of themes that were recurrent in public discourse but that the popular technical imagination rejected, resignified, or simply ignored. The area of military technology exemplifies this. Of more than five hundred letters considered by the secretariat, only three present designs or ideas concerning new weaponry, and two of those come from foreigners.[20] Other initiatives are aimed toward the defense sector, but rather than weapons, they concern artifacts that could be adapted from civilian use, such as mobile phones for troops on the move and an extendable tower for the observation of artillery practices. The author of

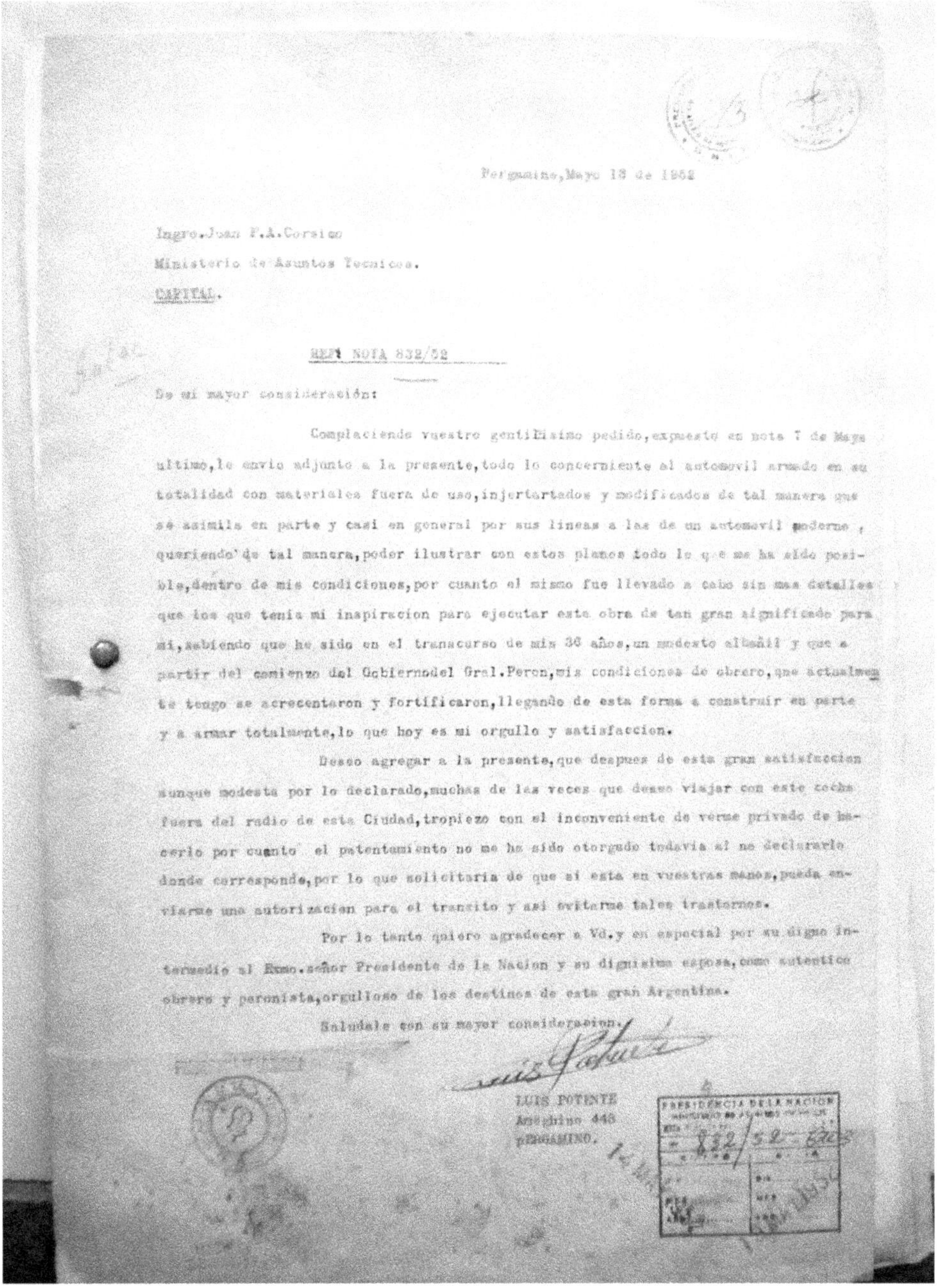

Pergamino, Mayo 18 de 1952

Ingro. Juan F. A. Corsico
Ministerio de Asuntos Tecnicos.
CAPITAL.

REF: NOTA 832/52

De mi mayor consideración:

Complaciendo vuestro gentilísimo pedido, expuesto en nota 7 de Mayo ultimo, le envio adjunto a la presente, todo lo concerniente al automovil armado en su totalidad con materiales fuera de uso, injertartados y modificados de tal manera que se asimila en parte y casi en general por sus lineas a las de un automovil moderno, queriendo de tal manera, poder ilustrar con estos planos todo lo que me ha sido posible, dentro de mis condiciones, por cuanto el mismo fue llevado a cabo sin mas detalles que los que tenia mi inspiracion para ejecutar esta obra de tan gran significado para mi, sabiendo que he sido en el transcurso de mis 36 años, un modesto albañil y que a partir del comienzo del Gobierno del Gral. Peron, mis condiciones de obrero, que actualmente tengo se acrecentaron y fortificaron, llegando de esta forma a construir en parte y a armar totalmente, lo que hoy es mi orgullo y satisfaccion.

Deseo agregar a la presente, que despues de esta gran satisfaccion aunque modesta por lo declarado, muchas de las veces que deseo viajar con este coche fuera del radio de esta Ciudad, tropiezo con el inconveniente de verme privado de hacerlo por cuanto el patentamiento no me ha sido otorgado todavia al no declararlo donde corresponde, por lo que solicitaria de que si esta en vuestras manos, pueda enviarme una autorizacion para el transito y asi evitarme tales trastornos.

Por lo tanto quiero agradecer a Vd. y en especial por su digno intermedio al Exmo. señor Presidente de la Nacion y su dignisima esposa, como autentico obrero y peronista, orgulloso de los destinos de esta gran Argentina.

Saludale con su mayor consideracion.

LUIS POTENTE
Amaghino 446
PERGAMINO.

Figure 9.1. Letter dated May 1952 from the inventor of a new automobile to the Argentine Technical Secretariat of the Presidency, Archivo General de la Nación, Buenos Aires, Secretaría Técnica de la Presidencia, Box 595, Initiative 832/52.

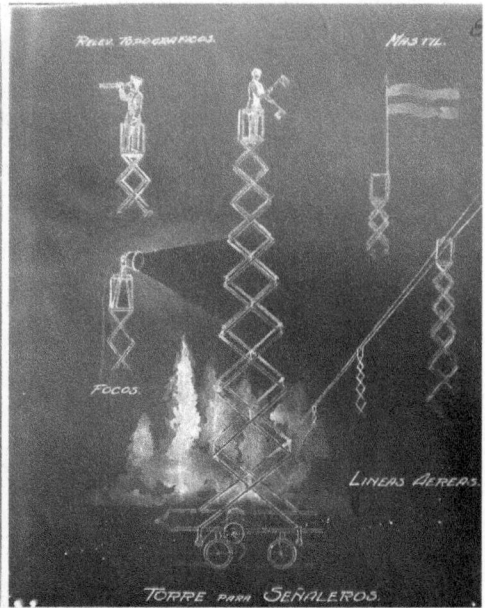

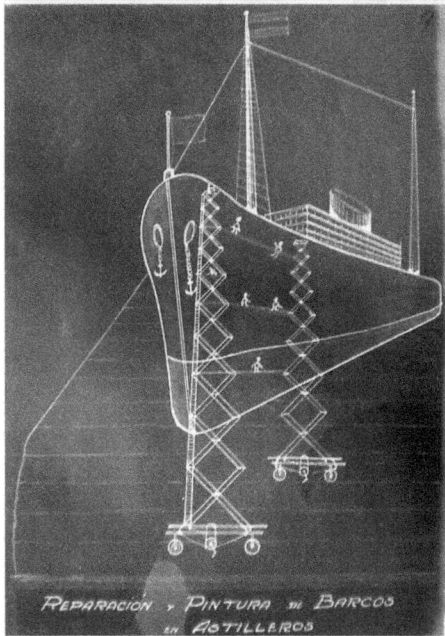

Figure 9.2. Drawings by the inventor of an extendable tower illustrating its multiple uses, 1952. Archivo General de la Nación, Secretaría Técnica de la Presidencia, Buenos Aires, Box 470, Initiative 4035.

the latter initiative, dated 1952, thought of various functions for his invention, only one of which was intended for the military sector; the possible civil applications of his invention included agricultural work, aeronautic signaling, and shipbuilding (figure 9.2).[21]

The recent experience of World War II was not absent in the correspondence, though the legacies of the war and its secret weapons are converted in the popular technical imagination in order to be placed at the service of production and socioeconomic development. To solve the problem of locust plagues in the agricultural interior of the country, one author proposed the manufacture of explosive cartridges that would kill insects through a wave blast; in a section entitled "Tactics" the same writer expands on the necessity of creating "forts" or "posts" in "strategic places," supplied with cartridges from factories in nearby towns and directed by specialists so that "cada provincia tendría su 'estado mayor'—comando—y una escuela, donde se enseñaría el manejo y táctica de combate" (each province would have its military staff—commando—and a school that would teach management and combat tactics) to be directed against the "falanges del enemigo" (phalanges of the enemy).[22] Another initiative was part of a supposed secret development in the German weapon industry, "un cañón, en el cual se hacía estallar 'ARGON,' recogiéndose las ondas sonoras en un espejo parabólico para formar una semi-onda dilatada de frecuencia capáz de dar muerte a un hombre" (a cannon in which 'ARGON' was exploded, collecting sound waves within a parabolic mirror in order to form a dilated semi-wave frequency capable of killing a man). But once again, it is not in war but in the fight against insects, "transmisores de gérmenes y enfermedades" (transmitters of germs and diseases), that the writer judges this revolutionary technology should be wielded.[23]

The relative absence of weapon technology within the letters sent to Perón is not consistent with the widespread presence of that subject in the newspapers and magazines of the epoch, making its absence more significant. It was the middle of the Cold War, atomic tests filled the front pages of newspapers, and in 1952 the Korean War was still going on. Perón himself—using the pseudonym Descartes in his columns in the newspaper *Democracia*—judged the Korean conflict to be the probable catalyst of a third world war for which Argentina needed to prepare. And yet, that same year, a man wrote to the secretariat claiming to have the calculations for "un dispositivo que multiplica revoluciones a gusto" (a device that multiplies revolutions with ease):

Bien mi primera intención es probar este dispositivo en un submarino tambien ideado por mi para el caso, pero como no hay apuros para armas belica, puesto que S. E. es la mejor garantia de nuestra paz pienso mejor seria por el momento provar en un coche de carrera: Asi los automovilista Argentinos cumplirían una vieja aspiración; correr con maquinas especiales propias y no extranjeras.[24]

(Well my first intention is to test this device in a submarine also devised by me for this case, but because there are no difficulties with military weapons, given that S.E. [Your Excellency] is the greatest guarantor of our peace, I think it best at the present moment to test it in a racing car: therein, Argentine motorists would fulfill an old aspiration, running with their own specialized machines and not foreign ones.)

Finally, I present the case of initiatives related to the pursuit of perpetual motion. These letters are significant because through them and beyond their specific historical context, mid-century Argentine amateur inventors inscribed themselves, consciously or unconsciously, into a tradition of mechanical experimentation dating back to at least the thirteenth century.[25] Within an extremely heterogeneous collection of letters, approximately one out of every ten proposals makes reference to some sort of perpetual motion device, hypothetical machines that depend on the possibility of an eternal and uninterrupted functioning following a single initial push; in other words, they are conceived as machines that once put into operation would not require the consumption of new energy to continue their movement and thus could function indefinitely. Many writers even hoped to be able to *produce* energy through these machines. The designs were based on a lack of knowledge, an erroneous reading, or simply a rejection of the principle of energy conservation; in any case, perpetual motion has been recognized by the scientific community as a "máquina imposible" (impossible machine) in that it would violate the second law of thermodynamics, whose most classical formulations date back to the middle of the nineteenth century.

Thus, if by 1870 perpetual motion had already moved from being a scientific exercise to something characterized by scientists as the product of ignorance, fifty years later that image would only be modified in relation to the expansion of potential inventors prompted by the development of the mass market. In October 1920, the cover of *Popular Science* featured a drawing by Norman Rockwell under the title "Perpetual Motion?" If the headline left any room for doubt, from the first line of the news story perpetual motion is characterized as an "ancient, ancient fallacy."[26] In fact, the only objective of the article is to demonstrate, once more, the impossibility of the machine on which an ever-growing public wasted its time. However, the same essay also suggests that the "problem" not only relates to a lack of knowledge of theoretical physics; many amateur inventors had heard the arguments and knew the theory but decided to reject or challenge them. In the face of the abstract nature of the laws of thermodynamics, the amateur opts for practical workshop experimentation and chooses to insert himself in the line of great thinkers who throughout history have encountered the same problem and dreamed of a solution that seems to be within reach of mechanical ingenuity.

Although the various initiatives aimed at resolving problems related to energy or to revolutionizing science, personal enjoyment also came into play. Bruno Jacomy describes it as the "instinto lúdico del mecánico" (ludic instinct of the mechanic) that leads one to "crear autómatas" (create automata) and transcends the particular historical context in which the creator is working.[27] Thus, versions of perpetual motion devices that could be found in the Renaissance are also apparent in the correspondences addressed to Perón in the mid-twentieth century, as machines that transmit their force through pulleys, gears, springs, counterweights, pistons, levers, moving parts, water, air pressure, magnets, and so forth.[28] As with the United States in the 1920s, mid-century Argentina already had a mass market that included magazines on mechanics and scientific dissemination; meanwhile, improvements in living and employment conditions among the working classes meant that resources were available to be poured into small domestic workshops and tools and that leisure time was available to be dedicated to the mechanic's "ludic instinct."

As a result of these circumstances, letter writers claiming to have unraveled the mysteries of machines that generate free energy or of engines that do not consume fuel multiplied in number. The technicians of the secretariat initially responded to such initiatives or even called the writers into the offices of the department to explain to them why the inventions could never work. Yet, faced with the growing number of initiatives aimed in this direction, they made it an internal policy not to consider correspondence referring to any kind of perpetual motion device. By 1953, when the influx of correspondence was greater, these letters were no longer sent to other offices for evaluation but simply dismissed with this note: "El Estado no presta su apoyo para ése tipo de realizaciones, por cuanto la imposibilidad de producir movimiento continuo está demostrada teórica y experimentalmente" (the state does not lend its support to these types of projects, since the impossibility of producing continuous motion has been demonstrated theoretically and experimentally).[29]

Conclusion

In the development of this brief study, reference has been made to a dialogue between public discourses relating to ST and the popular technical imagination. This does represent a dialogue in the strict sense; in the first place, it is clear from the sheer time and resources dedicated that state policy regarding the reception of citizens' initiatives cannot be dismissed as mere demagoguery. Second, this is the case because state policy responds to a social demand that precedes it. In fact, letters began to arrive at different offices of the state from the first months of 1946; thus, the call made by Perón in 1951 is not a starting point but a response to that social demand for recognition. Although I have not been able in the present inquiry to enter into a conceptual debate regarding the level of autonomy or heteronomy that characterized the popular technical imagination in the face of state discourses, the press, and the academy, I have demonstrated that a relative autonomy was very much possible; whether by rejecting, resignifying, or simply ignoring those discourses, the popular technical imagination constituted a dialogue with its historical context without necessarily conforming to an agenda marked by external preoccupations. Rather than being subordinate within a hierarchical structure, many of the "cartas a Perón" (letters to Perón) demonstrate instead the vitality of a heterogeneous and complex popular culture in the middle of a mid-twentieth-century Argentina traversed by profound social changes.

On the other hand, the same correspondence enables us to observe a sociopolitical reality that exceeds the technical adventures of amateur inventors. If the inventors were able to dream of the designs and inventions that would give form to a new Argentina that was scientifically and technologically modern, they did not act alone but proceeded in permanent dialogue with a state that also espoused a modernizing discourse. Thus, the epistolary practice presented here can also be read as an innovative mechanism of participatory democracy and as one of the spaces in which the popular classes were able to enter into contact with an active, accessible, and present state in a manner that went beyond the contact generated through the state's repressive or disciplinary activities such as the armed forces, police, and education. Even though the vast majority of the ideas sent to Perón were rejected, the letters provide evidence of the gratitude expressed by those who contacted the Technical Secretariat of the Presidency for the attention given by officials of the Peronist government. In reality, even the few initiatives that were evaluated favorably received only a recommendation to appear before the Industrial Credit Bank, apply for a loan, and run all the risks of bringing the idea to the market. In no case did the state take direct responsibility for any idea or invention received from the citizenry. And yet, the simple recognition of these popular inventors and self-taught thinkers, the predisposition to listen and dialogue with them, helped to forge new connections to the state and a lasting political bond and identification between the popular classes and Peronism that was able to survive even the overthrow of the Peronist government and the long exile of its leader.

Translated by Lucy Bollington.

Notes

1. For reasons of space, in the present essay I cannot dwell on the relationship these popular imaginaries had with the contemporary culture industry. For an analysis in this vein, see Comastri, "*Bull Rockett*," 239–257.

2. Gasparini, *Espectros de la ciencia*; Quereilhac, *Cuando la ciencia despertaba fantasías*.

3. Sarlo, *La imaginación técnica*.

4. Mariscotti, *El secreto atómico de Huemul*.

5. For example, Artopoulos, *Tecnología e innovación en países emergentes*.

6. Fontana, *La pugna antártica*.

7. Respectively, in Archivo General de la Nación (henceforth AGN): STP (Secretaría Técnica de la Presidencia), Box 450, Initiatives 6201 and 6365; Box 503, Initiative 2938; Box 582, Initiatives 658 and 777; Box 590, Initiative 1602; Box 587, Initiative 482/53; Box 470, Initiative 2891; Box 459, Initiative 1034; Box 463, Initiative 2548; Box 464, Initiative 2020, and Box 588, Initiative 4456.

8. STP, Box 472, Initiative 5828, and Box 332, Initiative 10116, AGN.

9. STP, Box 587, Initiative 2840; Box 598, Initiative 2579/47, AGN.

10. STP, Box 516, Initiative 4895, AGN. For other offers of service in the Antarctic, see, for example, Box 450, Initiative 1941.

11. An example would be expeditions to the center of the earth, as in STP, Box 583, Initiative 879/46, AGN.

12. STP, Box 458, Initiative 1794, AGN.

13. Comastri, "La apuesta por la energía atómica."

14. In AGN: STP, Box 592, Initiative 1278; Box 584, Initiative 1556/53; Box 579, Initiative 2300, and Box 464, Initiative 1269.

15. STP, Box 471, Initiative 71/50, AGN.

16. STP, Box 579, Initiative 2286, AGN.

17. STP, Box 597, Initiative 80/48, AGN.

18. In AGN: STP, Box 449, Initiative 3361; Box 450, Initiatives 522 and 2301; Box 461, Initiative 335/50; Box 464, Initiatives 1945 and 1528; Box 470, Initiative 2841/51; Box 474, Initiatives 4798 and 4775; Box 502, Initiatives 1991/47 and 1345; Box 503, Initiative 3310; Box 579, Initiative 2453; Box 582, Initiatives 2259, 1116 and 914; Box 586, Initiatives 6123 and 6114; Box 588, Initiative 2983; Box 590, Initiative 1477; Box 593, Initiatives 1525 and 1181; Box 598, Initiative 1828/47, and Box 332, Initiative 11964.

19. STP, Box 595, Initiative 832/52, AGN.

20. STP, Box 593, Initiative 1184, AGN, is from an Argentine; the two from foreigners are STP, Box 458, Initiative 1794, and Box 591, Initiative 254/54, AGN.

21. The mobile phone idea is STP, Box 471, Initiative 1/50, AGN; the extendable tower design is STP, Box 470, Initiative 4035, AGN.

22. STP, Box 599, Initiative 2262, AGN.

23. STP, Box 474, Initiative 2045, AGN.

24. STP, Box 503, Initiative 1131, AGN.

25. Dircks, *Perpetuum Mobile*.

26. Rowland, "Undying Lure of Perpetual Motion."

27. Jacomy, *Historia de las técnicas*, 86.

28. The letters on this topic are too numerous to be cited here in their totality. For examples in AGN, STP, see Box 388, Initiative 11232; Box 449, Initiatives 2920/51 and 2717; Box 450, Initiative 1226; Box 458, Initiative 1830; Box 459, Initiatives 869, 4552, 4594 and 950; Box 462, Initiatives 2111 and 2128; Box 463, Initiatives 5102 and 2374; Box 464, Initiatives 2677, 1265/54, 2603, 1894 and 2295, and Box 470, Initiatives 3094 and 3320.

29. For example, STP, Box 464, Initiative 1265/54, AGN.

References

Artopoulos, Alejandro. *Tecnología e innovación en países emergentes: La aventura del Pulqui II*. Buenos Aires: Lenguaje Claro, 2012.

Comastri, Hernán. "*Bull Rockett*, Héctor Germán Oesterheld y la imaginación técnica popular en la Argentina de mediados del siglo XX." *Anuario del Centro de Estudios Históricos Prof. Carlos S. A. Segreti* (Córdoba, Argentina) 14, no. 14 (2014): 239–257.

———. "La apuesta por la energía atómica: Guerra Fría, políticas de Estado e imaginación técnica popular en el primer peronismo (1946–1955)." In *Saberes desbordados: Historias de diálogos entre conocimientos científicos y sentido común (Argentina, siglos XIX y XX)*, edited by Jimena Caravaca, Caludia Daniel, and Mariano Plotkin. Buenos Aires: Libros del IDES, 2018.

Dircks, Henry. *Perpetuum Mobile; or, A History of the Search for Self-Motive Power, from the 13th to the 19th Century*. Vol. 2. London: E. and F. N. Spon, 1870.

Fontana, Pablo. *La pugna antártica: El conflicto por el sexto continente. 1939–1959*. Buenos Aires: Guazuvirá, 2014.

Gasparini, Sandra. *Espectros de la ciencia. Fantasías científicas de la Argentina del siglo XIX.* Buenos Aires: Santiago Arcos, 2012.

Jacomy, Bruno. *Historia de las técnicas.* Buenos Aires: Losada, 1992.

Mariscotti, Mario. *El secreto atómico de Huemul: Crónica del origen de la energía atómica en la Argentina.* Buenos Aires: Estudio Sigma, 2004.

Quereilhac, Soledad. *Cuando la ciencia despertaba fantasías: Prensa, literatura y ocultismo en la Argentina de entresiglos.* Buenos Aires: Siglo Veintiuno, 2016.

Rowland, Philip. "The Undying Lure of Perpetual Motion." *Popular Science* (October 1920): 26–29.

Sarlo, Beatriz. *La imaginación técnica. Sueños modernos de la cultura argentina.* Buenos Aires: Nueva Visión, 2004.

IV

Utopian Convergences between
Science and the Arts

Introduction to Section IV

MARÍA DEL PILAR BLANCO AND JOANNA PAGE

Writing in 1990, the Cuban-born poet Rafael Catalá contrasts Latin America's achievements in literature with its scientific underdevelopment.[1] He calls on those working in the arts and humanities to rectify this imbalance, arguing that they bear responsibility of making "viable" the creation and growth of other fields of knowledge that are disadvantaged.[2] He urges writers and critics to introduce science into their practice, as doing so will both foster a much-needed reflection on science from the perspective of ethics and enrich writing, according it new dimensions and textures.[3]

Although it is certainly true that, as Catalá asserts, literary criticism in Latin America has yet to integrate science into its "canon,"[4] there have been notable advances in this regard since the publication of his essay in 1990. Two areas of particular interest to recent scholars have been the evolution of the science-fiction genre in Latin America and the development of a scientific imaginary in periodicals from the late nineteenth century to the early twentieth century, particularly in Mexico and Argentina. Key works in these two fields have given insight into the extensive exchanges between arts and sciences that have shaped literary, artistic, and scientific culture in Latin America dating back to the colonial period and earlier.[5]

Convergences between scientific and literary imaginations are particularly prominent in magazines popular around the turn of the twentieth century. The peculiar blends of science, fantasy, theosophy, and other branches of empirical or esoteric knowledge offered to readers afford a privileged insight into how knowledge is synthesized and disseminated at a time of flux before the institutionalization of "mainstream" science. Soledad Quereilhac's chapter presents a selection of Argentine magazines that exemplify the feverish exchange of ideas that characterized this period. She focuses primarily on the interest in mystery and the occult shared by scientific writers and by *modernismo* in Latin America. *Modernista* writers such as Rubén Darío and Leopoldo Lugones explored areas of spiritual and supernatural experience excluded by positivist science while using the language of science; they did so, Darío affirms, in a bid to expand scientific methodologies to examine the "extraordinary" phenomena studied by the occult sciences, in anticipation of a time when science and religion would unite to forge a new "science of life."[6]

In addition to the extensive treatment of scientific and technological themes in literary and artistic movements since the beginning of the twentieth century in Latin America, these texts have often drawn on the potential in scientific ideas to inspire forms of literary innovation. This is the dynamic described by Julio Prieto in his chapter in this section, in which he suggests that in Severo Sarduy's poetry, "the laws and principles of mathematics and the natural sciences are engaged less in terms of their truth-value than in light of their aesthetic potentiality and/or textual productivity" (232, this volume). A long line of critics have found in Jorge Luis Borges's fictional labyrinths echoes of the key discoveries of Albert Einstein, Niels Böhr, and Werner Heisenberg, among other scientists whose work has changed our understanding of time and space.[7] More recently, the narrative experiments of Latin American Boom writers such as Julio Cortázar and Carlos Fuentes have also been read as literary entanglements with quantum mechanics, while theories of autopoiesis and complexity have been explored as figures for literary creativity in contemporary Argentine literature.[8]

The relation of influence between science and literature is not, of course, one-directional, as many Latin American writers have testified. Ricardo Piglia asserts that "los científicos son grandes lectores de novelas" (scientists are big readers of novels), and this ensures the transmission of the literary imagination of science throughout society as "los políticos les creen a los científicos . . . y los científicos les creen a los novelistas" (politicians believe scientists, . . . and scientists believe novelists).[9] Joanna Page's chapter in this section reflects on the work of Jorge Volpi, who draws on recent research in neuroscience to argue that fiction does not enter the world as a result of the flourishing of human society but may in fact have created the conditions in which that development became possible. In a similar way, cultural imaginaries of science do not merely register or critique specific advances that take place in another (separate) sphere; instead, they help to create the context in which those advances are received and understood. Volpi concedes that the popularization of discoveries by Einstein and Heisenberg led to a widespread misapplication of their ideas beyond their spheres of relevance; however, he considers that the names these scientists gave to their theories—relativity and uncertainty—were not unconnected to the cultural and social imaginaries of modernity and that this explains the resonance they gained in the social sphere.[10]

A strong materialist vein together with an attention to reading or artistic experience as embodied connects many of the writers and artists profiled here. This is seen first in Carlos Gutiérrez's 1895 account, cited by Quereilhac (218, this volume), of the physiological impact of artistic contemplation, in which the "conmoción" (shock) one experiences reaches "lo íntimo del cerebro, refluye en el sistema nervioso, altera nuestra corriente sanguínea, agita ó paraliza los músculos y hasta hay erizamientos en la piel y en el cabello: esto es una emoción" (the intimate core of the brain, flows back through the nervous system, disrupts blood flow, triggers or paralyzes muscles, and even makes the skin and hair bristle: this is an emotion).[11] Over a century later, Page shows, Marcelo Cohen also presents reading as an exchange of biochemical substances in the nervous system and the brain. As examined by Prieto, Jorge Eduardo Eielson's topological practices of reading and the literal knotting together of different traditions in his work also point to the importance of the material and the embodied in acts of intercultural communication and understanding.

Literary and critical reflections on the relation between science and art or literature often promote a vision of the universe as a single, interconnected system in which literature, like science, is a subsystem, not a separate realm somehow set apart from the natural world. For Catalá, his practice of "sciencepoetry" is an expression of this interconnectedness. He writes, "La cienciapoesía se sabe parte del sistema ecológico sociocultural. Es parte interdependiente del equilibrio homeostático de la humanidad" (Sciencepoetry knows that it is part of the sociocultural ecological system. It is an interdependent part of humanity's homeostatic equilibrium).[12] In Lugones's contributions to *La Quincena* and *Philadelphia*, Quereilhac observes in her chapter here "a curious spiritualist monism within which art does not find itself excluded but rather becomes the privileged channel for the perception of the analogies of the universe" (226, this volume). Likewise, Page finds in Cohen's texts an emphasis on the dynamics of codetermination and symbiosis that connects organisms with their environment and the act of reading with neurological and physiological processes.

The frequent representation of science and technology in Latin American literature as products imported from Europe or the United States means that their treatment is necessarily bound up with the broader geopolitics of (neo)colonialism. Jerry Hoeg identifies three "representative views" of technoscience in Latin American literature.[13] At one end of the scale, he places "the outright rejection of foreign, dominating technology" that he finds in Isabel Allende's *La casa de los espíritus* (1982), Laura Esquivel's *Como agua para chocolate* (1989) and Fernando Contreras Castro's *Los Peor* (1995), a Costa Rican novel that denounces the abuse of agrochemicals and champions indigenous knowledge over Western medicine.[14] In Carlos Diegues's *Bye Bye Brasil* (1980), Hoeg finds an intermediate position, as the film narrates a range of responses to technological change, including some that accommodate both the traditional and the new.[15] Catalá epitomizes for Hoeg a third alternative, a synthesis of literature and science that rejects any division between them as false.[16]

There is another important literary current, however, that imagines the fusion of science and magic and—increasingly—indigenous beliefs and knowledge. Hoeg would have needed to travel further back, to the late nineteenth and early twentieth centuries, to find examples of the intermixture of science and sorcery in the work of Eduardo Ladislao Holmberg or Horacio Quiroga; similarly, Hoeg's book predates the more recent flourishing of interest in interrogating the apparent divide between the scientific and the paranormal, or indigenous cosmologies. Here one could mention the synthesis of science fiction and indigenous shamanism in Jorge Baradit's *Ygdrasil* (2005), the exploration of image technologies and spectrality in Roque Larraquy's *Informe sobre ectoplasma animal* (2014), and the projection of indigenous ontologies into the futuristic worlds of several graphic novels, such as *Las playas del otro mundo* (Cristián Barros and Demetrio Babul, 2009) and two of Edgar Clement's works, *Operación Bolívar* (1990) and *Los perros salvajes* (2011-). The work of Eielson, explored here by Prieto, is closely associated with this move toward integration. It represents the "productive intertwining of three knowledge traditions": first, modern mathematics and physics; second, the avant-garde tradition (including concrete poetry, conceptualism, *arte povera*, and performance art); and third, the Andean practice of *khipu*, a writing and computing system that predates colonialism. If, as Bruno Latour asserts, Western modernity is built on the suppression of admixtures between the modern and the nonmodern,[17] the treatment of science and technology in Latin American literature has more often refused to separate the two.

The convergences between art and science imagined by the writers in this section gain a strongly utopian quality. Their work conceives "alternative forms of life and living together" (Prieto, 231) and attempts to "imagine other forms of society that are not based on self-interest and exclusionary paradigms of progress" (Page, 265); in a similar manner, the comprehensive science Emilio Becher imagines goes hand in hand with the "unity of all men" in the "universal collectivism," as envisioned in socialism (in Quereilhac, 225). What is also utopian about this work is the quest for a unified field of knowledge that transcends disciplinary and cultural divides. This search is manifest in the fictions and essays Quereilhac analyzes that epitomize efforts at the turn of the twentieth century to synthesize realms of knowledge and experience that many would consider antithetical—such as science and theosophy, positivism and spiritualism—within a truly monist conception of existence. Page reflects on "the crumbling distinctions between the physical and the nonphysical in many fields of knowledge and praxis" (264), as is evident in the work of Francisco Varela as well as those scientists—Erwin Schrödinger, for example—cited by Cohen in his fiction and critical essays. The interconnectedness of apparently opposed systems of knowledge also animates Eielson's hybrid poetics, which, as Prieto affirms, "is not a binary confrontation of European and American epistemologies but their *knotting*" (246). Prieto reads the work of Sarduy, Eielson, and Néstor Perlongher as a quest "to produce forms of epistemological pluralization," overcoming divides between the natural and human sciences, and between Western and non-Western knowledge, in response to what may be regarded as "one of the crucial challenges of our time" (248).

Many of these appeals to science could be read as an attempt to lend greater legitimacy to literary and artistic practices that have been sidelined as secondary in the evolution of human culture and society and inconsequential in relation to the real social or environmental problems that technology is attempting to overcome. What emerges from these essays, however, and from the works they discuss is a persuasive vision of literature, art, and poetry as powerful sites of knowledge production that should (and often do) shape our understanding of science, interrogate its social, cultural, and ethical implications, and even provide the conditions for the evolution of scientific thought and practice in society.

Notes

1. Catalá, "Para una teoría latinoamericana," 215.
2. Catalá, "Para una teoría latinoamericana," 216.
3. Catalá, "Para una teoría latinoamericana," 215–216, 218–219.

4. Catalá, "Para una teoría latinoamericana," 218.

5. See, among other works, Brown, *Cyborgs in Latin America*; Haywood Ferreira, *Emergence of Latin American Science Fiction*; Gasparini, *Espectros de la ciencia*; Ginway and Brown, *Latin American Science Fiction*; Hoeg and Larsen, *Science, Literature, and Film*; Page, *Science Fiction in Argentina*; Quereilhac, *Cuando la ciencia despertaba fantasías*; King, *Science Fiction and Digital Technologies*. Also see the two special issues of *Revista Iberoamericana* edited by Kurlat Ares and dedicated to Latin American science fiction, *La ciencia ficción en América Latina*, with the subtitles *Aproximaciones teóricas* and *Entre la mitología*. Since 2013 the University of South Florida in Tampa has been publishing the journal *Alambique: Revista Académica de Ciencia Ficción y Fantasía/Journal Acadêmico de Ficção Científica e Fantasía*.

6. Darío, "La esfinge," *La Nación*, March 16, 1895, quoted in Anderson Imbert, *La originalidad de Rubén Darío*, 203.

7. See, for example, Merrell, *Unthinking Thinking*; Mosher, "Atemporal Labyrinths in Time."

8. See, respectively, Rivero, "Heisenberg's Uncertainty Principle"; Page, *Creativity and Science*.

9. Piglia, *La ciudad ausente*, 141.

10. Volpi, "Las respuestas absolutas."

11. Gutiérrez, *La Quincena*, 228.

12. Catalá, "Para una teoría latinoamericana," 220.

13. Hoeg, *Science, Technology*, 109.

14. Hoeg, *Science, Technology*, 109–112.

15. Hoeg, *Science, Technology*, 112–116.

16. Hoeg, *Science, Technology*, 116–120.

17. Latour, *We Have Never Been Modern*.

References

Anderson Imbert, Enrique. *La originalidad de Rubén Darío*. Buenos Aires: Centro Editor de América Latina, 1967.

Brown, Andrew J. *Cyborgs in Latin America*. London: Palgrave, 2010.

Catalá, Rafael. "Para una teoría latinoamericana de las relaciones de la ciencia con la literatura: La cienciapoesía." *Revista de Filosofía de la Universidad de Costa Rica* 28, no. 67–68 (1990): 215–23.

Gasparini, Sandra. *Espectros de la ciencia: Fantasías científicas de la Argentina del siglo XIX*. Buenos Aires: Santiago Arcos, 2012.

Ginway, Elizabeth M., and J. Andrew Brown, eds. *Latin American Science Fiction: Theory and Practice*. London: Palgrave, 2012.

Gutiérrez, Carlos. "El arte desde el punto de vista fisiológico." *La Quincena: Revista de Letras* 2, no. 9–10 (January 1895): 222–230.

Haywood Ferreira, Rachel. *The Emergence of Latin American Science Fiction*. Middletown, CT: Wesleyan University Press, 2011.

Hoeg, Jerry. *Science, Technology, and Latin American Narrative in the Twentieth Century and Beyond*. Bethlehem, PA: Lehigh University Press, 2000.

Hoeg, Jerry, and Kevin Larsen. *Science, Literature, and Film in the Hispanic World*. London: Palgrave, 2006.

King, Edward. *Science Fiction and Digital Technologies in Argentine and Brazilian Culture*. New York: Palgrave Macmillan, 2013.

Kurlat Ares, Silvia, ed. *La ciencia ficción en América Latina: Aproximaciones teóricas al imaginario de la experimentación cultural*. Special issue of *Revista Iberoamericana* 83, no. 259–260 (April–September 2017).

———. *La ciencia-ficción en América Latina: Entre la mitología experimental y lo que vendrá*. Special issue of *Revista Iberoamericana* 78, no. 238–239 (January–June 2012).

Latour, Bruno. *We Have Never Been Modern*. Translation by Catherine Porter. Cambridge, MA: Harvard University Press, 1993.

Merrell, Floyd. *Unthinking Thinking: Jorge Luis Borges, Mathematics, and the New Physics*. West Lafayette, IN: Purdue University Press, 1991.

Mosher, Mark. "Atemporal Labyrinths in Time: J. L. Borges and the New Physicists." *Symposium: A Quarterly Journal in Modern Literatures* 48, no. 1 (1994): 51–61.

Page, Joanna. *Creativity and Science in Contemporary Argentine Literature: Between Romanticism and Formalism*. Calgary, Canada: University of Calgary Press, 2014.

———. *Science Fiction in Argentina: Technologies of the Text in a Material Multiverse*. Ann Arbor: University of Michigan Press, 2016.

Piglia, Ricardo. *La ciudad ausente*. Buenos Aires: Espasa Calpe/Seix Barral, 1995.

Rivero, Alicia. "Heisenberg's Uncertainty Principle in Contemporary Spanish American Fiction." In *Science and the Creative Imagination in Latin America*, edited by in Evelyn Fishburn and Eduardo L. Ortiz, 129–150. London: Institute for the Study of the Americas, 2005.

Volpi, Jorge. "Las respuestas absolutas siempre son mentiras." Interview by Joaquín María Aguirre Romero and Yolanda Delgado Batista, *Espéculo* 11 (March–June 2009). https://pendientedemigracion.ucm.es/info/especulo/numero11/volpi.html.

10

Modernismo, Spiritualism, and Science in Argentina at the Turn of the Twentieth Century

An Analysis of National Magazines

SOLEDAD QUEREILHAC

From a contemporary perspective, science and spiritualism appear to be antagonistic. However, the relation between the two was more permeable, and certainly less polar, at the turn of the twentieth century. In this essay, I will review some crossovers between spiritualism and scientific discourse as they are represented in Argentine magazines, in order to advance the hypothesis that the broad development of the sciences and the hegemony of positivist discourse not only had repercussions in specific disciplinary or academic arenas; these developments were also influential in cultural fields that existed at a distance from scientific work but were nevertheless deeply drawn to the scientific imaginary due to its futuristic projections and hypothetical games with fantasy.

I will analyze the appropriation and resignification of scientific discourse in Argentine magazines associated with literary *modernismo* on the one hand and with those forms of spiritualism that harbored scientific ambitions on the other, also known as the "occult sciences" in this period, such as spiritism and theosophy. I will focus on the complete collection of *La Quincena: Revista de Letras* (1893–1902), which provided a publication space for *modernista* writers as well as men and women of law, history, science, philosophy, and spiritualism. I will also consider some common ideas found in *Philadelphia* (1898–1902), the first theosophical magazine in Argentina, and the spiritualist magazine *Constancia* (1895–1902), which featured prominently the work of the young Emilio Becher.

In these magazines, which appear to be at a remove from "scientific culture,"[1] it is not only possible to discover numerous fiction stories, articles, news stories, and/or essays connected with the broad spectrum of sciences that existed at the time; it is also possible to find examples of the alternative ways in which scientific discourse was used. Moreover, the magazines contain the most fantastic, utopian, and darkest accounts of the potentials of scientific development; here, the historian and critic are able to encounter a wealth of literary material related to the history of the nascent genres of scientific fantasy and science fiction alongside nonfiction texts that express an interest in science and particularly science's capacity to resignify mysteries, pierce the veil of the occult, and reveal new wonders that are secular, tangible, and verifiable.

Throughout the almost nine-year time span of *La Quincena*'s publication, the scientific imaginary that characterized this period is evident in the short stories of Eduardo L. Holmberg, Carlos Monsalve, Miguel Cané, Eduardo Wilde, Emilio Godoy, Roman Pacheco, Alberto Ghiraldo, Julio Piquet, and Manuel Blancas, among others; it is evident as well in essays and articles that deal with questions related to astronomy, geology, X-rays, psychology, medicine, atavism, and other disciplines little discussed by other *modernista* magazines of the period. However, within this diverse mosaic of themes one encounters a central thread that explains how and why certain scientific discourses were selected and incorporated. This unifying thread derives from the magazine's inscription within *modernismo* and the guiding characteristics of this movement. The magazine's many and diverse contributors all displayed a belief in the value of literature and in the aesthetic value of decadence, mystery, and the occult in particular. The contributions were heterogeneous, written by Catholic and atheist authors, prestigious intellectuals and lesser-known young writers alike; when it came to scientific themes, however, there was clearly a common interest in those areas that had once been considered mysterious or supernatural.

The scientific imaginary of the period invaded *La Quincena* in a manner that was entirely consonant with the characteristic *modernista* repertoire of nymphs, exoticism, and palaces. This taste for the exotic, for reverie, for artificial paradises and decadent experiences also permeated many *modernista* approaches to the world of science. Far from prompting a rejection or an anti-scientific attitude, this sensibility molded a specific form of scientific reception, a deviant form somewhat hallucinatory in character that was prone to fantastical or nightmarish flights. In *Constancia* and *Philadelphia*, one also finds the widespread presence of scientific discourse, but in these cases science is summoned to support spiritualistic ambitions. In other words, the stimulus was neither an aesthetic sensibility nor a flight toward fantasy but rather an interest in articulating a total spiritual phenomenology supported by a discourse and a methodology and benefiting from a social legitimation that stemmed from the sciences. Spiritualists as well as theosophists were active receptors of news of scientific achievements in this epoch; each sought in their own way to extend the limits of what was scientifically observable and knowable in order to study the spiritual dimension of life, which was considered, without question, to be an empirical reality whose laws were still unknown.

In spite of their differences, within the *modernista* magazine as much as the spiritualist and theosophist magazines one finds evidence of the speculative reception and futuristic projection of the sciences in the epoch. These narratives represent deviant, hallucinatory, and fantastical readings of the same material that was heralded in other fields as the motor of progress.

Science among the *Modernistas*: The Case of *La Quincena*

La Quincena: Revista de Letras was one of the longer-lived of a number of magazines identified with a *modernista* aesthetic during the final years of the nineteenth century. The magazine was published on a biweekly basis at first, then monthly, and eventually, during certain periods, bimonthly. The managing editor, Guillermo Stock, a writer who was little known at the time and even less known today, was the primary translator in Argentina of Henrik Ibsen and other Norwegian writers. The editorial secretary listed in the first issues was Alberto Ghiraldo, who also used the pseudonyms Fur Heden and Marco Nereo. Additionally, Luis Berisso and Emilio Berisso participated actively; Luis was described in the magazine as a "patron of the arts," while Emilio was a young poet and critic and future dramatist who sometimes signed his name as "Azrael." The magazine was financed through annual subscriptions and advertisements for a variety of products.

This "tribuna modernista" (*modernista* forum), as it has been called by Héctor Lafleur, Sergio Provenzano, and Fernando Alonso,[2] demonstrated an avid interest in scientific questions. The wide spectrum of scientific discourses characteristic of the period, including pseudoscientific elements, can clearly be seen in fictional narratives as well as informative essays written by formally trained scientists or by self-taught scientists with a background in law.

Many contributions were attempts to think through how science might be articulated with other areas of knowledge or with legislation. In a book extract titled "El derecho penal ante la ciencia" (Criminal law in the light of science), the former anticlerical judge Juan Ángel Martínez claims, "El derecho debe salir de la región nebulosa de las abstracciones, y constituir sus puntos de partida en demostraciones *a posteriori* de las ciencias experimentales" (Law must leave behind the nebulous realm of abstractions, and form its points of departure in the a posteriori demonstrations of the experimental sciences).[3] A very different contribution is made by the curious essay "El arte desde el punto de vista fisiológico" (Art from a physiological perspective), in which Carlos Gutiérrez champions an understanding of art as a form of "vibrations" that can be captured and transmitted.[4]

In relation to literature, it is notable that in contrast to other magazines that devoted a significant amount of space to poetry, *La Quincena* afforded extensive space to short stories and to novels in installments, above all translations of works originally published in Norwegian or English. Those narratives include a considerable number belonging to the genres of scientific fantasy or general fantasy as well as stories that, without being fantastic in nature, address the world of corpses and dissections in autopsy amphitheaters, a world very well known to the group of medical narrators and medical students who participated in *La Quincena*, including Pacheco, Piquet, Blancas, and Wilde.

The interest in these types of narrative can be explained with reference to the first issue of the magazine (August 2, 1893). Julián Martel—the pseudonym of José María Miró, the celebrated author of *La Bolsa* (1891)—wrote an homage to Edgar Allan Poe in which he concludes,

> Tal fue la vida y la muerte del primer cuentista que ha habido en el mundo, del más original de los poetas, del que reunió en sí las más heterogéneas aptitudes, como que fue filósofo y poeta, matemático y novelista, creador de la novela científica. . . . Amalgamadas todas estas cualidades en una sola obra, resulta ser Poe el escritor más original del siglo.[5]

> (Such was the life and death of the first writer of stories in the world, the most original of poets, possessed with the most heterogeneous of aptitudes, such that he was a philosopher and a poet, a mathematician and novelist, a creator of the scientific novel. . . . Amalgamating all of these qualities in a single work, Poe ends up being the most original writer of the century.)

In the first issue, there were also two short stories, "Así," by Wilde, whose stories would occupy a privileged space in the magazine, and a strange story by Ghiraldo that takes place in a salon where doctors study corpses and are faced with the crude, material side of death. In this latter story, the narrator states,

> Estoy en el anfiteatro. En ese sangriento campo de batalla de la ciencia, en ese campo donde se lucha encarnizadamente, a arma blanca, con lo desconocido, con lo ignorado. Muchas veces he entrado a los cementerios. He pasado muchas horas contemplando sepulcros, contemplando muertos. Pero nunca ante las sombrías bóvedas, ante las sepulturas regias . . . he sentido la impresión de disgusto, de desagrado que experimento aquí.[6]

> (I'm in the amphitheater. In this bloody battlefield of science, in this field where one fights strenuously, with a white weapon, against the unknown, the overlooked. Many times I have entered cemeteries. I have spent many hours contemplating graves, contemplating the dead. But never, faced with dark vaults, with magnificent tombs . . . have I felt the sense of disgust and displeasure that I feel here.)

This new scientificist plot twist, which contrasts a scientific approach to death with the symbolic spiritual calm found in cemeteries, opened up a whole narrative direction within *La Quincena*. The high point of this trend was "La primera noche en el cementerio" by Eduardo Wilde, a shorter version of a story originally released in *El Sud Americano* in 1888.[7] Finally, toward the end of the first issue is a text about equality in capitalist societies, authored by the physician Eugenio Wasserzug and curiously subtitled "Un grano de ciencia" (A grain of science).[8] As these examples indicate, right from the first issue one sees the development of narratives that address scientific themes from surprising perspectives.

Indeed, three stories by the naturalist writer Eduardo Holmberg and at least two stories by the journalist and congressman Carlos Monsalve that were published in *La Quincena* would fall into the category of scientific fantasy or early science fiction.[9] These include two fragments of Holmberg's book *La casa endiablada*, presented as independent and previously unpublished stories, and his *Horacio Kalibang o los autómatas*, which had already been published in 1879.[10] Two stories by Monsalve are included that had been published in the 1880s: "De un mundo a otro" (From one world to another) and "Historia de un paraguas" (Story of an umbrella). In this way, the magazine editors seemed to want to insert into a new context scientific fantasies that had been forgotten or were unknown to younger readers. Similarly, with the publication of the two advances of *La casa endiablada*, which were both centered on the spiritualist experiences of Kasper and Otto, *La Quincena* introduced another strand of these early scientific fantasies in which contact with spirits—in this case a "peri-espíritu," the material nexus between the living organism and the beyond—is presented as another form of pseudoscientific experimentation. Not exempt from humor and irony, these stories narrate the cultural tendencies of the younger generation who defended Darwin and Newton but never fully discarded their curiosity for the afterlife.

For its part, Monsalve's "Historia de un paraguas" tells the story of an original voyage through space toward a satellite, made possible by a mysterious umbrella buried in Baltimore.[11] The protagonist, Nathaniel, receives a telegraphic message from a lightning bolt that gives him the clues he needs to undertake the voyage. The explicit reference to Poe (for example, the mention of the city where Poe died) and E. T. A. Hoffman (Nathaniel is the protagonist of "The Sandman") combine here in a story that does not merely draw on the tradition of the marvelous voyage but adds other aspects that also reappear in other texts in *La Quincena*. One is the theme of the criminal guilt of the insane, as when the narrator asks, "¿era criminal no siendo consciente?" (was it criminal if [the perpetrator] was not aware [of his act]?). Another is the strange and supernatural gaze of death introduced by the fear of catalepsy, the condition from which the protagonist suffers, and the question of the possible intelligence of the cosmos, captured in the "pensamiento inteligible" (intelligible thought) emitted by the storm. In a different vein, Monsalve's "De un mundo a otro" fantasizes about making contact with extraterrestrial life through a message written in Sanskrit that was discovered by a mad scientist named Dr. Pánax during his explorations in India.[12] The story mixes an antediluvian banquet of prehistoric meat and wine aged for thousands of years with a form of madness that borders on genius, manifested by the scientist in a way that recalls Holmberg's Burbullus in *El tipo más original* (1878–1879).[13]

Stories by Miguel Cané also form part of the magazine's mosaic, such as "Las armonías de la luz" (The harmonies of light) and "El canto de la sirena" (The siren's song), which address questions of science and fantasy.[14] Cané was much admired by the young people involved in the magazine for the quality of his prose and the fantastic pathways he opened for narrative exploration. In "Las armonías de la luz" he approaches the theme of synesthesia, which is important in *modernismo* and European symbolism, through scientific discourse; an old man claims to have constructed a "órgano de colores" (organ of colors) or a "clavicordio ocular" (ocular clavichord) for his sick daughter. Sandra Gasparini argues that the resolution of the plot comes about through the romantic device of the dream. Cané repeats this partial concession to scientific fantasy in "El canto de la sirena"; Broth, the musician-protagonist of the story, who reads Poe and brings to mind a Hoffman story, states that everything can be known, that behind every legend there is historical truth, and that he will therefore go in search of the sirens in order to reproduce their song. However, ultimately, he is fated to end up in a mental asylum. The failure of this quest for knowledge proceeds in the opposite direction to the corroboration of the supernatural that was present in earlier stories, and it perhaps illustrates Cané's distant or disenchanted perspective in relation to positivism and science, as Oscar Terán has indicated.[15]

Beyond the fantastic, though still animated by strange crossovers between naturalism, pathos, and gothic elements, lies the series of stories about the dead bodies that arrive in the amphitheater of the Faculty of Medicine, generally female corpses whom the narrators or protagonists—students or practicing medics—have loved and abandoned in life, such as suicidal women and single mothers. Pacheco in "Una autopsia" (An autopsy) and "Coquetería suprema" (Supreme coquetry), Piquet in "Margot: Boceto naturalista" (Margot: A naturalist sketch), and Ghiraldo (under the pseudonym Marco Nereo) in "El anfiteatro" (The ampitheater) all explore the sordid terror generated by "el carro de los pobres" (the cart of the poor), bringing in dead bodies for young people to tear open with scalpels. No drift toward the fantastic is evident here, only a pathetic and reflective perspective regarding the ominousness of dead flesh that imbues the stories with a sense of dark supernaturalism, of something perverse. The stories enunciate the sordid and melancholic side of medical practice while invoking a certain gothic sensibility in representing the return of the known in a terrifying form, like the body of a female lover—a necrophilic reversal of the diurnal erotic experience, now lost. While the formal features of these works are simple and they function more as anecdotes than as full stories, they are testimonies of an experience behind the scenes of scientific activity, an experience that is only apprehensible through literary language. In these stories, the doctor-writers of the new generation express the horror to which their own profession awakens them. Thus, for instance, in "Margot," the narrator claims to be going mad after witnessing how another practitioner ripped the eyes from the corpse of a child born by Caesarean section.[16]

In the stories published by *La Quincena* one finds the blueprints of a literary perspective on the sciences that takes full form in later years in literary works by Leopoldo Lugones, Atilio Chiáppori, Ricardo Rojas, and Horacio Quiroga, among others. This perspective is neatly summarized in a text authored by "Dr. Moorne" (the pseudonym of the Spanish writer Francisco Teodosio Moreno) entitled "La fantasía y la ciencia," as follows:

> Las demostraciones científicas y las comprobaciones experimentales muestran bajo su aparente aridez un caudal inventivo e imaginativo extraordinarios siendo tan variados como múltiples los ejemplos que podrían presentarse fácilmente en prueba de tal aserción. . . . Es en vano, pues, que alguien se afane en afirmar que la ciencia influye de una manera desfavorable sobre la fantasía.[17]

> (Scientific demonstrations and experimental tests show beneath their apparent aridity an extraordinarily inventive and imaginative wealth; diverse and manifold examples could easily prove this assertion. . . . It is thus in vain, then, that anyone would bother to declare that science has an unfavorable influence on fantasy.)

This affiliation between science and fantasy, and the drawing of an equivalence between literary and scientific creativity to which Eduardo Holmberg had already referred,[18] are proposed by an author who, curiously, had lashed out against *modernista* aesthetics and the tutelary figure of Rubén Darío as part of a defense of social art in another contemporary magazine, *La Revista Literaria*, edited by Manuel Ugarte. Moreno, as Dr. Moorne, did not pursue a dalliance with "las excelencias de la mitología, de los reyes y las princesas" (the excellence of mythology, kings, and princesses), considering such mythology to be an evasion of reality,[19] although he did share an affinity for another area that attracted modernists: the occult sciences and their possible overlaps with positivist or "materialist" sciences, as they were called at the time. Indeed, years later, from 1901 to 1905, Dr. Moorne published four volumes in Spain on the *Vulgarización de las ciencias ocultas* (Popularization of the occult sciences) in addition to *Los maravillosos secretos de los naipes: Arte de echar las cartas* (The marvelous secrets of cards: The art of reading cards), among other titles.[20] It is not by chance that a fondness for the occult is present here, considering that his vision of the close bond between science and fantasy is related to the understanding developed in spiritualism and theosophy.

Alongside Dr. Moorne, spiritualists such as Cosme Mariño, Felipe Senillosa, and Emilio Becher contributed to *La Quincena*.[21] In their contributions, these spiritualists sought to convey their vision of a science reconciled with the spiritual world, a subject that was extensively developed in the magazine *Constancia*.

The Spiritualization of the Sciences in *Constancia* and *Philadelphia*

In the pages of *Constancia,* which was first published in 1877 and enjoyed a long life during the twentieth century, one finds an original and well-developed enterprise with the aim of disseminating, appropriating, and modifying scientific discourse in order to bring spiritualist practices within the broad field of "the scientific."[22] The magazine was published fortnightly at the turn of the century and financed by subscriptions and through the support of the magazine's wealthiest partners. *Constancia* frequently reviewed the latest scientific discoveries and technological innovations: new rays, the invention of the wireless telegraph and the telephone, gas liquefaction, and the isolation of microbes, among many others. Alongside this coverage, the magazine advanced the argument that if what was once considered impossible or nonexistent was now coming into view thanks to scientific investigation, then soon the "materialidad" (materiality) of the spirit, the "fuerza" (force) of the mind, the transmission of thought, or the transmigration of souls to other planets would become a matter for scientific revelation and explanation.

Constancia's editors and contributors were convinced that spiritualistic phenomena constituted authentic objects of scientific study but that for that end to be fully realized, positivist science needed to relax its methodological prerogatives. Far from ascribing to an antiscientific dogmatism, the spiritualists looked to embody the foundations of a future science that would overcome the pitfalls and limitations of positivism. Thus, one of the editorial notes of 1901 states the following:

> El espiritismo y la ciencia positiva conocida hasta hoy no son dos sistemas de conocimiento opuesto el uno al otro: es la misma ciencia que se alecciona y se complementa. . . . El espiritismo fue, en un principio, desechado en absoluto, ridiculizado y tenido como un delirio, una fantasía de mente desequilibrada; una nueva forma del fanatismo religioso, un resurgimiento de lo sobrenatural y maravilloso, que la ciencia positiva creyó haber enterrado para siempre. Más tarde . . . empezó a verse que el

espiritismo era la ciencia misma positiva, proyectándose más allá de los límites de los sentidos corporales; era una nueva faz de la ciencia misma, que aparecía en el momento en que, atrapada en un callejón sin salida, le era necesario tomar otra dirección o permanecer estacionada.[23]

(Spiritualism and positivist science as they are known to date are not opposing systems of knowledge: it is the same science that is taught and complemented. . . . At first, spiritualism was dismissed completely, ridiculed and treated as a delirium, a fantasy of the unbalanced mind; a new form of religious fanaticism, a resurgence of the supernatural and the marvelous that positivist science believed had been buried forever. Later . . . it began to be seen that spiritualism was the same positivist science, projecting itself beyond the limits of bodily sensations; it was a new face of the same science that appeared at the moment when, trapped in a dead end, it became necessary to take a new direction or remain deadlocked.)

An attempt is made here to trace a continuity between scientific work and spiritualist interests. One can also find in *Constancia* a proposal similar to the one made earlier in *La Quincena* by the positivist Víctor Mercante in his "Moral y religión positivas" (Positivist morality and religion),[24] albeit with a subtle difference. If Mercante proposed science as the unique discipline capable of providing the new beliefs that would replace those of religion, Becher saw in spiritualism a perfect synthesis that would supplant religious dogmas insofar as it would be able to prove the existence of the beyond.

Becher began writing for the spiritual press at the age of seventeen, contributing sporadically to *La Quincena* and maintaining a strong presence in *Constancia* between 1898 and 1902. A reader of Herbert Spencer and Max Nordau, Becher imagined the culmination of the emancipatory power of the sciences, once they had succeeded in revitalizing religion, in the following terms:

Sería admirable y verdaderamente digno de la ciencia que, después de haber arrancado de nuestras almas hasta la última esperanza, fuera ella la que restaurara la fe y la que nos devolviera la certidumbre de una existencia inagotable y la superior alegría de la inmortalidad.[25]

(It would be admirable and truly worthy of science if, after having torn from our souls the last hope, it was science itself that restored faith and gave us back the certitude of an inexhaustible existence and the superior happiness of immortality.)

Becher imagined a comprehensive science, a "síntesis de todas las conquistas dispersas del positivismo moderno" (synthesis of all the scattered conquests of modern positivism) that would also correspond on an economic plane with "la unidad de todos los hombres en un colectivismo universal de fuerzas" (the unity of all men in a universal collectivism of forces), that is, with socialism. Becher dreamed of a total science that in accordance with the spiritualist dream would be able to overcome obscurantism, religion, and irrationalism and include the spiritual dimension as an object of study.

The theosophists of Buenos Aires associated with the magazine *Philadelphia* articulated a similar idea with respect to the pertinence of science to theosophy, though their ambitions were greater than those of the spiritualists. In *Philadelphia*, whether in the essays of Argentine authors such as Leopoldo Lugones and Alejandro Sorondo or in the reproduction of lectures and parts of the movement's reference books such as those by Helena P. Blavatsky and Annie Besant, science invariably appears in capital letters. This was so because it did not refer to modern disciplines but to the accumulation of esoteric and exotic knowledge from the East and the West in which positivist science was only a small and questionable chapter. By contrast to the spiritualists, the theosophists did not seek to embody a science of the future or to replace positivism. Instead, they posited an inverse temporal logic: positivist or materialist science of the nineteenth century came to corroborate empirically that which theosophy already knew through revelation, clairvoyant intuition, the knowledge encoded in mythology, and the esoteric tradition of Western and Eastern religions, among other sources. Along these lines, in 1899–1900 *Philadelphia* published numerous installments of "Corroboraciones científicas de la teosofía" (Scientific corroborations of theosophy) by the Brazilian João Benedicto de Azevedo Marques. The essays address each of the scientific advances achieved in previous years in the areas of bacteriology, astronomy, radiation, and others while declaring that these discoveries were prefigured in Blavatsky's *Secret Doctrine* (1888).

Theosophy was conceived of as the synthesis of all religions, sciences, and philosophies; one of its principal objectives was the scientific study of spiritual phenomena. Its proponents sought a level of legitimation for theosophy as a producer of scientific knowledge in a manner comparable to spiritualism but also to absorb science under the banner of "Scientific Theosophy." Its proposed objective, in short, was the utopian task of spiritualizing science, of giving science a philosophical beginning and end, of providing it with a framework rooted in cosmogony. Theosophists dreamed of a strange form of monism, an idea that was actually taken from a certain current of Comtean positivism. Thus, Arthur Arnould, the French theosophist and director of the magazine *Lotus Bleu*, declares in "Ciencia teosófica," published in *Philadelphia*,

> La Teosofía . . . es una ciencia, o más bien dicho, es la ciencia, toda la ciencia, la sola y única ciencia, es decir, la síntesis de la *sola,* única *y eterna* verdad . . . poniendo de acuerdo todas las religiones, todas las filosofías y todas las ciencias, absorbiendo y disolviendo a la vez el deísmo, el ate-ísmo, el materialismo y el espiritualismo.[26]

> (Theosophy . . . is a science, or rather, it is science, all science, the only, unique science, that is, the synthesis of the only, unique and eternal truth . . . bringing into agreement all religions, philosophies, and sciences, absorbing and dissolving at the same time deism, atheism, materialism, and spiritualism.)

Leopoldo Lugones contributed to both *Philadelphia* and *La Quincena*. As well as several poems, he published an essay entitled "Los climas del arte" (The climates of art) in *La Quincena*; in addition to contemplating decadence as the ideal domain for *modernismo* and positing "mysteries" as the privileged material of art, he describes theosophy as providing the perfect method for knowing art in depth.[27] In *La Quincena* and *Philadelphia*, Lugones, acting both as poet and theosophist, sought to argue for a premodern and monist articulation of art with recourse to philosophy and science. In the first essay he wrote for *Philadelphia*, "Acción de la teosofía" (The effect of theosophy),[28] he announces the advance of spiritualism in scientific disciplines in the place of obsolete material-ist theories. Free from submission to any traditional church, spiritualism was slowly driving science toward the formulation of questions that were previously forbidden due to metaphysical phobias: the why and what for of the phenomena of life. The model of this new science was to be found, paradoxically, in antiq-uity, in a science that "sabía más, aunque conociera menos" (knew more, even while it knew less), thanks to having always been preceded by a philosophical system. In "Nuestras ideas estéticas," Lugones suggests that the common point shared between this science of the past and the science that would necessarily impose itself in the future was the search for a single law, for a law synthesizing the operation of all the phenomena in the universe. Here one finds a curious spiritualist monism within which art is not excluded but rather becomes the privileged channel for the perception of the analogies of the universe.[29]

Conclusion

Within *modernismo*, modern spiritualism, and theosophy it is possible to find projections of the scientific imaginary that break the limits of strict positivism and that look to bring science into the terrains of the spiritual, the fantastic, and the utopian. Across these three areas are diverse exercises in imaginative reasoning about the potential of science. In *La Quincena*, this occurs through the publication of stories with scientific themes on the one hand and the dissemination of discoveries on the other hand. In *Philadelphia* and *Constancia*, these exercises in imaginative reasoning are seen in the articulation of scientific discourse in relation to the spirit world and the occult. Although in these two journals fiction was practically nonexistent, narrative material was offered through the presentation of real cases, that is, through the compilation of testimonies about the phenomena of apparitions, telepathy, and communication beyond the grave. The anecdotes found in such testimonies are curiously similar to the arguments foregrounded in many literary stories of the era.[30] Thus, elements of the scientific imaginary can be seen to slip into the palaces, gallant coquetry, eroticism, and dreams of the *modernistas* and among the tripods, sensitivities, and cosmogonies of the spiritualists. These fantastical imaginaries continue to ring with the echoes of the affective reception of the new world revealed by science.

This use of science in nonscientific contexts represents an important cultural phenomenon at the turn of the twentieth century in Argentina and Latin America. In these speculative, fantastical uses, the layperson drew on the imaginary resources of science to construct his own dreams. Writers, doctors, self-taught scholars, spiritists, and theosophists forged new and heterogeneous routes through this scientific imaginary, fusing elements of expert scientific knowledge with concerns of an aesthetic, spiritual, or mystical nature. These excursions form a crucial part of the cultural history of the sciences in Latin America.

Translated by Lucy Bollington.

Notes

1. Terán, *Vida intelectual*, 9.
2. Lafleur et al., *Las revistas literarias argentinas*, 42.
3. *La Quincena* 1, no. 5–6 (October 18, 1893): 10.
4. *La Quincena* 2, no. 9–10 (January 1895): 229.

5. Martel, "Edgar Allan Poe," *La Quincena: Revista de Letras*, no. 1 (August 2, 1893): 3.

6. Marco Nereo (pseudonym for Ghiraldo), "En el anfiteatro," *La Quincena*, no. 1 (August 2, 1893): 12.

7. Cf. Korn, "Estudio preliminar," 11–32.

8. *La Quincena*, no. 1 (August 2, 1893): 14.

9. For definitions of these genres, please see Abraham, *La literatura fantástica argentina*; Gasparini, "La fantasía científica"; Quereilhac, *Cuando la ciencia despertaba fantasías*; Page, *Science Fiction in Argentina*.

10. E. L. Holmberg, *Horacio Kalibang*, then included in *La Quincena* 1, no. 15–16 (1893–1894). E. L. Holmberg, *La casa endiablada*, two fragments then included in *La Quincena*: "La perí," no. 2 (August 18, 1893): 23–24; and "Kasper," no. 11–12 (January 1894): 218–219.

11. *La Quincena* 7, no. 1–4 (March–April 1899), 25–32.

12. *La Quincena* 6, no. 13–16 (September–October 1898): 410–418.

13. Abraham, "Carlos Monsalve," in *La literatura fantástica argentina*, 336–347.

14. Both stories were originally published in 1877 in Cané, *Ensayos*, then, respectively, in *La Quincena* 3 (September 1895–August 1896): 101–118 and no. 13–14 (March 1895): 313–320.

15. Terán, *Vida intelectual*, 13–82.

16. *La Quincena* 2, no. 19–20 (June 1895): 428–430.

17. *La Quincena* 4, (September 1896–August 1897): 25–26.

18. E. L. Holmberg, "Olga," *La Nación*, September 1, 1878.

19. Galasso, *Del vasallaje a la liberación nacional*, x.

20. Buil Pueyo, *Gregorio Pueyo*, 126.

21. Some examples are Felipe Senillosa, "Al espiritualismo por la ciencia," *La Quincena* 6, no. 17–20 (November–December 1898): 470–472; Cosme Mariño, "Nuevos rumbos de la filosofía y la ciencia," *La Quincena* 7, no. 5–6 (May–June 1899): 155–167 and no. 11–14 (November–December 1900): 350–362; Emilio Becher, "Un drama espiritista," *La Quincena* 7, no. 5–6 (May–June 1899): 195–199.

22. I have worked extensively on the development of spiritualism and theosophy in Argentina in *Cuando la ciencia despertaba fantasías*, which is why less attention is devoted here to these magazines in comparison with *La Quincena*, a magazine that has barely been studied by critics.

23. "El espiritismo y la ciencia oficial," *Constancia*, November 24, 1901, n.p.

24. *La Quincena* 6, no. 21–24 (January–February 1899): 576–583.

25. "El siglo XX," *Constancia*, January 15, 1901, 419.

26. *Philadelphia*, August 7, 1898, 18.

27. *La Quincena* 5 (March 1897–February 1898): 310–314, 443–446. For a detailed analysis of this and other essays by Lugones, see Salazar Anglada, "Modernismo y teosofía."

28. *Philadelphia*, December 7, 1898, 167–176.

29. Lugones, "Nuestras ideas estéticas," *Philadelphia*, November–December 1901, 151–161.

30. One example of many is Camile Flammarion's text "Las fuerzas desconocidas," *Constancia*, March 13, 1898, 82–83.

References

Abraham, Carlos. *La literatura fantástica argentina en el siglo XIX*. Buenos Aires: Ciccus, 2015.

Buil Pueyo, Miguel Ángel. *Gregorio Pueyo (1860–1913): Librero y editor*. Madrid: Instituto de Estudios Madrileños, 2010.

Cané, Miguel, *Ensayos*. Buenos Aires: Imprenta de La Tribuna, 1877.

Galasso, Norberto. *Del vasallaje a la liberación nacional*. Vol. 1 of *Manuel Ugarte*. Buenos Aires: Eudeba, 2014.

Gasparini, Sandra. "La fantasía científica: Un género moderno." In *El brote de los géneros*. Vol. 3 of *Historia crítica de la literatura argentina*, edited by Alejandra Laera, 119–147. Buenos Aires: Emecé, 2010.

Holmberg, E. L. *Horacio Kalibang o los autómatas*. Buenos Aires: Álbum del Hogar, 1879.

———. *La casa endiablada*. Buenos Aires: Compañía Sudamericana de Billetes de Banco, 1896.

Korn, Guillermo. "Estudio preliminar." In *Prometeo y Cía*, by Eduardo Wilde, 11–32. Buenos Aires: Biblioteca Nacional, 2005.

Lafleur, Héctor, Sergio Provenzano, and Fernando Alonso. *Las revistas literarias argentinas (1893–1967)*. Buenos Aires: El Octavo Loco, 2006.

Page, Joanna. *Science Fiction in Argentina: Technologies of the Text in a Material Multiverse*. Ann Arbor: University of Michigan Press, 2016.

Quereilhac, Soledad. *Cuando la ciencia despertaba fantasías: Prensa, literatura y ocultismo en la Argentina de entresiglos*. Buenos Aires: Siglo XXI, 2016.

Salazar Anglada, Aníbal. "Modernismo y teosofía: La visión poética de Lugones a la luz de 'Nuestras ideas estéticas.'" *Anuario de Estudios Americanos* 62, no. 2 (2000): 601–626.

Terán, Oscar. *Vida intelectual en el Buenos Aires fin-de-siglo (1880–1910): Derivas de la cultura científica*. Buenos Aires: FCE (Fondo de Cultura Económica), 2000.

11

Doing Poetry with Science

Unthinking Knowledge in Sarduy, Perlongher, and Eielson

JULIO PRIETO

Galileo famously claimed in *Il saggiatore* (1623) that "the book of nature is written in the language of mathematics."[1] This emblematic statement hints at a conflict that pervades modernity since its inception: the divide between the natural and the human sciences, or what the British chemist and novelist C. P. Snow called in an influential essay the "two cultures" of the West,[2] that is, the divide between knowledge based on mathematics and knowledge based on natural language. Since the 1960s, however, several powerful critiques have challenged the assumptions of modern science and its accompanying divides. These critiques, more or less associated with the emerging field of science studies, have questioned the universality of scientific truth claims, showing how they are socially and/or culturally determined and depend for their articulation on historical sets of defining metaphors, in other words, showing how science in one way or another always *tells a story*.[3]

Bearing in mind these critiques and focusing on the twentieth-century Latin American tradition, in this essay I explore three modes of relation between science and poetry that undermine the modern divide of the sciences and its concurrent ideology. That divide appears in the hierarchical distinction between a universal or pure type of knowledge, largely associated with the natural or "hard" sciences, and a contingent, mixed, and all-too-human type of knowledge, the supposed area of expertise of the "soft" sciences known as the humanities. In examining Latin American modes of mixing poetry with science it might be useful to recall Julio Ramos's well-known thesis that the uneven process of literary autonomization in Latin America has fostered since the nineteenth century the proliferation of "mixed forms."[4] Beyond this thesis I would argue that the exploration of mixed and open forms—in Latin America and elsewhere—became a key trait of artistic modernity in the twentieth century and that, particularly in the case of poetry, both transgeneric hybridization and the exploration of *Unbegrifflichkeit* (inconceptuability)[5] came to be essential elements of modern poetics,[6] by virtue of the exilic and heterologic vocation of modern poetry.[7]

In Latin America, the historical constitution of cultural fields as spaces of heightened discursive hybridity dovetails throughout the twentieth century with modern Western traditions of interartistic and *transscientific* practice. Hence, the productivity of modern and contemporary Latin American poetry as a field of experimentation and reimagination of science, discourse, and culture. To sum up my argument, I propose to read Latin American poetry as a laboratory of the "inconceivable," a rich archive of *experimentos del impensar*, experiments with not-knowing and the unknown, through which delegitimized knowledge practices are mobilized and projected toward the task of reimagining the present, that is, toward the imagination of alternative forms of life and living together.

Sarduy's *Big Bang*: Neo-Baroque *Retombée* and Concretist Astrophysics

In a very schematic way, three models of relation can be distinguished between science and poetry in Latin America during the twentieth century: appropriation, infiltration, and mediation or *entre-capture* (intercapture).[8] The first model can be described as the exploration of the poetic productivity of scientific discourse, as seen, for instance, in Severo Sarduy's first poetry volume, *Big Bang* (1974). In his playful "transcreation" of modern astrophysics,[9] Sarduy combines the insights of two influential Latin American precursors, both eminent models of "science diction" (if I may be allowed a little *calembour*), one distant, the other very close. The distant precursor is Sor Juana's *Primero sueño* (1692), arguably the highest accomplishment of the Spanish American Baroque, a lyrical and epistemological tour de force that offers an astounding synthesis as well as a poetic transmutation of the seventeenth-century natural and human sciences (scholastic theology, logic, physiology, geography, pre-Enlightenment natural philosophy, and so forth).[10] The not-so-distant precursor, perhaps unsurprisingly, is Jorge Luis Borges,[11] whose practice of fictional appropriation of the theories of modern mathematics and physics is taken forward in Sarduy's poetry and essay writing as well as in his novels' parodic reworking of Telquelian critical discourse: a potpourri of new trends and emerging disciplines in the humanities in the 1960s (poststructuralist linguistics, semiotics, Lacanian psychoanalysis, neo-Marxist political philosophy), which in Sarduy's work comingle with the vocabulary of mathematics, astrophysics, and biology.

As in the nightmarish folds of Riemannian geometry in Borges's labyrinths, in Sarduy's concretist rendition of key notions of modern astrophysics and biology—the drifting away of galaxies, the typology of stars, space-time curvature, bilateral symmetry—the laws and principles of mathematics and the natural sciences are engaged less in terms of their truth-value than in light of their aesthetic potentiality and/or textual productivity, thus generating a deviant species of cognition: the lateral form of knowing enabled by textual gnosis. In a prose poem entitled "Vagabundas azules" (Blue stragglers), Sarduy performs a sardonic cross-dressing of astronomic metaphorology (in Hans Blumenberg's sense) to describe a group of Spanish drag queens:

> Todas galácticas, nubladas de pies y manos, dejando un remolino de estrellitas de strass, las Cosméticas salieron de Toledo. . . . la Tutsi, tan estrellada y doble y cubierta de emulsiones sensibles al infrarrojo, que era un homenaje vivo al astrónomo italiano Paolo Maffei.
>
> Así, microcósmicas—querían citar textualmente el universo—, partieron, digo, de Toledo.[12]

> (All of them galactic, overcast from head to toe, leaving behind a whirlpool of little stars of strass, the Cosmetics left Toledo. . . . Tutsi, so starry and double and covered in infrared-sensible emulsions that she was a walking homage to Italian astronomer Paolo Maffei.
>
> Thus, microcosmic—they wanted to quote the universe, textually— they headed off, I say, from Toledo.)

In this type of dissonant alliance of *poiesis* and *mathesis*, the theorems and basic tenets of the natural sciences are treated as ruins. In a way, one could see in this type of poetic reworking of scientific fragments a significant trait of the Baroque imagination, as theorized by Benjamin in his reflection on the Baroque allegory as melancholic discourse, which, as Sor Juana would say, "las glorias deletrea / entre los caracteres del estrago" (spells glories / among the characters of wreckage).[13] As Irlemar Chiampi keenly puts it, "The Baroque is a recycling of forms, an energization of discarded materials, that had its first moment of evidence as cultural fact in the seventeenth century, and that will occur whenever literary discourse reproduces the imaginary of science, handling its statements (fragments) as if they were metaphors."[14]

Perlongher's Ethnography of the Margins: Queering Science and the Poetics of Drift

The second model proposes an inverse operation: an infiltration of scientific discourse by the excess of poetic vision or of an idiosyncratic authorial figure. Such is the strategy found in Néstor Perlongher's *O negócio do michê* (1987), a study of male prostitution in São Paulo that was presented as a master's thesis in the Faculty of Social Anthropology at the State University of Campinas. In this anthropological study, described in the first chapter as an "etnografia das margens" (ethnography of the margins),[15] Perlongher inscribes several lines of flight toward a rhizomatic poetic practice—a neo-Baroque poetics he calls *neobarroso transplatino*[16]—as well as toward what would now be called queer discourse.[17] In this model, the ascendancy of the Latin American essayistic tradition should be taken into consideration if not, more generally, the modern tradition of the essay as an "open form."[18] This double tradition, I would surmise, generates a field of transdiscursive irradiation that favors the drift of works produced within very specific disciplinary frameworks toward fictional modulations and poetic intensities that exceed the conventions of scientific discourse.

In Perlongher's case, it might be argued that the anthropological study is per se a "mixed form": a fringe form of scientific writing, always bordering on the literary, largely partaking of the strategies of narrative fiction. This quality is pointed out by Perlongher himself in his definition of anthropology as "ciência do sutil," a "subtle science" that does not seek to establish objective truths so much as productive connections with certain "practices" and "populations."[19] Perlongher indeed points out how anthropological accounts of the nocturnal life of modern cities—what the sociologist Manuel Castells (interestingly invoking a cinematographic reference) calls the "luzes da cidade" (city lights)[20]—tend to exceed sociology and mutate into "a literary genre halfway between the lyric and science-fiction."[21]

Yet it is precisely in Perlongher's exacerbation of the lyrical intensities of the "city lights," in his vindication of a specific epistemic productivity of the drift— what he calls "as imprevisibilidades do trottoir" (the unforeseeables of the sidewalk)[22]—and in his view of anthropological writing as "um lugar de experimen-tação conceitual" (a site of conceptual experimentation), rather than as a plane of demonstration of "rigorously pre-established hypothesis,"[23] that Perlongher deviates from the disciplinary propriety of the anthropological study, even if this is defined as an always already hybrid form. Going beyond the fictional propensities of that scientific genre,[24] Perlongher actually mobilizes a queer-ing poetic practice that multiplies lines of flight across generic, subjective, and linguistic definitions, in its exploration of *portuñol* and in its poetics of drift, or in what he calls "uma 'linha de fuga' emaranhada em certa paixão de abolição" (a "line of flight" tangled up in a certain passion for abolishment).[25] Unthinking pre-scripted knowledges in terms of genre, gender, language, and sexuality, his ethnography of the margins deploys an activist poiesis and a theoretical prac-tice,[26] simultaneously uniting and dismantling poetry and science in a veritable "anthropology of ecstasy."[27]

Eielson: Knot-Poetics, Not-Objects, and Intercultural Topology

A third model of science diction could be described as a symbiotic intercap-ture,[28] not only between number- and letter-based sciences (C. P. Snow's two cultures of the West) but also between Western and non-Western sciences, be-tween hegemonic and delegitimized knowledges. This model upsets the two great divides of the modern world system: the "internal" divide of Western culture between nature and society and the "external" divide between Western and non-Western cultures, or rather, between Western purported universal-ity and cultures of the rest of the world. It does so by a procedure akin to what Bruno Latour calls "symmetrical anthropology."[29] What is at stake in the poetic practices linked to this model is an operation of redefinition of knowledge, in line with Foucault's description in the opening lecture of his 1976 course at the Collège de France: "In fact it is a question of bringing into play local knowledges—discontinuous, unqualified, not legitimized—against the unitary theoretical instance that tries to filter them, to rank them, to order them in the name of true knowledge and the rights of a science that is held by few."[30] In other words, these poetic practices seek to generate alternative modes of *partage du sensible*:[31] other modes of partitioning/sharing the sen-sible than those prescribed by the hegemonic knowledge distribution dictated from the centers of power of global capitalism. In order to illustrate this third model, I will focus on the work of the Peruvian poet and visual artist Jorge

Eduardo Eielson, although it would be equally productive to analyze from this perspective the work of other Latin American authors, from José María Arguedas's bilingual poetry to Cecilia Vicuña's poetics of the "precarious," or from Nicanor Parra's "anti-poetry" to the transdiscursive experiments developed in the context of the historical avant-gardes and their various "ethnographic surrealisms."[32]

In what follows, I will examine Eielson's practice as poetic-scientific "mediator,"[33] focusing on the productive intertwining of three knowledge traditions explored by his work: modern mathematics and physics (quantum mechanics, topology, string theory), the twentieth-century avant-garde tradition (concrete poetry, conceptualism, *arte povera*, performance art), and the ancient Andean practice of *khipu*, a nonalphabetical writing and computing technique based on the correlation of colored knots tied on woolen threads. Eielson's iconotextual and performative experiments involving the ancient knowledge of *khipu* are particularly interesting as a reactivation of a simultaneously computational and poetic practice that not only undoes the number/letter binarism (or, for that matter, the divide between natural and human sciences) but also, and perhaps more crucially, resignifies the Western notions of "language," "writing," and "science," producing hybrid assemblages that mobilize poetry and artistic practice toward the development of alternative epistemologies.

One possible entry to Eielson's work is through *Leonardo's Codex on the Flight of Birds and on Knots* (1993) (figure 11.1). This is a work that is many works, a description that may be applied to the whole of Eielson's production, to the extent that it does not propose discrete and finished works so much as series of open experiments that dialogue with one other and involve different media, disciplines, and traditions.

In its dual condition of illegible text and plastic artwork, *Leonardo's Codex* somehow condenses the two main veins of Eielson's work, which he himself describes as a combination of "written" and "visual" poetry.[34] *Leonardo's Codex on the Flight of Birds and on Knots* is indeed, first, the quaint title of a strange intermedial object that perhaps should be called a "hybrid," a quasi-object/quasi-subject.[35] It is a knot obtained by twisting a cloth printed with Leonardo's text (the "codex" of the title) and displayed in 1993 as part of an installation at the Galleria delle Stellini in Milan. Second, it is also the title of this installation, consisting of forty of these textual and textile "knots" suspended in midair through golden threads attached to the ceiling (figure 11.2).

Figure 11.1. Jorge Eduardo Eielson, *Leonardo's Codex on the Flight of Birds and on Knots* (1993). Copyright Archivio Jorge Eielson, Saronno and Centro Studi Jorge Eielson, Florence. Reproduced with permission.

In turn, this installation is part of a double *Homage to Leonardo*, consisting of two adjoining rooms that evoke the continuous dialogue between Leonardo da Vinci's painting and writing, between his theoretical-scientific research and his artistic practice. Thus, the space in which Leonardo's codex is "knotted" and put "in flight" directly communicates with an adjoining installation that revisits one of his most famous paintings, *The Last Supper*.

At the same time, these installations and hybrid objects are integrated into several intermedial series that continually link Eielson's written and visual po-

Figure 11.2. Jorge Eduardo Eielson, *Homage to Leonardo. Leonardo's Codex on the Flight of Birds and on Knots* (1993). Copyright Archivio Jorge Eielson, Saronno and Centro Studi Jorge Eielson, Florence. Reproduced with permission.

etry. Thus, *Homage to Leonardo* and *Leonardo's Codex* are taken up in 1996 in two variations on this subject that oscillate between the canvas and the *objet d'art* as well as between the word and the image (figure 11.3); the knotting that retrieves Leonardo's text while twisting it directly connects with the *Khipu* series (figure 11.4). The *Khipu* series is a sequence of abstract paintings begun in the 1960s in which a similar technique of dyeing, tightening, and twisting of fabrics is used to interrogate the relation between the canvas as classic support of European painting and *khipu* as nonverbal writing system of native Andean cultures; a rare form of "semasiography,"[36] *khipu* was largely forced into oblivion by the destructive impact of the Conquest.[37]

Figure 11.3. Jorge Eduardo Eielson, *Leonardo's Codex on the Flight of Birds and on Knots* (1996). Copyright Archivio Jorge Eielson, Saronno and Centro Studi Jorge Eielson, Florence. Reproduced with permission.

In a 1997 essay, "La escalera infinita" (The endless staircase), Eielson relates his work with knots not only with the textual-textile knowledge of Andean *khipu* but also with information theory and mathematical topology. Describing his art as "an encounter between cultures," he zeroes in on the central notion of "network":

La metáfora de la red, al igual que la del nudo (y, evidentemente, no hay red sin nudos), es también la metáfora de la existencia. Desde la cadena de nudos en forma de espiral que constituye el ADN primordial de la vida, hasta el insondable paquete de nervios y neuronas que conforman ese milagro de la evolución que es el cerebro humano, toda nuestra existencia

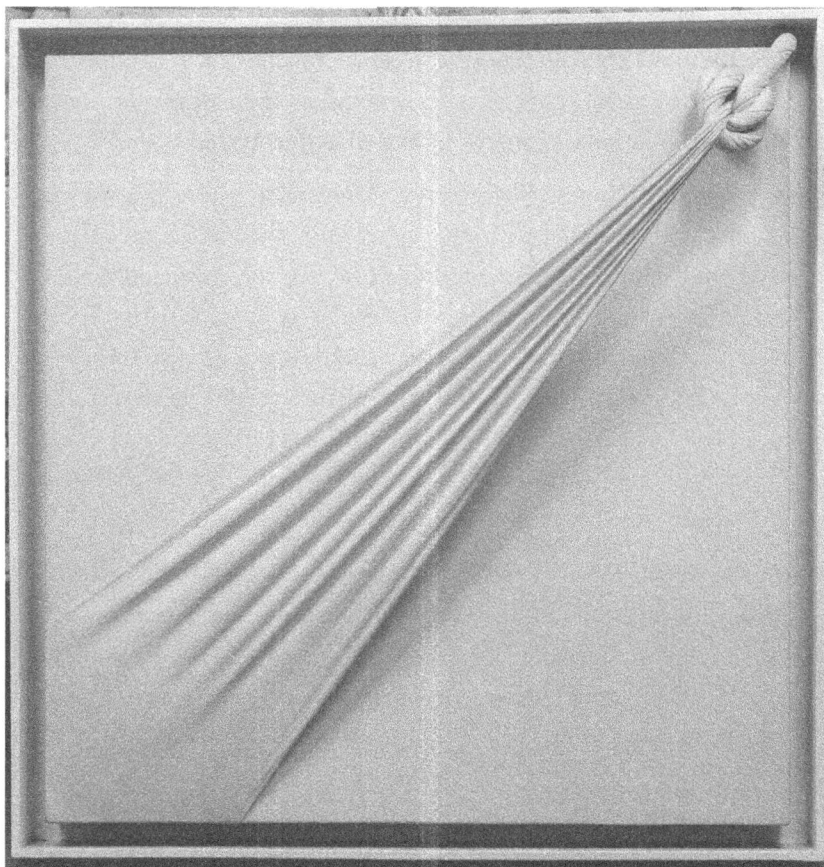

Figure 11.4. Jorge Eduardo Eielson, *Khipu 29 A* (1972). Copyright Archivio Jorge Eielson, Saronno and Centro Studi Jorge Eielson, Florence. Reproduced with permission.

es la historia de una estructura que, para sobrevivir, debe continuamente inventar para sí una infinita red de informaciones y de relaciones interactivas que amplíen su horizonte vital.[38]

(The metaphor of the network, like that of the knot (and, evidently, there is no network without knots), is also the metaphor of existence. From the spiral chain of knots that constitutes the primordial DNA of life, to the unfathomable bundle of nerves and neurons that conform that miracle of evolution that is the human brain, our entire existence is the history of a structure that must continually invent for itself an infinite network of information and interactive links in order to survive and be able to expand its vital horizon.)

Interestingly, he goes on to describe his conception of the knot as mediator between the world of "words" and the world of "numbers," as exemplified by his own written and visual poetic practice (figure 11.5), in a passage permeated by the vocabulary of quantum mechanics and chaos theory:

> Las palabras, que son también nudos de significación, y los nudos, que son también palabras, es decir, objetos significantes, se manifiestan de forma simultánea en un universo misteriosamente caótico y ordenado, regidos sólo por el número y por el caso, en el cual el ser humano participa con soberano y libre albedrío. Un ejemplo visual de esta concepción es mi serie "Nudos como estrellas, estrellas como nudos" de 1990–93, en la que los núcleos de colores (los nudos) proliferan sobre el espacio bidimensional, fuera y dentro de las coordinadas celestes, creando una "caótica" lluvia de valores cromáticos y dinámicos que conforma una auténtica galaxia.[39]

> (Words, which are also signifying knots, and knots, which are also words, that is, signifying objects, simultaneously appear in a mysteriously ordered and chaotic universe, ruled only by the number and the case, in which human beings participate with free and sovereign will. A visual example of this conception is my series "Knots as Stars, Stars as Knots" of 1990–93, in which the nuclei of colors (the knots) proliferate in the bidimensional space, in and out of the celestial coordinates, creating a "chaotic" rain of chromatic and dynamic values that conform a true galaxy.)

Also tellingly, he describes his poetic practice as a form of cognition beyond logical-verbal communication that is "casi en los límites del silencio" (almost on the verge of silence), equating it to "abstract forms" such as music and mathematics.[40]

But I want to go back to that bizarre "signifying object" named *Leonardo's Codex on the Flight of Birds and Knots*, a (k)not-object "beyond words, sounds, and forms."[41] William Rowe observes, regarding Eielson's poetry, that "there are two main paths to enter and leave Eielson's work: through cultural history and through the topology of reading."[42] What I would like to consider here is how these two paths in fact continuously crisscross. A topological analysis of the visible and the readable in Eielson's knots would be seriously impaired were it not to consider the cultural history in which they are inscribed. That history has partly to do with what Rowe elsewhere calls "the historical violence of the letter" in Peru.[43] Eielson's "topology of reading" is of interest in that sense in that

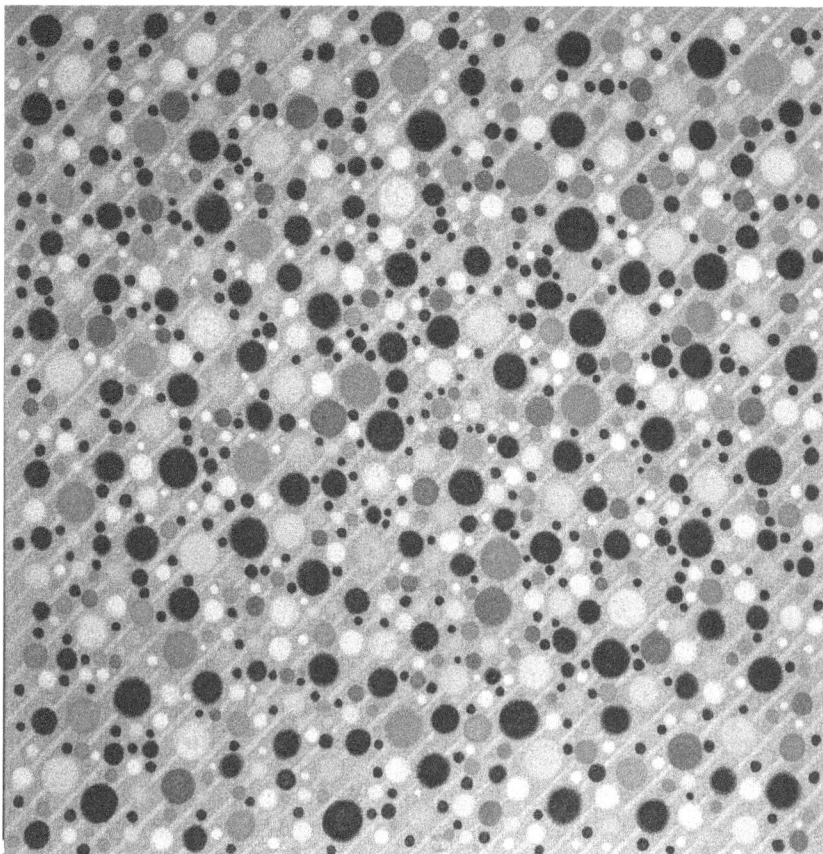

Figure 11.5. Jorge Eduardo Eielson, *Knots as Stars, Stars as Knots* (1990). Copyright Archivio Jorge Eielson, Saronno and Centro Studi Jorge Eielson, Florence. Reproduced with permission.

it functions as a critical figure that intervenes in a historical field teeming with visions and theories of intercultural relation as well as with visions and theories of writing. This historical field, far from being limited to the Peru of the 1990s, pertains to the extended history of Latin American modernity. It was shaped throughout five centuries of conflictive interplay of European and American knowledge cultures, in an uneven global field marked by the divide between the colonizing "sciences" of the West and the subalternized "epistemologies of the South."[44]

How, then, do Eielson's knots intervene in this history? In order to approach this question, one must look closer at the transdisciplinary dimension of Eielson's "topology of reading." "Topology" refers here to the mathematical theory developed in the mid-nineteenth century in the pioneering studies of Gauss,

Tait, and Lord Kelvin, a theory with contemporary ramifications in numerous disciplines (computer science, theoretical physics, biochemistry, and so forth). "Knot theory" and "network topology," specific branches of this theory, can be described as the algebraic study of multiply connected regions and their possibilities of mutation and permutation. Tait's table of "Tenfold knottiness" (figure 11.6) offers, for example, a calculation of possible knots with ten crossing points. For the sake of this discussion, I will call them "points of relation." As epistemological figures, or "meta-images,"[45] enabling a specific way of seeing/thinking (through) intercultural relation, Eielson's knots retain the topological dimensions of multiconnectivity and continuous mutability and thus undermine the binarisms and linear logics that often lurk in the metaphors of intercultural relation privileged in the tradition of Latin Americanism, metaphors such as Vasconcelos's *mestizaje*, Ortiz's and Rama's "transculturation," and Cornejo Polar's "heterogeneity."

Eielson's knots suggest a different figure, in which the dynamic of intercultural relation remains open to different positionalities, alternately euphoric and dysphoric.[46] Far from the determinism inherent in the ethnobiological metaphor of *mestizaje*, Eielson's knots highlight the dimensions of desire and performativity associated with (inter)cultural projections and imaginations. If the knot signals the tension and violence that often go along with the intercultural bind, it also makes visible the contingency and mutability of that bind. However complex a knot may be, it can be undone or reworked in other ways. In "La escalera infinita," he emphasizes precisely the dynamic dimension of "la topología de los nudos [que] se modifican con extrema flexibilidad" (the topology of knots [that] are modified with extreme flexibility), opposing it to "el viejo modelo de la máquina y del fácil determinismo" (the old model of the machine and of simple determinism).[47]

Thus, if Eielson's knot-poetics clearly refers to the historical violence of cultural exchange in Latin America, it does not, however, exclude a specific mode of topological enunciation that short-circuits binarisms, circumventing the extreme pole of euphoric synthesis (as in Vasconcelos's *mestizaje* or in Rama's *transculturación*) as much as the dysphoric pathos of eternally antagonistic actors (as in Cornejo Polar's *heterogeneidad*). In Eielson's "knottiness" there is a measure of naughtiness; his topology of reading brings forth a vein of irony that has something of the an-aesthetic mode of Duchamp's "ready-mades" and something of Oswald de Andrade's "anthropophagic" gesture, as when he declares in his *Manifesto antropófago* (1928), "Tupi or not tupi, that is the question."[48]

Figure 11.6. Peter Guthrie Tait, *Tenfold Knottiness* (1885).

In Eielson's *Codex* the historical "violence of the letter" in Peru and through-out Latin America becomes strikingly visible in the sheer action of twisting and tightening that produces the knot, to begin with. But that violence is articulated in a mode of enunciation that appeals less to the pathos of heterogeneity than to the irony and the polylogical complexity of a nomadic vision.[49] The play of tensions between seeing and reading, as well as between writing and painting, exposed by the hybrid object is framed by a specific colonial history of negation of non-European writing. Eielson's knot-poetics at once remembers and resignifies the historical violence of the gesture of not seeing the other's writing. To that effect, he deploys a strategy involving two operations: first, deactivating or displacing European writing (Leonardo's codex) and second, making visible the *khipu* as a form of writing while affirming its value as textual practice.

As to the latter, Eielson questions the Eurocentric ideology that historically denied that peoples seen as "primitive" could have writing, that supposed lack being precisely what allowed them to be classified as "backward" or devoid of "civilization," thus providing a convenient argument to justify colonization. Explicitly, Eielson treats *khipu* as forms of synesthetic, nonalphabetic writing, seeking to reactivate the ancient practice of Andean peoples, as is evident in his piece *Alphabet* (figure 11.7) and in many of his essays on Andean art and culture. In a text entitled "Luz y transparencia en los tejidos del antiguo Perú" (Light and transparency in the textile crafts of ancient Peru, 1988), he describes pre-Inka textile art as an "alfabeto visual inagotable" (inexhaustible visual alphabet) and as "escritura de algodón" (cotton writing).[50] In this, Eielson seems to concur with Derrida's wry observation, "In fact, the peoples said to be 'without writing,' lack only a certain type of writing."[51]

As to the first operation performed by Eielson's knots, the reimagination of Andean writing is simultaneous to the gesture of displacing the European text: by twisting and knotting Leonardo's codex, reducing it to the materiality of a stained cloth, Eielson performs a violent manipulation that reinscribes in an aggressive reply an entire history of suppression of Andean textuality. However, Eielson's poetic gesture is not confined to the binary confrontation of the European and the Andean or to the inscription of a pathos linked to the vision of the vanquished. Rather, faithful to the key figure of the knot, he puts together a system of folds and ironic contortions, a complex interplay of perspectives that would be best described in terms of what the art historian Hans Belting calls *Blickwechsel*, an "exchange of looks."[52] After all, Eielson's *Codex* not only offers a challenging replica—a twisting and debunking replication as much as an aggressive response—but also a tribute to Leonardo; in that sense one could point out the various levels of identification that Eielson establishes as interdisciplinary artist with the multifaceted figure of the Italian painter, writer, scientist, engineer, anatomist, and inventor, among other vocations.

Of course, Eielson's vindication of *khipu* as textual practice does not entail the reconstruction of a lost language but rather its poetic reinvention, that is, the

Figure 11.7. Jorge Eduardo Eielson, *Alphabet* (1973). Copyright Archivio Jorge Eielson, Saronno and Centro Studi Jorge Eielson, Florence. Reproduced with permission.

activation of its potentiality as alternative textuality or as a way of reading otherwise. Thus, at stake in Eielson's topological poetics is the production of a dynamic illegibility in the interval between the always already read and the desire or the incalculable event of a legibility to come: the creative event of a future vision.[53] His knots are continuously "in flight" between the Andean and the European as well as between the textual and the visual. In this sense, the complex oscillation between the (un)readable and the (in)visible articulated by Eielson's knot-objects evokes the iconotextual dynamic of some emblematic works of the Peruvian tradition—Guaman Poma's *Nueva coronica e buen gobierno* (ca. 1615, figure 11.8)

Figure 11.8. Sketch of a *khipu* reader, Guamán Poma de Ayala, "Contador Maior i Tezorero," chapter 18, *Nueva corónica e buen gobierno* (1615).

or José María Arguedas's posthumous novel *El zorro de arriba y el zorro de abajo* (1971)—as much as the tension between word and image that defines Leonardo's own writing, insofar as Leonardo always writes *through drawing*.

What Eielson's topological poetics envisions, then, is not a binary confrontation of European and American epistemologies but their *knotting*: the visualization of a complex system of approaches and deviations. In this poetic assembly and political assemblage of not-objects/not-subjects, Andean and European knowledges and writing practices are knotted, that is, they relate to one another asymmetrically and multidirectionally. A work like *Leonardo's Codex on the Flight of Birds and on Knots* is multiply ironic; the poignant exchange of glances that it mobilizes has something of the Mona Lisa's ambivalent smile and something of the fundamental ambiguity that Eielson appreciated in the pre-Inka art of the Chavín culture.[54] Or perhaps it has something of what he describes elsewhere, referring to the art of the Chancay as a culture of resistance, as "esta sonrisa textil que atraviesa la entera costa central del Perú" (this textile smile that crosses the entire central coast of Peru).[55]

To Be Continued: Latin American Poetry in the Global Ecology of Knowledges

As I have shown, Sarduy, Perlongher, and Eielson explored different ways of traversing knowledges, producing distinct modes of science diction: three ways of doing poetic things with science as much as of doing scientific things with poetry. This, come to think of it, is not as far-fetched as it might appear *prima facie*. After all, poetry and science have something fundamental in common; if Coleridge famously defined fiction as "suspension of disbelief," poetry and science on the other hand are driven by a basic "suspension of lived belief."[56]

In their experimenting with poetic-scientific hybrids that interrupt long-standing lived beliefs, these Latin American authors sought to create a paradoxical life environment, an alternative "ecology of knowledge practices,"[57] thus joining in the efforts of a broad constellation of artistic trends and political movements that have been moving in that direction in the past decades. Beyond this epochal scenario, one might consider the affinities of these poetic experiments with the "minor" knowledge practices vindicated by Isabelle Stengers in *La Guerre des sciences*, practices that accompany the history of modern science like a dissonant or barely heard *basso continuo*: abandoned hypotheses, scientific paths not taken (the practice of hypnosis, for example, discarded by Freudian psychoanalysis, or psychoanalysis itself today, increasingly embattled in the era of pharmaceutical psychiatry), eccentric theories and experiments validated in retrospect (precursors in physics and chemistry to the contemporary sciences of irreversibility and self-organization), magic and occultism, alchemy, witchcraft, shamanism.[58]

From this hybrid junction, a program of transversal investigation can be extrapolated: to study poetry—and Latin American poetry in particular—in terms of its productivity as a practice of potential knowledges. In this vein, poetry can be read as a future science or "virtual ecology,"[59] or as practice of the "not yet," in Ernst Bloch's sense.[60] This would be a practice that stresses the virtuality of *conocimientos en devenir* (a knowledge in the making, a knowledge yet to come), projecting lines of flight onto the historicity of "sciences" as accumulative formations of knowledge/power, since (as Eielson would say) a word is a moving knot, an unstill knot, a not yet.

These poets, in their experiments with hybrids, sought to produce forms of epistemological pluralization, which may be regarded as one of the crucial challenges of our time, for "there is no global social justice without global cognitive justice."[61] It is a challenge that in the past decades has inspired some of the most daring thinking in the arts and in the sciences (natural, human, social, and various mixtures thereof), generating innovative intellectual currents, many of which—from chaos theory, complexity studies, and science studies to cultural studies, gender and queer studies, affect theory, and the material and digital turns in the humanities—can be seen as attempts to overcome the two great divides of the modern world system: the divide between natural and human sciences and that between Western sciences and non-Western knowledges and epistemologies.

To what extent that double divide has been eroded in the past years through the concerted efforts of artists, poets, scientists, philosophers, and scholars working from different disciplines and perspectives remains an open question, its deactivation still being (and likely to continue to be in the foreseeable future) an ongoing task. Insofar as that double divide is still operative, despite recent critiques, one might say that Latin American poetry, as a practice doubly marginalized—being geopolitically and epistemologically on the losing side of those divides—is critically concerned with them and thus in an optimal position to tackle the task of undoing them, to keep taking creative risks in unthinking knowledges. For as Hölderlin would say, "Wo aber Gefahr ist, wächst / das Rettende auch" (where danger is, there also grows the saving power).[62]

Notes

1. In Popkin, *Philosophy of the Sixteenth and Seventeenth Centuries*, 65.
2. Snow, *Two Cultures*.
3. See Blumenberg, *Paradigmen zu einer Metaphorologie*; Foucault, *La Archéologie du*

savoir; Kuhn, *Structure of Scientific Revolutions*; Serres, *La Traduction*; Latour, *Science in Action*; Stengers, *L'Invention des sciences modernes* and *La Guerre des sciences*.

4. Ramos, *Desencuentros de la modernidad*, 169.

5. Blumenberg, *Theorie der Unbegrifflichkeit*.

6. On poetry as transrational event and experience of "un-knowing," see Casado, *La palabra sabe*, and Nancy, *Corpus*.

7. Milán, *Un ensayo sobre poesía*; Prieto, "Hilos transversales."

8. Stengers, *La Guerre des sciences*, 64–68.

9. "Transcriação" (transcreation) is the term coined by the Brazilian poet Haroldo de Campos to name his theory and practice of creative translation ("Da tradução como criação e como crítica"). Here it should be understood also in the sense of "cultural translation" developed by the philosophers of science Michel Serres and Bruno Latour, either as translation between different accounts of science (Serres, *La Traduction*) or between scientific and sociopolitical representations (Latour, *We Have Never Been Modern*).

10. For an illuminating reading of Sor Juana's *Primero sueño* as compendium of pre-Enlightenment European learning, see Hill, *Scepters and Sciences*.

11. See Merrell, *Unthinking Thinking* for a thorough study of the scientific underpinnings of Borges's writing.

12. Sarduy, *Obra completa*, 170.

13. de la Cruz, *Obra completa*, 198.

14. Chiampi, *Barroco y modernidad*, 33.

15. Perlongher, *O negócio do michê*, 63.

16. Perlongher, "Caribe transplatino."

17. Balderston and Matute-Castro, *Cartografías queer*.

18. Adorno, "Essay as Form."

19. Perlongher, *O negócio do michê*, 60.

20. Castells, *City and the Grassroots* (quoted in Perlongher, *O negócio do michê*, 70).

21. Perlongher, *O negócio do michê*, 70.

22. Perlongher, *O negócio do michê*, 60.

23. Perlongher, *O negócio do michê*, 50.

24. On the proximity and frequent comingling of ethnography, literature, and the arts, see Clifford, *Predicament of Culture*, and Foster, "The Artist as Ethnographer," in his *Return of the Real*, 171–204.

25. Perlongher, *O negócio do michê*, 252.

26. On the need to unthink rather than rethink science, see Wallerstein, *Unthinking Social Science*. For a vindication of "theoretical practices" from the perspective of non-representational theory, see Thrift, *Non-Representational Theory*.

27. Ferrer, "Escamas de un ensayista," 182.

28. Stengers, *La Guerre des sciences*, 64–66.

29. Latour, *We Have Never Been Modern*, 97–103.

30. Foucault, *Il faut défendre la société*, 19.

31. Rancière, *Le Partage du sensible*.

32. Clifford, "On Ethnographic Surrealism."

33. Latour, *We Have Never Been Modern*, 79–82.

34. *Poesía escrita* (Written poetry) is the title Eielson gave to the successive editions compiling his poetic work published in book form (Lima, 1976; Mexico City, 1989; Bogotá, 1998). "Visual poetry," on the other hand, is the term with which he usually refers to his plastic and performative works.

35. Serres, *Parasite*; Latour, *We Have Never Been Modern*.

36. Salomon, *Cord-Keepers*.

37. Most of the existing *khipu* (a Quechua word meaning "knot" or "link") are from the Inka period (ca. 1400–1532 AD), although the oldest ones date from as far back as 2500 BC. Some *khipu* are still given ritual uses by contemporary Andean communities, but their exact functioning as mnemonic, accounting, or narrative devices remains unknown, as it did not survive the violence of European colonization. For a thorough discussion of *khipu* as a nonalphabetical writing system, see Urton, *Signs of the Inka Khipu*; Mignolo and Boone, *Writing without Words*.

38. Eielson, "La escalera infinita," 19–20.

39. Eielson, "La escalera infinita," 21.

40. Eielson, "La escalera infinita," 21.

41. Eielson, "La escalera infinita," 21.

42. Rowe, "J. E. Eielson," 94.

43. Rowe, "Sobre la heterogeneidad," 223.

44. de Sousa Santos and Meneses, *Epistemologías del Sur*.

45. Mitchell, *Picture Theory*.

46. For a detailed discussion of this point, see Prieto, "De nómadas y sujetos migrantes."

47. Eielson, "La escalera infinita," 21.

48. de Andrade, *De Pau-Brasil*, 171. On the an-aesthetic dimension of Duchamp's ready-mades see Oyarzún.

49. What Eielson himself describes as "my cultural nomadism" ("Defensa de la palabra," 443) is akin in that sense to Edouard Glissant's *Poetics of Relation,* a theory and a poetics of intercultural relation that emphasizes the creative dimension of nomadism as much as the "contaminating violence" of the relation.

50. Eielson, "Luz y transparencia," 320.

51. Derrida, *Of Grammatology*, 83.

52. Belting, "Blickwechsel mit Bildern," 66–67.

53. For a comprehensive study of illegibility in Latin American literature and culture, see Prieto, *La escritura errante*.

54. Eielson, "La religión y el arte chavín," 335.

55. Eielson, "Luz y transparencia," 320.

56. Massumi, *Parables for the Virtual*, 232.

57. Stengers, *La Guerre des sciences*.

58. See Massumi, *Parables for the Virtual*, 292–293, for a discussion of this aspect of Stengers's work.

59. Guattari, *Chaosmosis*.

60. Bloch, *Principle of Hope*.

61. de Sousa Santos, *Descolonizar el saber*, 46.

62. Hölderlin, *Gedichte nach 1800*, 173.

References

Adorno, Theodor W. "The Essay as Form." In *Notes to Literature*, vol. 1, edited by Rolf Tiedemann, 3–23. New York: Columbia University Press, 1991 [1958].

Balderston, Daniel, and Arturo Matute-Castro, eds. *Cartografías* queer: *Sexualidades y activismo LGBT en América Latina*. Pittsburgh, PA: Instituto Internacional de Literatura Iberoamericana, 2011.

Belting, Hans. "Blickwechsel mit Bildern: Die Bildfrage als Körperfrage." In *Bilderfragen: Die Bildwissenschaft im Aufbruch*, edited by Belting, 49–76. Munich: Fink, 2007.

Benjamin, Walter. *The Origin of German Tragic Drama*. Translation by John Osborne. London: Verso, 2003 [1925].

Bloch, Ernst. *The Principle of Hope*. Cambridge, MA: MIT Press, 1995 [1947].

Blumenberg, Hans. *Paradigmen zu einer Metaphorologie*. Frankfurt: Suhrkamp, 1997.

———. *Theorie der Unbegrifflichkeit*. Frankfurt: Suhrkamp, 2006.

Casado, Miguel. *La palabra sabe, y otros ensayos sobre poesía*. Madrid: Libros de la Resistencia, 2012.

Castells, Manuel. *The City and the Grassroots*. Berkeley: University of California Press, 1984.

Chiampi, Irlemar. *Barroco y modernidad*. Mexico City: Fondo de Cultura Económica, 2000.

Clifford, James. "On Ethnographic Surrealism." *Comparative Studies in Society and History* 23, no. 4 (1981): 539–564.

———. *The Predicament of Culture: Twentieth-Century Ethnography, Literature and Art*. Cambridge, MA: Harvard University Press, 1988.

de Andrade, Oswald. *De Pau-Brasil* à *Antropofagia e* às *Utopias*. Vol. 6 of *Obras completas*. Rio de Janeiro: Civilização Brasileira, 1978.

de Campos, Haroldo. "Da tradução como criação e como crítica." In *Metalinguagem: Ensaios de teoria e crítica litéraria*, 21–38. Rio de Janeiro: Vozes, 1967.

de la Cruz, Sor Juana Inés. *Obra completa*. Mexico City: Porrúa, 1996.

de Sousa Santos, Boaventura. *Descolonizar el saber, reinventar el poder*. Montevideo: Trilce, 2010.

de Sousa Santos, Boaventura, and María Paula Meneses, eds. *Epistemologías del Sur: Perspectivas*. Madrid: Akal, 2014.

Derrida, Jacques. *Of Grammatology*. Baltimore, MD: Johns Hopkins University Press, 1997 [1967].

Eielson, Jorge Eduardo. *El cuerpo de Giulia-no*. Mexico City: Joaquín Mortiz, 1971.

———. "Defensa de la palabra: A propósito de 'El diálogo infinito.'" In *nu / do: Homenaje a J. E. Eielson*, edited by José Ignacio Padilla, 438–43. Lima: Pontificia Universidad Católica del Perú, 2002.

———. *El diálogo infinito: Una conversación con Martha Canfield*. Mexico City: Universidad Iberoamericana, 1995.

———. "La escalera infinita." In *nu / do*, edited by Padilla, 19–22.

———. "Luz y transparencia en los tejidos del antiguo Perú." In *nu / do*, edited by Padilla, 317–321.

———. *Poesía escrita*. Bogotá: Norma, 1998.

———. "La religión y el arte chavín." In *nu / do*, edited by Padilla, 332–336.

Ferrer, Christian. "Escamas de un ensayista." In *Lúmpenes peregrinaciones: Ensayos sobre*

Néstor Perlongher, edited by Adrián Cangi and Paula Siganevich, 181–193. Rosario, Argentina: Beatriz Viterbo, 2006.

Foster, Hal. *The Return of the Real: The Avant-Garde at the End of the Century*. Cambridge, MA: MIT Press, 1996.

Foucault, Michel. *La Archéologie du savoir*. Paris: Gallimard, 1969.

———. *Il faut défendre la société: Cours au Collège de France (1975–1976)*, edited by Mauro Bertani and Alessandro Fontana. Paris: Seuil, 1997.

Glissant, Édouard. *Poetics of Relation*. Translation by Betsy Wing. Ann Arbor: University of Michigan Press, 1997.

Guattari, Felix. *Chaosmosis: An Ethico-Aesthetic Paradigm*. Sydney, Australia: Power Institute, 1995.

Hill, Ruth. *Scepters and Sciences in the Spains: Four Humanists and the New Philosophy (ca. 1680–1740)*. Liverpool, England: Liverpool University Press, 2000.

Hölderlin, Friedrich. *Gedichte nach 1800: Sämtliche Werke*. Vol. 2, edited by Friedrich Beissner. Stuttgart: Kohlhammer, 1953.

Kuhn, Thomas. *The Structure of Scientific Revolutions*. Chicago: University of Chicago Press, 1962.

Latour, Bruno. *Science in Action: How to Follow Scientists and Engineers through Society*. Cambridge, MA: Harvard University Press, 1987.

———. *We Have Never Been Modern*. Translation by Catherine Porter. Cambridge, MA: Harvard University Press, 1993.

Massumi, Brian. *Parables for the Virtual: Movement, Affect, Sensation*. Durham, NC: Duke University Press, 2002.

Merrell, Floyd. *Unthinking Thinking: Borges, Mathematics and the New Physics*. West Lafayette, IN: Purdue University Press, 1991.

Mitchell, William J. T. *Picture Theory: Essays on Verbal and Visual Representation*. Chicago: University of Chicago Press, 1994.

Mignolo, Walter, and Elizabeth Hill Boone, eds. *Writing without Words: Alternative Literacies in Mesoamerica and the Andes*. Durham, NC: Duke University Press, 1994.

Milán, Eduardo. *Un ensayo sobre poesía*. Mexico City: Libros del Umbral, 2006.

Nancy, Jean-Luc. *Corpus*. Madrid: Arena Libros, 2003.

Oyarzún, Pablo. *Anestética del ready-made*. Santiago, Chile: LOM, 2000.

Perlongher, Néstor. "Caribe transplatino." In *Prosa plebeya*, edited by Christian Ferrer and Osvaldo Baigorria, 93–102. Buenos Aires: Colihue, 1997.

———. *O negócio do michê: a prostituição viril em São Paulo*. São Paulo: Fundação Perseu Abramo, 2008.

Padilla, José Ignacio, ed. *nu / do: Homenaje a J. E. Eielson*. Lima: Pontificia Universidad Católica del Perú, 2002.

Popkin, Richard, ed. *The Philosophy of the Sixteenth and Seventeenth Centuries*. New York: Free Press, 1966.

Prieto, Julio. "De nómadas y sujetos migrantes: Arguedas, Cornejo Polar, Eielson (Un ensayo de arqueología crítica)." In *De nómades y migrantes: Desplazamientos en la literatura, el cine y el arte hispanoamericanos*, edited by Andrea Castro and Anna Forné, 15–59. Rosario, Argentina: Beatriz Viterbo, 2015.

———. *La escritura errante: Ilegibilidad y políticas del estilo en Latinoamérica*. Madrid: Iberoamericana, 2016.

———. "Hilos transversales: Nomadismos en la poesía de Cecilia Vicuña." In *Poéticas del presente: Perspectivas críticas sobre poesía hispánica contemporánea*, edited by Ottmar Ette and Julio Prieto, 237–258. Madrid: Iberoamericana, 2016.

Ramos, Julio. *Desencuentros de la modernidad en América Latina: Literatura y política en el siglo XIX*. Mexico City: Fondo de Cultura Económica, 1989.

Rancière, Jacques. *Le Partage du sensible: Estétique et politique*. Paris: La Fabrique, 2000.

Rowe, William. "J. E. Eielson: Palabra, imagen, espacio." In *nu / do. Homenaje a J. E. Eielson*, edited by José Ignacio Padilla, 408–414. Lima: Pontificia Universidad Católica del Perú, 2002.

———. "Sobre la heterogeneidad de la letra en *Los ríos profundos*: Una crítica a la oposición polar escritura/oralidad." In *Heterogeneidad y literatura en el Perú*, edited by James Higgins, 223–252. Lima: Centro de Estudios Literarios A. Cornejo Polar, 2003.

Salomon, Frank. *The Cord-Keepers: Khipus and Cultural Life in a Peruvian Village*. Durham, NC: Duke University Press, 2004.

Sarduy, Severo. *Obra completa*. Edited by Gustavo Guerrero and François Wahl. Madrid: ALLCA XX, 1999.

Serres, Michel. *The Parasite*. Baltimore, MD: Johns Hopkins University Press, 1982.

———. *La Traduction (Hermès III)*. Paris: Minuit, 1974.

Snow, Charles Percy. *The Two Cultures and the Scientific Revolution*. London: Cambridge University Press, 1959.

Stengers, Isabelle. *La Guerre des sciences (Cosmopolitiques I)*. Paris: La Découverte, 1997.

———. *L'Invention des sciences modernes*. Paris: La Découverte, 1993.

Thrift, Nigel. *Non-Representational Theory: Space, Politics, Affect*. London: Routledge, 2007.

Urton, Gary. *Signs of the Inka Khipu: Binary Coding in the Andean Knotted-String Records*. Austin: University of Texas Press, 2003.

Wallerstein, Immanuel. *Unthinking Social Science: The Limits of Nineteenth-Century Paradigms*. Philadelphia: Temple University Press, 2001.

12

The Science of Reading Fiction

New (Post-Darwinian) Metaphors to Live By

JOANNA PAGE

The prominent role that Latin American literature has played in the critique of positivism since the late nineteenth century has generated a series of dystopian images of science. In the narratives of Leopoldo Lugones, Horacio Quiroga, Bioy Casares, Edmundo Paz Soldán, and others, science has justified dehumanizing experiments, delivered new techniques of surveillance and torture, obfuscated the masses, deepened inequalities, and generally served the interests of the ruling elite. Literature became in this way a crucial site for the political and philosophical critique of science's role in political repression and socioeconomic exclusion and of the increasing dominance of scientific epistemologies. Idelber Avelar observes a twist on this relationship during the Boom period of the 1960s and 1970s in Latin America when instead of pointing to the cost of scientific advance, the innovations of literary modernism "imaginarily compensated" for the region's scientific and socioeconomic backwardness.[1] The representation of science in more recent Latin American fiction bears witness to the continued presence of two contradictory imaginaries: the apocalyptic, on the one hand, and the (often comically) underdeveloped, on the other. In this essay, however, I focus on a new imagined reconciliation between science and literature in a selection of contemporary Latin American narratives. Underpinning this convergence is a shared interest in exploring a post-Darwinian understanding of human evolution.

I will explore a recent rapprochement between literature and science evident in the narrative projects of Marcelo Cohen and Jorge Volpi. Both writers have drawn on research in cognitive neuroscience and evolutionary biology and on theories of autopoiesis and emergence in complex systems to illuminate and extend their literary and critical explorations of human perception, causality, and human-nonhuman interactions. Cohen finds in the physics of uncertainty a stimulating source of inspiration for a new kind of literary realism; inversely, he even suggests that in its embrace of chaos, science becomes "more literary."[2] Following Jonah Lehrer in *Proust Was a Neuroscientist*, Volpi similarly emphasizes that the insights of artists, novelists, and philosophers have often ended up being "confirmed by science."[3] Jorge Luis Borges—as always—got there first: the Argentine neuroscientist Rodrigo Quian Quiroga, who is credited with an important discovery in research on mirror neurons, devotes a whole book to explaining how Borges's narratives anticipate many recent findings in the neuroscience of perception and memory.[4]

Cohen and Volpi engage with scientific research in a much more explicit manner than Borges ever did, however. Volpi explores at some length recent discoveries in cognitive neuroscience with regard to the unique role played by literature in the evolution of human society and culture. In particular, he focuses on how the study of reading, and especially the reading of fiction, has begun to make important contributions to the current understanding of brain function and development. For Cohen, similarly, text and world are bound together in a "continuidad inconsútil" (seamless continuity);[5] in many of his fictions and essays, theories of emergence and complexity are shown to illuminate the ways in which our perceptions, our sense of self, and our acts of reading and writing arise from our cognitive and chemical entanglements with the nonhuman world.

As I will show, both Volpi and Cohen explore post-Darwinian approaches to evolution in which symbiosis and forms of social cooperation become more important than the competitive forces of natural selection. Many of the concepts they employ originate in the pioneering work of Chilean biologists Francisco Varela and Humberto Maturana on autopoiesis, structural coupling, and the neurobiology of cognition. In their theory of "natural drift," Varela and Maturana propose that evolution is not about the survival of the fittest at all, but arises as a result of the aleatory variations that occur through continuous structural couplings between organisms and their environment such that, as they explain, "there is no progress or optimization of the use of the environment, only conservation of adaptation and autopoiesis."[6] Evolution—in their definition—is "somewhat like a sculptor with wanderlust: he goes through the world collecting a thread here, a hunk of tin there, a piece of wood here, and he combines them in a way that their structure and circumstances allow, with no reason other than what is *able* to combine them."[7] Natural drift therefore provides an alternative way of understanding the diversity of life that rejects both the teleology of Darwinian evolution and its emphasis on competition. The processes of coevolution and mutualism described by Varela and Maturana, among others, become important metaphors for Volpi and Cohen in understanding the role of fiction in human culture and society. As I will suggest, their interest in these metaphors lies at least in part in the possible alternatives to capitalist individualism they allow their readers to imagine.

Volpi's *Leer la mente: El cerebro y el arte de la ficción* (2011) draws on recent developments in cognitive neuroscience to voice a passionate defense of the vital importance of art, and especially literary fiction, in creating and sustaining what is uniquely human. A brief bibliography at the end of the book lists Volpi's sources in studies of consciousness, the relation between cognitive science and fiction, meme theory, Gödel's incompleteness theorems, artificial intelligence, and mirror neurons. This last subject is the one to which *Leer la mente* most often returns. Intense discussion and debate has followed the discovery of mirror neurons by Giacomo Rizzolatti, Vittorio Gallese, and other neurophysiologists based at the University of Parma in the 1980s and 1990s. In their work on macaque monkeys, Rizzolatti's team found that when monkeys observed other individuals' actions and emotions, those perceptions activated precisely the same areas of the cerebral cortex that were involved in their own experience of those actions and emotions.[8] Marco Iacoboni, whose research has extended this earlier work to consider specifically human brain functions, affirms that "by helping us to recognize the actions of other people, mirror neurons also help us to recognize and understand the deepest motives behind those actions, the intentions of other individuals."[9] In a more recent synthesis, Rizzolatti and Corrado Sinigaglia, in agreement with many other researchers in the field, emphasize the evolutionary advantage that mirror neurons appear to have conferred both in diagnosing potential threats to the individual and in enabling the social bonding that is characteristic of human societies.[10]

These benefits of mirror neurons for survival and social bonding become the key focus of Volpi's exposition in *Leer la mente*. In this involuntary response in the brain originates the ability to imitate and therefore also to understand others. As Volpi puts it, "Cuando te miro llorar, en mi cerebro *tú* lloras y *yo* lloro al mismo tiempo: las dos personas gramaticales se confunden. En las neuronas espejo, el *yo* y el *otro* se traslapan, se trenzan, se enmarañan" (When I see you cry, in my brain *you* cry and *I* cry at the same time: the two grammatical persons are fused. In mirror neurons, the self and the other overlap, are woven together, tangled up).[11] Yet Volpi's major interest in mirror neurons lies in their capacity to shed light on the determining role played by aesthetic experience, and fiction in particular, in human evolution. For him, the existence of mirror neurons provides hard biological evidence to support the age-old intuition that to read a novel is to inhabit a world.[12]

The fact that the human brain reacts to a novel in the same way as it does to the real world—evaluating every situation by comparing it to existing patterns in order to gauge possible outcomes[13]—strongly suggests for Volpi (as for many of the neuroscientists he cites) that fiction does not simply arise as a result of the flourishing of human culture and society; instead, it may be understood to *produce* those advances. If inventing possible futures helps humans to react faster and better to external threats and aids social cooperation, then what might be dismissed as "un pasatiempo inútil" (a useless pastime) may now be seen as a significant motor of human evolution.[14] Art does not simply offer a uniquely intense kind of pleasure, like the "sensual wallop" of the cheesecake in Steven Pinker's famous analogy.[15] Nor is it merely a by-product of the evolution toward greater cognitive sophistication. Instead, Volpi flips the equation: "El arte no sólo es una prueba de nuestra humanidad: somos humanos gracias al arte" (Art is not just a proof of our humanity: we are human because of art).[16]

Volpi and the researchers he cites clearly move beyond the strictures of classical theories of evolution based on genetic determinism and the rigid laws of natural selection. This is evident in the weight they give to cooperation rather than competition as a key motor of human evolution. There has been a noticeable shift since the 1980s away from an emphasis on the "selfishness" of genes and the disguised self-interest of human acts of altruism,[17] and toward a more pluralistic approach to evolution in which social cooperation plays an important role in new theories of group selection.[18] Interestingly, *En busca de Klingsor* (1999)—the novel in which Volpi undertakes his most significant exploration to date of the imbrication of science and social life—is markedly less confident than *Leer la mente* in the capacity of humans to decode the intentions of others. *En busca de Klingsor* points instead to the social and moral complexities of human interactions and the mistakes to which intuitions can lead. These works remain united, nevertheless, by two themes that are highly relevant to the particular convergence Volpi imagines between science and literature: human exceptionality and the increasing mutual encroachment of the biological and the social.

En busca de Klingsor presents itself as a translation into narrative fiction of Kurt Gödel's incompleteness theorem. In the frenzy of denazification in postwar Germany, Bacon, a promising mathematician from Princeton, is charged with the task of uncovering the identity of the shadowy Klingsor, Hitler's chief scientific adviser. Bacon fails in this mission. What makes Klingsor's identity undecidable within the parameters of the novel, *pace* Gödel's theorem, is Volpi's decision to embed the entire narrative, including those sections told from Bacon's own perspective, in an end-of-life confession volunteered by a physicist, Gustav Links.[19] Links is, however, one of the chief suspects. In a version of the Cretan paradox, Bacon is warned early on that all physicists are liars,[20] and Links strongly intimates at various points that he will be lying. Is Links lying when he suggests that he will be lying (in which case he is telling the truth), or telling the truth (in which case he is lying)? This kind of self-reference propels the reader into a logical paradox, a Gödelian "strange loop," as Douglas Hofstadter shows.[21]

The novel often warns against the use of mathematical and scientific theories to throw light on real-life choices and moral dilemmas. Bacon states clearly that Einstein's theory of relativity has nothing to do with moral relativism.[22] The fictional Erwin Schrödinger is equally adamant that the indeterminacy of quantum mechanics does not extend to the human exercise of free will and morality.[23] Bacon's youthful mistakes in love are only aggravated by his attempts to frame his choices with the help of Von Neumann's game theory. Volpi suggests that the moral complexity of the world cannot be explained or solved by an understanding of its physical laws, although these have certainly provided attractive metaphors for thinking about causality, choices, uncertainty, and unpredictability. The reader witnesses instead in *En busca de Klingsor* the extent to which science may be advanced, halted, or diverted by the ambitions, intimacies, and betrayals of social and political life. The novel therefore acts as a sobering counterpoint to the theories of art and reading that Volpi develops in *Leer la mente*, showing that the scientization of social life may lead not to greater knowledge and understanding but to the abominations and abuses of science for which the Nazi regime was so notorious.

Thus for Volpi, the complexity of human social life is—as Guillermo Martínez also suggests in his novel *Crímenes imperceptibles* (2003)—irreducible to the laws of mathematical logic or rational game theories. If this appears to contradict Volpi's enthusiastic embrace of scientific theories to explain human culture in *Leer la mente*, the two texts do subscribe to a common idea of some significance. In both, the human response to other individuals is not based on rational calculation but on the involuntary firing of neurons and a learned capacity to read intention through a perception of others' emotions that one cannot, biologically, separate from one's own emotional response. In this way, both texts are inscribed into the broader shift in neurobiology that leads away from understanding the human brain as a computational machine and toward an appreciation of the importance of neuronal plasticity.

Volpi's perspective on evolution does not ultimately challenge the principle of natural selection but enlarges it to include cooperation as a kind of competitive advantage. Cohen, in comparison, moves further away from the Darwinist consensus in his emphasis on the importance of symbiosis rather than competition. His interest in human-nonhuman couplings chimes instead with the insights of contemporary ecological developmental biologists. As I will show, Cohen also draws on theories of autopoiesis and dissipative structures to imagine a new kind of literary realism that does not mirror the world but recognizes its relationship with the world as one of codetermination and coevolution. In Volpi's work, the use of scientific theories to cast light on human experience and the work of literature leads him to a humanist reassertion of the uniqueness of human cognition and culture. In contrast, Cohen's broader interest in autopoiesis and emergent structures in nature affords a distinctively posthumanist vision, sketched out within a radical philosophy of immanence.

The theory of autopoiesis developed by Varela and Maturana in the 1970s has had far-reaching implications in the fields of systems theory, evolutionary biology, cognitive science, and phenomenology, among others.[24] Autopoietic systems produce the components they need to maintain their own organization by means of exchanges with their environment, with which they are bound in a relationship of structural coupling. Ilya Prigogine's "dissipative structures," which also include nonliving systems, are thermodynamically open systems that exchange energy and matter with their environment in such a manner that a steady state is maintained.[25] In Cohen's own definition, "Se define como estructura disipativa al sistema capaz de mantenerse estable a condición de abrirse permanentemente a los flujos del medio" (A dissipative structure is defined as a system that maintains its stability by opening itself up permanently to the flows of its environment).[26] Cohen returns repeatedly to the language and dynamics of such systems in his critical essays and narrative fiction in order to combat fixed ontologies of the self and to explore the potential in such accounts for a more accurate understanding of how literature interacts with the world.

In Cohen's *Donde yo no estaba* (2006), the protagonist Aliano learns from his guru that humans have no identity, character, or morality beyond the "chemical processes" sparked by the substances that enter their bodies: "Líquido amniótico, leche de teta: con eso empieza todo. Después comemos carne o papas y la química más compleja, la del cerebro, produce una friolera de ilusiones, por ejemplo la del ser" (Amniotic liquid, breast milk: everything begins with that. After that we eat meat or potatoes and a more complex chemistry, that of the brain, produces a series of mere illusions, for example that of the self).[27] Consciousness in Cohen's fiction is often presented in this way as an illusion produced by the constant biochemical exchanges that take place across the ostensible boundaries of the body and its organs. Even the act of reading, apparently an exercise of constructing mental images, produces a physiological effect; the excitement Aliano feels when reading Lumel's poems causes a flood of endorphins that soothe his headache (182).

Aliano's quest is to live and write in such a way that reflects the plurality and porousness of his self. He aims to "depersonalize" himself in the manner recommended by his guru, who suggests—with recourse to biological metaphors of ingestion and digestion—that "un buen procedimiento de despersonalización es ingerir lo que de sus personas suelten otros, y en el mismo acto evacuar parte de uno" (13) (a good method for depersonalization is to ingest what others let go from themselves, and in the same process evacuate part of oneself). In the world of *Donde yo no estaba*, forms of multiple or distributed consciousness are common and are often expressed in the language of mutualism or parasitism. "Mutual parasitism" is how Aliano describes his relationship with Yónder, a disturbed youth (355), and he begins to refer to "el Yónder-en-mí" (the Yónder-in-me) to describe those parts of his consciousness that originally belonged to Yónder or have fused in some way with his. Yónder himself has been mentally "colonized" (355), his mind hijacked by another person.

If such invasions are often unwelcome, *Donde yo no estaba* also presents a more harmless and pleasurable version of such intermingling of consciousnesses in the form of the "Panconciencia," a kind of virtual network that allows individuals to enter the minds of others and intercept their sensory experiences. In Cohen's fiction, the Panconciencia is "el apogeo evolutivo" (the evolutionary apogee) of the human need to communicate (382), and it takes on many of the functions of reading. As Aliano listens to "el vocerío del multiverso interior" (271) (the clamor of the interior multiverse), he understands that his individual past is no longer a solitary one. Individuals participate in a kind of expanded consciousness in which their encounters with the world extend beyond their own personal perceptions and experiences.

Cohen's characters not only engage in consciousness transfers with other humans but also form dynamic assemblages with cyborgs and bionic animals, mechanical or virtual prostheses, media technologies, and neuropharmaceuticals. These induce mental states and memories to such a degree that it becomes meaningless to try to identify to whom they might originally have belonged. Indeed, the living organisms of Cohen's fiction appear more like ecosystems than individuals. The illusion of biological individuality is expressly countered in recent work in ecological development biology, and the importance of symbiotic processes has taken center stage with the discovery of the vital function of bacteria in normal animal development. In an essay, Cohen also draws attention to the significant roles played by bacteria in the early stages of evolution, recombining damaged genes, sharing molecular chains with their neighbors, and invading other cells as parasites, thereby aiding their growth and reproduction.[28]

Cohen's vision, here and elsewhere, resonates strongly with the emphasis placed by ecological development biologists on the "codetermining" function of organism and environment. Alongside nuclear fission, the microchip, and chaos theory, he observes, twentieth-century science has brought an understanding that humans are nonfoundational beings, as "con nuestra acción nos hacemos y hacemos de lo real un mundo o un medio para nosotros" (with our actions we make ourselves and make of reality a world or an environment for ourselves).[29] This is a kind of codetermination in which neither organism nor environment exists independently from each other. The evolutionary biologist Richard Lewontin affirms, "The environment is not a structure imposed on living beings from the outside but is in fact a creation of those beings. . . . Just as there is no organism without an environment, so there is no environment without an organism."[30]

Cohen's fiction dramatizes this dynamic, by which beings continually form and are formed by other organisms and their environment in a series of complex interactions. These exchanges challenge the classical precepts of natural selection. Cohen suggests, instead, that evolution should be understood, in a way that recalls Maturana and Varela's concept of natural drift, as "la acción ciega de un conjunto de mutaciones genéticas aleatorias" (the blind action of a combination of aleatory genetic mutations).[31] What is more, he explains, a number of biologists have come to believe that the cell from which all plants and animals originally derived did not arise from a genetic mutation but from symbiosis.[32] The pioneering work by Lynn Margulis from the 1960s onward has indeed brought a significant challenge to the supposed centrality of natural selection in evolution by proposing symbiosis instead as the major driver in evolutionary change.[33]

In ¡*Realmente fantástico!* Cohen proposes a new kind of literary realism that would leave behind the paradigms of natural selection—mirrored in a literature governed by the exploration of causality and determinism[34]—and open to "el ritmo de la incertidumbre" (the rhythm of uncertainty) (144). He cites chaos theory as a "hipótesis de trabajo" (working hypothesis) for the inventions of literature (144). This hypothesis would reveal a literature that is engaged in a process of actively shaping itself and its environment through myriad exchanges, in which a novel becomes a "zone of relations" (10). Unlike classical realism, what Cohen calls "la narración de lo real incierto" (the narrative of uncertain reality) abandons neat plot resolutions, linearity, simple causality, and exhaustive revelation of motives and instead embraces digressions, analogies, multiple references, and "el poder transformador de la resonancia" (149) (the transforming power of resonance). Like a Borgesian garden of forking paths rather than expressing self-sufficiency and completion, "cada relato anuncia la existencia de varios mundos" (145) (each story announces the existence of several worlds). This is the kind of relationship between book and world imagined by Deleuze and Guattari, "an aparallel evolution" in which the book is not an image of the world but forms a rhizome with it, just as the orchid and the wasp have evolved for their mutual benefit.[35] To call this operation one of mere mimicry, they hold, would be to ignore the "transversal communications" that "scramble the genealogical trees" and the many molecular and even submolecular ways in which organisms are allied with each other.[36]

Of particular interest in Cohen's fiction is the way he articulates the dynamics of resonance, codetermination, and symbiosis through experiments with narrative voice and syntax; such experiments become, in the case of *Donde yo no estaba*, part of Aliano's search for a new, "depersonalized" writing style. These techniques include the unexpected switching between first- and third-person narrative voices, which suggests a distribution of experiences and perspectives beyond the bounded self. He also confounds the distinction between subject and object to express reciprocity and exchange, in phrases such as "Lo que vemos ve. Lo que oímos oye. . . . Digo y me oyen y oigo lo que me dicen y lo que dicen las cosas, y toco y muevo y soy movido porque toda la experiencia es de un mismo cuerpo" (What we see, sees, and what we hear, hears. I speak and they hear me and I hear what they say to me and what things say, and I touch and move and am moved because all experience is of the same body).[37] Infinitives are used to suggest actions that transcend individual subjects: "Que la vida pase por los verbos. Ocasear. Aguar. Verdear. Trinar. Enlunecer. Acompañar. Espaciar. Durar. Estañar" (May life take place through verbs. To sunset. To water. To green. To chirp. To moonize. To accompany. To spread out. To last. To solder).[38] Such techniques underscore the radical philosophy of immanence explored throughout Cohen's fiction, in which the world and the self are mutually enfolded in multiple ways. "Enlunecer" brings to mind Borges's idealist creation, "Tlön, Uqbar, Orbis Tertius," in which impersonal verbs and adjectives replace all nouns. By contrast, Cohen's unconjugated verbs of becoming usher in a world that is resolutely materialist in its account of the embodied interminglings in which the human and the nonhuman participate.

While in Volpi's humanist approach, reading and writing fiction allow us to participate in experiences that are both unique and universal with respect to the human, for Cohen, such activities more frequently open up to the posthuman and to a Deleuzean "becoming-other" in which the notion of the human is continually expanded. Yet, both writers draw on the powerful metaphors of symbiosis and cooperation that have emerged in recent work on evolution in order to question a classical Darwinist narrative in which individual organisms enter into competition with each other within a rigorous process of natural selection. Indeed, they find a similar dynamic of symbiosis at work in the relation between science and literature itself; this relation becomes yet another example of mutualism and coevolution rather than a battle for survival between two distinct and competing fields of knowledge.

While these new encounters between two old antagonists, literature and science, are most welcome, we should not neglect to ask what epistemological or political projects this apparent confluence might serve. To what extent might this appropriation of scientific theories represent an attempt to legitimize the apparently unproductive activity of reading fiction? One could fully understand the incentive to find scientific evidence for the evolutionary benefits of art, particularly in a society that has forgotten how to value practices whose outcomes cannot be quantified. Other less salutary objectives might, even unwittingly, be served by this imagined coupling. When culture and biology converge, determinism and the naturalization of difference are ever-present dangers. What new determinisms might lie in literature's embrace of the science of complexity or evolution as privileged tools explain the workings of human consciousness and society? Does the appeal to neurophysiological constants in the evolution of human brain function and culture repair to the universalism of humanist thought and its claims to human exceptionalism? This is clearly a question that may be directed toward those texts by Volpi that expound upon the unique qualities of human society, while in others he rejects altogether the scientization of social life. Only in Cohen do we glimpse an expansion of the social itself to include those nonhuman others (other species, environments, and technologies) with which humans continually interact and coevolve and others that may play a much greater role in human evolution than previously thought.

This imagined convergence between the scientific and the literary certainly becomes an expression of the crumbling distinctions between the physical and the nonphysical in many fields of knowledge and praxis, including several branches of science. Cohen shares with Varela a deep interest in the potential for dialogue between theories of animal cognition and self-organization in the physical world on one hand and Buddhist ontology and psychology on the other. Alongside a number of other cognitive neuroscientists, Varela and his colleagues also trace important links with phenomenology, with its emphasis on experience and embodiment.[39] Arguably, this imagined rapprochement between science and philosophy may resist the more problematic kinds of naturalization and determinism that have often accompanied the biologization of culture and social life. This is due to its emphasis on plasticity and experience rather than genetic determinism, and on the possible and the viable rather than the optimal, according to the much broader model of adaptation offered by natural drift as compared to natural selection.

That metaphors of symbiosis and rhizomatic entanglement are taking root so powerfully across the disciplines attests, at least in part, to the urgency of calls to understand human culture and society in ways that do not depend on Darwinist paradigms of competition and progress, paradigms that have so thoroughly naturalized laissez-faire capitalism. A critique of capitalism subtends several works by Cohen and Volpi. Volpi extends his investigation into the human capacity to read others' intentions in another novel, *Memorial del engaño* (2014), a critique of the deceit and impunity of capitalism that climaxed in the 2008 crisis. Many of Cohen's works, including *El oído absoluto* (1989), *El fin de lo mismo* (1992), and *El testamento de O'Jaral* (1995), construct a neoliberal scenario of hyperinflation and homogenization in which the state has been virtually subsumed into vast, unregulated corporations. Like Deleuze and Guattari, Cohen and Volpi find in the dynamics of symbiosis and social cooperation a way of thinking about society that is not entirely dominated by capitalist logic. Their work has the effect of denaturalizing capitalism as, in Volpi's words, "una ideología tan poderosa que nos hizo creer que no tenía ideología" (an ideology so powerful that it made us believe it was not one).[40] The entanglements they explore return the reader again and again to the ethics of coexistence with human and nonhuman others.

In both science and literature, the metaphors used to imagine human evolution and social life cannot be other than deeply political. The neuroscientist and philosopher Catherine Malabou points to the political importance of being conscious of the brain's *plasticity* and not confusing it with its "mistaken cognate," *flexibility*.[41] While the former reveals the power of the brain to give form as well as to receive it, to offer resistance and resilience, the latter aligns us with the "flexible" working practices of late capitalism and its ceaseless demand for workers to adapt.[42] She warns that there is much about the decentered, connectionist character of "the new neuronal paradigm" that bears suspicious structural resemblances to contemporary neoliberal capitalism: "The biological and the social mirror in each other this new figure of command."[43] The mirroring of the biological and the social will always, perhaps, incur a risk of naturalization. The contradictory power that lies in metaphors drawn from the biology of symbiosis—both potentially to resist and to reinforce techniques of control in neoliberal societies—becomes an explicit concern in another recent Argentine novel, Pola Oloixarac's *Las constelaciones oscuras* (2015), which is particularly alert to the dangers and the affordances of such metaphorical borrowings. While philosophy and the arts continue to play crucial roles in the critique of the naturalization of science, this should not preclude a strategic adoption of new paradigms that obey what the developmental biologists Scott F. Gilbert and David Epel call a "moral imperative" to "tell better stories" about the course of human evolution.[44] For Volpi and Cohen, like Gilbert and Epel, those stories allow us to imagine other forms of society that are not based on self-interest and exclusionary paradigms of progress.

Notes

1. Avelar, *Untimely Present*, 35.

2. Cohen, *¡Realmente fantástico!*, 143.

3. Volpi, "Ciencia-Ficción." Also see Lehrer, *Proust Was a Neuroscientist*.

4. Quian Quiroga, *Borges and Memory*.

5. Cohen, *¡Realmente fantástico!*, 135.

6. Maturana and Varela, *Tree of Knowledge*, 115.

7. Maturana and Varela, *Tree of Knowledge*, 117.

8. Rizzolatti and Sinigaglia, *Mirrors in the Brain*, xii.

9. Iacoboni, *Mirroring People*, 6.

10. Rizzolatti and Sinigaglia, *Mirrors in the Brain*, 177.

11. Volpi, *Leer la mente*, 118.

12. Volpi, *Leer la mente*, 121.

13. Volpi, *Leer la mente*, 26.

14. Volpi, *Leer la mente*, 23.

15. Pinker, *How the Mind Works*, 525.

16. Volpi, *Leer la mente*, 15.

17. See, for example, Ghiselin, *Economy of Nature*; Dawkins, *Selfish Gene*.

18. See Meloni, "How Biology Became Social"; Sober and Wilson, *Unto Others*.

19. The technique of free indirect discourse used in these sections presents an insider's perspective on Bacon's world that could only be the result of supposition on Links's part.

20. Volpi, *En busca de Klingsor*, 289.

21. Hofstadter, *Gödel, Escher, Bach*, 17–18.

22. Volpi, *En busca de Klingsor*, 103.

23. Volpi, *En busca de Klingsor*, 279.

24. For an account of some of these, see Mingers, *Self-Producing Systems*.

25. See Ilya Prigogine, *End of Certainty*.

26. Cohen, *¡Realmente fantástico!*, 146.

27. Cohen, *Donde yo no estaba*, 35.

28. Cohen, *¡Realmente fantástico!*, 215.

29. Cohen, *¡Realmente fantástico!*, 173.

30. Lewontin, "Organism as the Subject and Object of Evolution," cited in Varela, Thompson, and Rosch, *Embodied Mind*, 198.

31. Cohen, *¡Realmente fantástico!*, 147.

32. Cohen, *¡Realmente fantástico!*, 147.

33. Margulis, *Origin of Eukaryotic Cells*.

34. Cohen, *¡Realmente fantástico!*, 136.

35. Deleuze and Guattari, *A Thousand Plateaus*, 11–12.

36. Deleuze and Guattari, *A Thousand Plateaus*, 12.

37. Cohen, *Donde yo no estaba*, 150.

38. Cohen, *Donde yo no estaba*, 530.

39. Varela, Thompson, and Rosch consider their research to be "a modern continuation" of the phenomenological philosophy of Merleau-Ponty; they also explore in significant depth the relation between their approach to cognition as embodied action and

Buddhist meditative psychology (*Embodied Mind*, xv, 21–33). Scott F. Gilbert and David Epel's tome on ecological development biology contains a coda dealing with the "philosophical concerns" raised by their perspective on developmental biology and locating this within the tradition of "interdependence, harmony, and integration" bequeathed by Kant and Goethe rather than the "autonomous, competitive" philosophy of Locke and Hobbes (*Ecological Developmental Biology*, 406).

40. Volpi, "Jorge Volpi, el engaño, el capitalismo y el poder."
41. Malabou, *What Should We Do with Our Brain?*, 5, 12.
42. Malabou, *What Should We Do with Our Brain?*, 9–12.
43. Malabou, *What Should We Do with Our Brain?*, 41–42, 33.
44. Gilbert and Epel, *Ecological Developmental Biology*, 416.

References

Avelar, Idelber. *The Untimely Present: Postdictatorial Latin American Fiction and the Task of Mourning*. Durham, NC: Duke University Press, 1999.

Cohen, Marcelo. *Donde yo no estaba*. Buenos Aires: Norma, 2006.

———. *¡Realmente fantástico! y otros ensayos*. Buenos Aires: Norma, 2003.

Dawkins, Richard. *The Selfish Gene*. Oxford: Oxford University Press, 1976.

Deleuze, Gilles, and Félix Guattari. *A Thousand Plateaus: Capitalism and Schizophrenia*. Translation by Brian Massumi. London: Continuum, 2004.

Ghiselin, Michael T. *The Economy of Nature and the Evolution of Sex*. Berkeley: University of California Press, 1974.

Gilbert, Scott F., and David Epel. *Ecological Developmental Biology: Integrating Epigenetics, Medicine, and Evolution*. Sunderland, MA: Sinauer, 2009.

Hofstadter, Douglas R. *Gödel, Escher, Bach: An Eternal Golden Braid*. London: Penguin, 1979.

Iacoboni, Marco. *Mirroring People: The New Science of How We Connect with Others*. New York: Picador, 2009.

Lehrer, Jonah. *Proust Was a Neuroscientist*. Boston: Houghton Mifflin, 2007.

Lewontin, Richard C. "The Organism as the Subject and Object of Evolution." *Scientia* 77, no. 18 (1983): 63–82.

Malabou, Catherine. *What Should We Do with Our Brain?* Translation by Sebastian Rand. New York: Fordham University Press, 2008.

Margulis, Lynn. *Origin of Eukaryotic Cells: Evidence and Research Implications for a Theory of the Origin and Evolution of Microbial, Plant, and Animal Cells on the Precambrian Earth*. New Haven, CT: Yale University Press, 1970.

Martínez, Guillermo. *Crímenes imperceptibles*. Buenos Aires: Planeta, 2003.

Maturana, Humberto R., and Francisco J. Varela. *The Tree of Knowledge: The Biological Roots of Human Understanding*. Translation by J. Z. Young. Boston: Shambhala, 1991.

Meloni, Maurizio. "How Biology Became Social, and What It Means for Social Theory." *Sociological Review* 62, no. 3 (August 1, 2014): 593–614.

Mingers, John. *Self-Producing Systems: Implications and Applications of Autopoiesis*. New York: Springer Science and Business Media, 2013.

Pinker, Steven. *How the Mind Works*. Harmondsworth, England: Penguin, 1997.

Prigogine, Ilya. *The End of Certainty: Time, Chaos, and the New Laws of Nature*. New York: Free Press, 1997.

Quian Quiroga, Rodrigo. *Borges and Memory: Encounters with the Human Brain*. Cambridge, MA: MIT Press, 2012.

Rizzolatti, Giacomo, and Corrado Sinigaglia. *Mirrors in the Brain: How Our Minds Share Actions and Emotions*. Translation by Frances Anderson. Oxford: Oxford University Press, 2008.

Sober, Elliott, and David Sloan Wilson. *Unto Others: The Evolution and Psychology of Unselfish Behavior*. Cambridge, MA: Harvard University Press, 1998.

Varela, Francisco J., Evan Thompson, and Eleanor Rosch. *The Embodied Mind: Cognitive Science and Human Experience*. Cambridge, MA: MIT Press, 1991.

Volpi, Jorge. "Ciencia-Ficción: Dos modos de entender el mundo. Entrevista a Jorge Volpi." Interview by Mónica Prandi. *Letra Urbana*, October 25, 2013. http://letraurbana.com/articulos/ciencia-ficcion-dos-modos-de-entender-el-mundo-entrevista-a-jorge-volpi/.

———. *En busca de Klingsor*. Barcelona: Seix Barral, 1999.

———. "Jorge Volpi, el engaño, el capitalismo y el poder: 'Uno de mis combates privados es contra la religión.'" Interview by Ana Rodríguez. *Clinic Online*, May 22, 2014. http://www.theclinic.cl/2014/05/22/jorge-volpi-el-engano-el-capitalismo-y-el-poder-uno-de-mis-combates-privados-es-contra-la-religion/.

———. *Leer la mente: El cerebro y el arte de la ficción*. Mexico City: Punto de Lectura, 2015.

V

Science, Epistemology,
and the Critique of Modernity

Introduction to Section V

MARÍA DEL PILAR BLANCO AND JOANNA PAGE

In Latin America, the construction of science as an analogue of modernity was cemented in the late nineteenth century with the appropriation of positivism by ruling liberal elites to promote their modernizing agendas. Positivism—itself a reaction against the conservative, colonial, and Catholic consensus that had previously dominated intellectual life in Latin America—provided the justification for Comtean doctrines of "order and progress" that advocated economic liberalism while shoring up social hierarchies. The belief that human society could be perfected through the application of scientific methods underpinned a series of modernizing projects around the turn of the century, particularly in Argentina, Brazil, Chile, and Mexico. Science and modernity patently created the conditions for their mutual advancement; even the speed with which scientific theories were radiating across the globe seemed in itself to create a new vision of the interconnectedness of the modern world.

If positivism was embraced with enthusiasm in many parts of Latin America during the nineteenth and early twentieth centuries, the quest to prove its limits was marked with equal vigor. Prominent antipositivists have included Alejandro Korn (Argentina, 1860–1936), José Enrique Rodó (Uruguay, 1872–1917), and José Vasconcelos (Mexico, 1882–1959). These thinkers argued against the new dogma of a scientific rationalism that did not interrogate its own assumptions and that rejected, out of hand, alternative routes to knowledge, including the metaphysical, the spiritual, and the aesthetic. Their objections were also rooted in a repudiation of the utilitarian politics fueled by positivism, perhaps most overtly during the Porfiriato in Mexico (1876–1911). Other critiques were launched from positions much closer to positivism by thinkers such as Carlos Vaz Ferreira (Uruguay, 1872–1958) and José Ingenieros (Argentina, 1877–1925), who argued for the importance of a metaphysical dimension to scientific thought.

Latin American literature has often become a site for the critique of (positivist) science and its inscription in liberal ideologies of progress. The texts discussed by Carlos Fonseca in his chapter are exemplary in their focus on universalism and linearity as part of a broader critique of the geopolitics of scientific knowledge and modernity. The apparently out-of-sequence development of science and technology in Latin America relativizes the particular course of history such development has followed in Europe or North America. It introduces folds, cleavages, and reversals, allowing a glimpse of a global history of scientific knowledge that does not proceed in a linear fashion but is marked by spatial and temporal discontinuities. These are perhaps most evident in the unfathomable coincidence of discoveries made in unconnected places. Such is the mystery offered by the plainsman discovered by Humboldt in Fonseca's essay in this section whose "reinvention" of the electric machine seems to posit a dissemination of knowledge that does not require contact.

The famous case of the Indian mathematician Srinivasa Ramanujan, whose lack of formal training did not impede him from solving problems that had stumped his contemporaries in Cambridge, lent weight to claims that mathematics operates as a kind of natural, universal language. Latin American literature's solitary geniuses embody similar desires for universality. To the inventors presented in Fonseca's essay in this section—Borges's Carlos Argentino Daneri, García Márquez's José Arcadio Buendía, and Piglia's Luca Belladona—one could add Arlt's Silvio Astier and Remo Erdosain and the protagonists of several novels by Bioy Casares. These figures inspire belief in the possibility that scientific innovation can take place far from its presumed home in the well-equipped laboratories of North American and European institutions and commercial corporations or that it could leap over centuries of technological development, catapulting Latin America into the modern world. This is a belief haunted by unbelief, however, as these innovators are often deranged or desperate men. They hold out the promise of brilliance but sink into madness, violence, or obscurity. It is remarkable that Latin America's literary scientists repeatedly fail where its historical scientists, even the self-taught ones, have more often succeeded. These characters must bear the contradictory weight of a desire for modernity and a simultaneous rejection of its illusions, among which is the notion of scientific truth as universal and absolute.

Since Einsteinian science challenged the clean causalities and absolutes of time and space in classical mechanics, the concepts of relativity and complexity have become privileged frameworks for thinking across multiple fields in Latin America, from cybernetics to literary metafiction, education, environmental studies, and public policy. The complex entwining of human and nonhuman agencies in Latin America—quickly destroying any illusion of autonomy that might be attributed to the political, the ecological, the cultural, or the economic—seems to provide a particular insight into complexity. The condition of dependency might be seen to magnify the impact of socioeconomic and natural processes that interact in a complex manner, acquiring a dynamic of their own that far outstrips the ability of human planners to control causality.

The two very different Latin American thinkers whose work is explored in this section—the Peruvian intellectual José Carlos Mariátegui and the Argentine physicist and theorist Rolando García—are excellent examples of the use of post-Newtonian science to inform political thought in the region. The appropriation of scientific theories for political ends has a long and infamous history in Latin America. The difference here is that science is used precisely to counter the universalism and determinism that have underpinned racial discrimination and other forms of marginalization in the region. Brais Outes-León suggests here that Mariátegui "subverts the role played by science within the coloniality of power" (299–300, this volume), attempting to decolonize science by exposing its role in hegemonic European projects of progress, civilization, and capitalism, and harnessing its power to strengthen a (Marxist) project of political emancipation in Latin America.

García's epistemological project shares something of the decolonizing thrust of Mariátegui's work. It challenges the nature/culture divide that structures European modernity and reframes the politics of knowledge in ways that emphasize the significance of the local and the historically contingent within dynamic processes of global reach. García's work has been extremely influential in the interdisciplinary fields of complexity theory, systems theory, and cybernetics in Latin America. Mara Polgovsky Ezcurra argues that his understanding of the imbrication of the scientific and the political does not give rise to overly simplistic dogmas, but the reverse: an appreciation of the nonlinear relations of causality that link different spheres of knowledge and action together in ways that challenge our capacity to measure or predict their consequences. García belonged to a circle of *rioplatense* philosophers of science from the 1950s onward that included Mario Bunge, Gregorio Klimovsky, and a number of others, whose work traveled beyond the region, becoming well known in the English-speaking world.[1] Alberto Cordero affirms that no other branch of philosophy in Latin America has enjoyed the same prestige as the philosophy of science; despite the obstacles they faced, major Latin American philosophers of science have succeeded in producing work "of the highest international level."[2] One could speculate that the antipositivist critiques of the early twentieth century onward may have laid the foundations for a metaphysical approach to science that would reflect critically on its underlying assumptions, account for the importance of the political, and understand empirical epistemologies as severely limited in their capacity to account for complexity.

Taken together, the chapters in this section emphasize the extent to which the scientific is always already bound up with the political, perhaps more manifestly so in a region construed as peripheral to the powerhouses of scientific advance. Their writers, much like the figures whose work they explore, question accounts of universalism and causality in the production and dissemination of knowledge. What operations result in the occlusion of ideas originating in the "periphery," also thereby concealing the geopolitical forces at work in the circulation of "global" knowledge? Can the flow of information between "center" and "periphery" be remapped to account for a transmission of knowledge that is by no means unidirectional or linear? To write about science in Latin America is always to confront universalist epistemologies with the force of the local and the historical and to question the relations of power that mark out a global history of knowledge and modernization.

Notes

1. See Cordero, "Philosophy of Science," 371–373.
2. Cordero, "Philosophy of Science," 380.

References

Cordero, Alberto. "Philosophy of Science." In *A Companion to Latin American Philosophy*, edited by Susana Nuccetelli, Ofelia Schutte, and Otávio Bueno, 370–382. Oxford: Blackwell, 2013.

13

Laboratories of Universality

A Genealogy of Solitary Latin American Inventors

CARLOS FONSECA SUÁREZ

In the mesmerizing persona of Alexander von Humboldt and his intrepid scientific travels, the emerging Latin American public imagination devised a foundational figure capable of personifying the complex historical changes the continent endured during the nineteenth century, a century that marked its independence from metropolitan Spain as well as a refashioning of the relation between nature and culture that coincided with the birth of modern science and the demise of the old paradigm of natural history. Traversing the shaky frontiers that were soon to be erased and redrawn by the nascent nation-states, carrying the notebooks, chronometers, telescopes, sextants, pendulums, and other scientific instruments with which they aimed to measure the exuberant nature of the equinoctial regions, Humboldt and his colleague Aimé Bonpland unexpectedly became witnesses to the changing political landscape of a continent they had romantically imagined as an Edenic natural paradise.

The revolutionary wind of universal history, which under the flags of equality and freedom had already guided the French Revolution and the American Revolution, would soon sweep through Spanish America, linking science and politics in unexpected ways. One can then imagine the surprise Humboldt must have felt when, as he traversed in 1801 what would soon be the Venezuelan plains, he found an impressive electrical machine lost amid the American wilderness. It was "an apparatus nearly as complete as our first scientific men in Europe possess," he would later recount in his *Personal Narrative of a Journey to the Equinoc-*

tial Regions of the New Continent.[1] As Humboldt soon realized, this machine had not been imported from Europe. It had been assembled by a simple man from the plains, an inventor who had no person to consult and whose only acquaintance with the phenomenon of electricity came from reading the treatises of de la Fond and Franklin. Despite his absolute seclusion, this solitary plainsman, Carlos del Pozo, had been able to participate in a debate that proved to be central within the Enlightenment: that concerning the nature of electricity.

In the image of this solitary plainsman seemingly lost amid the Venezuelan *llanos*, devoid of direct contact with the metropolitan centers of knowledge production, Humboldt must have found an uncanny mirror image of himself and his scientific enterprise. In that foundational scene, Latin America's aporetic role within the history of the Enlightenment is sketched out in relation to the language of science and technology. One must not forget that Carlos del Pozo's technological machine was not just any sort of machine, but an electrical one, an invention that became a symbol for the Enlightenment itself. As has been explored by James Delbourgo in "The Electric Machine in the American Garden," the electric machine, made possible by Franklin's discovery of a scientific language capable of explaining electricity, came to represent the new ways in which reason had been able to explain an otherwise miraculous phenomenon: *action at a distance.* I would like to stress this phenomenon—action at a distance—since what remains fascinating about the scene Humboldt describes is precisely its capacity to reenact, in cartographic and political terms, another aspect of the phenomenon. Constructing an electric machine without recourse to metropolitan materials or models, the plainsman had enacted a new modality of action at a distance; his "reinvention" of the electric machine forces us to think of knowledge as a phenomenon whose dissemination works beyond the notion of contact. Humboldt is quick to emphasize, "Up to then he [Carlos del Pozo] had enjoyed astonishing uneducated people with his experiments, and had never travelled out of the llanos. . . . The names of Galvani and Volta had not yet echoed in these vast solitudes."[2]

Like electricity, this foundational scene refashions a paradoxical connection between center and periphery—the imperial distance that then separated Europe and America—in ways that sketch a thwarted cartographic causality that acts not by proximity but rather by other means. In fact, as Delbourgo has noted, the electrical machine quickly became a spectacular attraction that traveled the world, tracing a global network that included Havana, Mexico City, Manila, China, Persia, and Turkey.[3] The solitude of Humboldt's plainsman, his paradoxical entrance into modern scientific debates, instantiates, following the work of Bruno Latour, a "metrological network" of diffusion that problematizes the model of center and periphery, of global and local, of universal and contingent, that characterized the scientific enterprise of modernity. Latour has explained in books such as *We Have Never Been Modern* that within a landscape that was slowly constructing a complex and often problematic relationship between locality and globality, science was given the task of filling the holes left within the universalizing task of metrology, understood as the project that was intended to render the outside world measurable, factual, a mirror image of the modern laboratory. Latour argues that

> science . . . always renews and totalizes and fills the gaping holes left by the networks in order to turn them into sleek, unified surfaces that are absolutely universal. Only the idea that we have had of science up to now rendered absolute a dominion that might have remained relative. All the subtle pathways leading continuously from circumstance to universals have been broken off by the epistemologists, and we have found ourselves with pitiful contingencies on one side and necessary Laws on the other—without, of course, being able to conceptualize their relations.[4]

The question regarding science becomes, then, a question regarding the ways in which universality has been imagined and constructed within modernity. What Humboldt's plainsman brings into question is the modern enterprise to establish science as the universal language of an enlightened modernity.

I believe one could sketch—around this foundational scene—a genealogy of fictional Latin American inventors who, in their solitude and their failures, will help us reflect upon the impasses of the model of scientific universality and progressivist modernity proposed by the image of Humboldt's plainsman while simultaneously exploring the critical potential of such an illuminating conceptual figure. A new concept of universality might arise from such a genealogical exploration. Taking Humboldt's scene as a starting point I will read Jorge Luis Borges's short story "El Aleph" as a reflection upon the limits of technological universalism as well as a reconfiguration of modern cosmopolitanism. I will then go on to explore the figure of José Arcadio Buendía—founder of Macondo in *Cien años de soledad* (1967)—who in his obsession with scientific innovation takes Borges's exploration of technological modernity and the impasses of modern progressivism even further, proposing instead a new dialectical model of universalism. Finally, I will conclude by adding a final star to this constellation by exploring how the character of Luca Belladona in Ricardo Piglia's 2010 novel *Blanco nocturno* allows for a rereading of this foundational scene in the contemporary socioeconomic context, where the relation between the global and the local, center and periphery, becomes intertwined in the elusive informational networks of global capital.

Before moving on to the Latin American texts, however, I want to return one last time to the scene itself, to the strange causality that is there disclosed. What the plainsman's electrical machine conjures is a causality that seems to escape the classical logic of knowledge dissemination, that is, a tendency to think of scientific invention as functioning within the logic of model and copy, of center and periphery. How could this be? In returning to the scene, one encounters a hint: according to Humboldt, Carlos del Pozo had been able to "reinvent" the electric machine from having only *read* two books, Sigaud de la Fond's *Traité* and Franklin's *Mémoires*. Namely, Humboldt's passage encrypts, at its very heart, a *scene of reading*, and moreover, not only a scene of reading, but also—considering the books themselves—a *scene of translation*. What I hope will emerge from the genealogy I am about to sketch are the political stakes at play in this act of reading that is simultaneously an act of translation. Following the work of Dipesh Chakrabarty, Gayatri Spivak, Alain Badiou, and Étienne Balibar, among others, I would argue that what is at stake in this scene of reading and translation is the cultural politics of competing models of universality and therefore of alternative models of modernity. By staging the limits of science as a universal metalanguage, each of these authors—Borges, García Márquez, and Piglia—proposes a critical model of cosmopolitanism based on what later on, following Balibar, I will call the negative dialectics of universality. Within this critical paradigm, what is encountered is a reading of modernity against the grain, in which Latin America does not copy Europe but rather sketches ways that the relationship itself between model and copy, between center and periphery, is questioned by the reformulation of classical paradigms of writing, reading, and translating.

Carlos Argentino Daneri: A Forgotten Universe on Garay Street

Every time I reread the passage where Humboldt describes encountering the figure of Carlos del Pozo and his electric machine in the Venezuelan *llanos*, I cannot avoid thinking that the image—in its mocking beauty and its conceptual complexity—could have very well been written by Borges. Borges, who in his short story "Pierre Menard, autor del Quijote" thwarts causality by proposing a copy of the original novel that proved infinitely richer than the original, could have easily imagined the preeminence of del Pozo's apocryphal machine over Franklin's. Borges, who in his essay "Kafka y sus precursores" reimagines the notion of influence by positing that each writer retrospectively reconstructs the history of literature, could have perfectly understood that perhaps it was only through the plainsman's act of reinventing the electrical machine after reading merely two books that Franklin's invention truly gained its historical density. I then realize that somehow his short story "El Aleph" could be read as a rewriting of Humboldt's plainsman anecdote, with the character of del Pozo disguised as the megalomaniac Carlos Argentino Daneri, who from the solitary grounds of his house on Garay Street claims to be writing a universal poem pretentiously entitled *La Tierra*.

Published in 1945, with the rise of nationalisms and World War II as its immediate historical background, the short story can be read as simultaneously a defense and a critique of modern universalism in the context of technological modernity. When Borges, the narrator, is introduced by Daneri to his universalist project, its pretensions are sketched in relation to the possibilities of modern science:

> Carlos Argentino lo probó, lo juzgó interesante y emprendió, al cabo de unas copas, una vindicación del hombre moderno.—Lo evoco—dijo con una animación algo inexplicable—en su gabinete de estudio, como si dijéramos en la torre albarrana de una ciudad, provisto de teléfonos, de telégrafos, de fonógrafos, de aparatos de radiotelefonía, de cinematógrafos, de linternas mágicas, de glosarios, de horarios, de prontuarios, de boletines. . . . Observó que para un hombre así facultado el acto de viajar era inútil: nuestro siglo XX había transformado la fábula de Mahoma y de la montaña: las montañas, ahora, convergían sobre el moderno Mahoma.[5]

> (Carlos Argentino tasted it, pronounced it "interesting," and, after a few drinks, launched into a glorification of modern man. "I view him," he said with a certain unaccountable excitement, "in his inner sanctum, as though in his castle tower, supplied with telephones, telegraphs, phonographs, wireless sets, motion-picture screens, slide projectors, glossaries, timetables, handbooks, bulletins . . ." He remarked that for a man so equipped, actual travel was superfluous. Our twentieth century had inverted the story of Mohammed and the mountain; nowadays, the mountain came to the modern Mohammed.)[6]

With this "vindication of modern man," Carlos Argentino Daneri provides a scientific prologue to his poetic enterprise, leading the reader to understand that the universalist pretensions of this minor librarian from "una biblioteca ilegible de los arrabales del sur"[7] ("an unreadable library out on the edge of the Southside of Buenos Aires")[8] are grounded upon the possibilities granted to him by modern technology. Modern science has made the local global and the global local, Daneri seems to suggest. Borges, who one decade later would state in his 1953 seminal essay "The Argentine Writer and Tradition" that as Argentine writers "nuestro patrimonio es el universo"[9] ("our patrimony is the universe"),[10] here seems to suggest that this founding cosmopolitanism is a result of the progress of technology itself. The allegorical demonym that Borges grants his character as a middle name—Carlos *Argentino* Daneri—further stresses the tension between his local identity as an Argentine and the global pretensions that end up endowing his character with an aura of false grandeur. This air of pathetic delirium hints at the idea that "El Aleph" is not a mere story about the triumph of the cosmopolitan progressivism made possible by modern technology but rather a melancholic exploration of the limits of such a universalist project. If, as Piglia has suggested, every short story tells two stories, "El Aleph" is as much the story of the narrator's discovery of the secret aleph hidden in the basement of Daneri's house as it is the story of the narrator's mourning after the death of Daneri's cousin, his beloved Beatriz Viterbo. Namely, it is both a story about the possibility of universality and an allegory regarding its limits, its impasses; it is a story, I would venture to say, about the internal dialectical tension that keeps the project of universality going.

In the same essay in which he proclaims the universe as the legitimate heritage and playground for Argentine writers, Borges explains his thesis by describing Latin American writers as akin to the Jews and the Irish, in their liminal position as insiders and outsiders of Western tradition:

> Creo que los argentinos, los sudamericanos en general, estamos en una situación análoga: podemos manejar todos los temas europeos, manejarlos sin supersticiones, con una irreverencia que puede tener, y ya tiene, consecuencias afortunadas.[11]

> (I believe that we Argentines, we South Americans in general, are in an analogous situation; we can handle all European themes, handle them without superstition, with an irreverence which can have, and already does have, fortunate consequences.)[12]

"El Aleph," like most of Borges's oeuvre, plays upon this irreverent liminal position, positing the great symbol for universality and then denying the possibility of gaining any experience from it before going on to suggest that another model of the universal might subsist in different marginal locations. In the short story, the narrator's encounter with the aleph is accompanied by a great agony, that of having seen a traumatic vision whose painful memory only goes away after long nights of insomnia allow for its forgetting. The narrator's act of mourning is now double; learning to lose Beatriz is also learning to lose the aleph's universal vision, both of whose memories are washed away once the house on Garay Street is destroyed. The postscript of the short story is, in this sense, illuminating; there Borges narrates how, after the events recounted, he came to understand—through the reading of several texts—that the aleph on Garay Street might not actually have been the only one: "Por increíble que parezca, yo creo que hay (o que hubo) otro Aleph, yo creo que el Aleph de la calle Garay era un falso Aleph"[13] ("incredible as it may seem, I believe that the Aleph of Garay Street was a false Aleph").[14] He then goes on to hypothesize about these other possible alephs, providing a list that includes the mirror of Iskandar Zu al-Karnayn upon whose surface the whole universe is reflected, the goblet of Kai Josrú, the spear that in Capella's *Satyricon* belonged to Jupiter, and the stone columns of the Amr mosque in Cairo upon whose surface the ears of the attentive pilgrims can hear the poignant resonance of the whole universe. Paradoxically, by depriving us of the singular, "truthful" aleph, Borges ends up giving us a multitude of them.

This last gesture, which both critiques and reinstates the possibility of universality, gives rise to a further reflection on the scene of Humboldt's plainsman and the model of universality that is staged there. Basing himself on a world in which what remains primordial is the act of reading and translating—and therefore erasing the distinction between model and copy, original and replica—Borges is able to put on, within the struggle for universality itself, a daring performance. Removing the universal from the realm of the true and placing it within the realm of the written, he hints at a "negative dialectics" that, following the work of Étienne Balibar, one could claim lies at the very heart of the struggle for universality as ongoing process:

> And it is because I want to incorporate some of its conditions (including the *negative conditions*, or the "conditions of impossibility") *within* the discourse of universalism, or to put it more philosophically, because I want to outline a discourse of universalism that opens up the possibility

of incorporating within itself its contradictory conditions, that I adopt a critical *and* dialectical point of view.[15]

Balibar has proposed in books such as *Saeculum* that the battle for universality, which is always a struggle with the violence of universality itself, finds in the project of cultural translation a model for its infinite dialectical task. Namely, the universal must always be pursued with an awareness of its exclusions and its failures.[16] One must never settle for a "true" universality but rather explore the dialectical path of pluralizing the universal without therefore reducing it to a mere sum of particularisms. Taking Judith Butler's idea of a "conflictual universalism" as his point of departure, Balibar suggests that the quest toward universality is always a dialectical process of ongoing critique of the project of the Enlightenment, with its enthroning of scientific reason as the universal metalanguage. Following his work, one could then say that, like Carlos Argentino Daneri or Humboldt's plainsman, what the author of *Historia universal de la infamia* performs within a secluded basement on Garay Street is the dialectical freeing up of the universal from the fixed teleology of center and periphery that could only imagine Latin America as a peripheral suburb in the Eurocentric landscape of modernity. If there are only apocryphal alephs in Borges's world, if the false Argentine aleph, as Borges explains in a postscript to the story, is so easily replaced by the mirror found by Tárik Benzeyad, it is precisely because the concept of originality is here bracketed, and what remains is the image of a world where the struggle for universality—imagined as a game of readings and translations—can be found hidden in the heart of any stone. As for Balibar, for Borges the quest toward universality is constantly being questioned by the contingencies of locality and particularity.

José Arcadio Buendía: Coding the Impasses of Globalization

In the tumultuous 1940s amid the rise of fascism, Nazism, and Peronism, Borges's plea for a critical universalism could be read as an attack on nationalism as well as a critique of a certain history of technocratic modernity that had—as Adorno and Horkheimer famously state in their seminal *Dialectic of the Enlightenment*—ended up naturalizing as well as mythologizing its own progressivism. From then on, the founding image of the aleph would remain the archetype that Latin American writers would have to *rewrite* in order to ascertain new models of universalism and cosmopolitanism. As Borges himself has shown in the postscript, once the aleph on Garay Street had been destroyed, what remained was the incessant search for new alephs. This is undoubtedly the case of Gabriel García Márquez's *Cien años de soledad*, a novel in which the foundation and eventual destruction of its mythical Macondo is grounded upon the advances and impasses of science and technology. Writing in the 1960s, a central protagonist of what later gained the name "the Latin American Boom," García Márquez was part of the new struggle between what Jean Franco has called the "conflicting universals" of the Cold War.[17] At once deeply local and universal, thoroughly grounded in Colombia's national history and biblical in its planetary reach, the novel stages the tensions inherent in the internationalization of the Latin American novel in the aftermath of the Cuban Revolution, the rise of mass culture, and the growth of a neoliberal market that included the publishing industry. In the novel's initial pages, the foundation of Macondo is recounted in terms of its relation to science and, moreover, with respect to a particular series of technological discoveries that involve—in a gesture that takes us back to Humboldt's plainsman—the phenomenon of action at a distance. The inventions that the gypsy Melquíades brings from his exotic travels to far-off lands and that quickly become the obsession of Macondo's founder José Arcadio Buendía are precisely those scientific artifacts that are able to relate distant objects. From the discovery of the magnet to that of the magnifying glass, from cartographic instruments to the daguerreotype, these instruments conjure a new cartography in which, as Melquíades himself proclaims, "La ciencia ha eliminado las distancias"[18] ("science has eliminated distance").[19]

Locked in the laboratory he has constructed for himself, Buendía is consumed by the potential of these inventions without realizing that the reason for his obsession lies in the solitude that surrounds him. Macondo, like Daneri's house on Garay Street, is a place secluded from the world. Science becomes, for José Arcadio, the route for a possible contact with the centers of knowledge, a dream that will end the day on which, guided by the compass, the astrolabe, and the sextant gifted to him by Melquíades, he makes his way north, only to wretchedly discover the sea and to hear his wife Úrsula proclaim in dismay: "Nunca llegaremos a ninguna parte. . . . Aquí nos hemos de pudrir en vida sin recibir los beneficios de la ciencia"[20] ("we will never get anywhere. . . . We're going to rot our lives away here without receiving the benefits of science").[21] This image of a secluded, island-like Macondo that nevertheless receives through Melquíades the wonders of science—reproduced in the isolation of José Arcadio's laboratory—returns us to the image of Carlos del Pozo and his reinvention of the electrical machine. Like Borges before him, García Márquez imagines his novel as a laboratory where questions regarding universality, knowledge, and modernity are staged in relation to a cartography of knowledge that defies the classical notion of connectedness.

Local and global, secluded yet founded upon science, marked by the sporadic arrival of gypsies, Macondo reflects what Bruno Latour has called "actor-networks," networks that help "lift the tyranny of geographers in defining space and offers us a notion which neither social nor "real" space, but simply associations. . . . Instead of having to choose between the local and the global view, the notion of networks allows us to think of a global entity—a highly connected one—which remains nevertheless continuously local."[22] Melquíades, the traveling gypsy, the polyglot, produces a network of scientific knowledge that defies geographical proximity and instead proposes a world of interacting languages and codes in which the possibility of universality is explored in relation to a continual act of reading and translating. The history of Macondo, in a way, is the history of these network interactions as well as the history of the decipherment of the secret language that encrypts Melquíades's posthumous manuscript, the one that describes in detail the catastrophic destruction of the town and the tragic destiny of the Buendía family.

Like Borges's "El Aleph," *Cien años de soledad* is also a Janus-faced story; García Márquez's achievement lies in his capacity to highlight the dangers of technological progressivism while at the same time hinting at a model of universality that bypasses the teleological univocity of the metalanguage of science. This is done, aesthetically, through an intertwining of epistemological codes that could be best described in terms of translation. Indeed, magical realism could be said to be an aesthetic model predicated upon the possibility, sketched out by Dipesh Chakrabarty in *Provincializing Europe: Postcolonial Thought and Historical Difference*, of a model of "cross-cultural and cross-categorical translations that do not take a universal middle term for granted. The Hindi *pani* may be translated into the English 'water' without having to go through the superior positivity of H_2O."[23] Like this cross-cultural translation, García Márquez's magical realist aesthetic could be seen as a search for a universality without assuming the existence of a metalanguage. Namely, within García Márquez's magic-realist poetic it is never a matter of translating Latin American myths into the universal language of realism as it was conceived by the European Enlightenment but rather of exposing how the universalism of such linguistic code works through the violent exclusion of other epistemological codes. Magical realism allows for a world where both codes coexist without either being required to relinquish its specificity; neither is forced to bear the burden of truth.

Science—as the assumed metalanguage of reason—becomes, throughout the novel, the site of an ongoing critique of the enlightened project for a progressive modernity. The history of Macondo traces a critique of the Eurocentric teleological metanarratives of progress predicated upon a metalanguage of value. Science and reason, in their incessant search for development and progress, end up destroying the town they had helped build. As Balibar states, this take on modernity is doomed to fail precisely because it ends up contradicting its equalitarian claims: "Indeed, there is no metalanguage of universality, or the surest way to destroy the universality of a universalistic discourse is to claim that it provides the metalanguage of universality, as Hegel already knew it."[24]

While aesthetically sketching—through the construction of a magical realist poetic—a way out of this dilemma, *Cien años de soledad* ends up with a powerful reminder of the dangers of searching for such a metalanguage. The great enigma of the novel, after all, is Melquíades's manuscript, written in a secret code that hides the truth of the town. Like Borges's aleph, like enlightened reason, Melquiades's manuscript hides the universal code, the key for unlocking and understanding the universe. However, as in the case of the aleph, its decipherment unleashes the worst catastrophe. The history of the secrets hidden in Melquiades's manuscript ends on the day Aureliano Babilonia decodes its secret language and, in reading its pages, realizes that he has unleashed the tragic destiny of his family; Macondo is destined to be swept away by a biblical hurricane that begins to destroy the town as he starts reading. By unlocking the universal metalanguage of the town's history, he has unknowingly provoked its destruction. Like Carlos Argentino Daneri, destroyed by his conviction that science would finally bring him in contact with the world, the last of the Buendía family understands in that final instant that his search for the universal key will be his damnation. The novel's ending, where Macondo is catastrophically destroyed by this hurricane, stages—in its apocalyptic imagery—a striking statement on the dangers of progressivist history and technological universalism. Written in the mid-1960s, the novel remains an emblematic text that underlines and critiques a new wave of universalism that was then beginning to sweep through the world. Perhaps unknowingly, García Márquez was hinting at the dangers of a globalizing process that would soon envelop the literary world within a new universal metalanguage, that of the free market, where the flow of capital would soon become the new universal value of equivalence.

Luca Belladona: Dialectical Fictions of Value

Not only is money, as Marx would have it, the universal equivalent but also, the Argentine writer Ricardo Piglia notes, its greatest storyteller: "El dinero—podría decir Arlt—es el mejor novelista del mundo" (Money—Arlt could have said—is the world's best novelist).[25] Few writers have understood the complex intertwining of fiction and money as well as the author of *Plata quemada*. In a literary career that spanned the last three decades of the twentieth century and the beginning of the twenty-first, Piglia wove a series of conceptual fictions in which the alignment between the rise of the neoliberal market and the circulation of state-sponsored fictions was subjected to a thorough critique that took as its basis an insightful theory of reading. Grounded in Borges's intuition regarding the primacy of the copy over the original, Piglia's work explores the ways in which value is fictionally constructed through circulation. Abolishing the distinction between the true and the false—taking as one of its basic symbols the notion of a "fake coin"—Piglia's analysis of the economics of storytelling remains an exploration into those apocryphal alephs first disclosed by Borges in his postscript to "El Aleph." Piglia writes,

> Arlt no asocia—como podría pensarse—el poder del dinero con la verdad, sino con la mentira, el crimen y la falsificación: por de pronto el dinero, signo del oro, obligado a circular sin reposo, no es más que la ficción, el simulacro—o como diría Marx: el enigma—del valor. Al mismo tiempo, en una sociedad que sostiene la ilusión de enriquecerse en el mito de *hacer dinero*, la falsificación aparece como la metáfora misma del trabajo productivo.[26]

> (Arlt does not relate—as one might think—the power of money to that of truth, but rather with the world of lies, crime, and falsifications: money, the sign for gold, forced to circulate without pause, is nothing but the fiction, the simulacrum—or as Marx would say: the enigma—of value. At the same time, in a society that maintains the illusion of enriching itself through the myth of *making money*, falsification appears as the metaphor of productive labor.)

The circulation of capital, Piglia seems to suggest, is the origin of fiction understood as the birth of value; it lies, therefore, at the very center of the question regarding universality. Borges's second great short story about universality is precisely a story about a coin. In "El Zahir," Borges, in yet another story about

mourning, explores the impossibility of forgetting the coin that in its representation of an entire economy comes to hypostatize the universe. Like the aleph,
the coin is the principle of equivalence and therefore of translation; precisely
because it does not stand for anything concrete, it stands for the possibility of
everything. It should not come as a surprise, then, that in his 2010 novel *Blanco
nocturno* Piglia decides to weave, around the death of a gambler and his enigmatic suitcase full of money, a detective fiction that explores the ways in which
the relation between the global and the local, center and periphery, is caught up
in the elusive informational networks of global capital. Published when the Argentine government's 2008 conflict with the agricultural sector was still a lively
memory, the novel explores, through its retelling of the story of the Belladona
family, the ways the peripheral Argentine countryside is simultaneously cast
off and woven into the larger cartography of global capitalism and its claims of
universality.

As Croce, the detective leading the case, starts to uncover the facts around
the death of Tony Durán, the Puerto Rican playboy whose death triggers the
plot, the novel begins to unravel the story of the decline of the family's business as well as the mad project of its last member to keep it alive. And so,
alongside the detective and Emilio Renzi, the journalist covering the case, we
are introduced to the fascinating character of Luca Belladona. He is a madman who, from within the ruins of the old car factory, is engaged in a mega-
lomaniacal project through which he wishes to salvage the economic future of
the factory and of his family. Like the electrical machine lost in the Venezuelan plains, the Belladona factory remains a paradoxical mirage of rationalism
lost in the desert: "Una construcción increíble, a diez kilómetros del pueblo,
entre los cerros, con una arquitectura racionalista, que impresionaba aislada
en medio del campo, como una fortaleza en el desierto"[27] (an unbelievable
building of architectural rationalism, ten kilometers outside the town, in the
hills, especially striking in the middle of the countryside, like a fortress in the
desert).[28] And it is precisely from that rationalist fortress that Luca Belladona
puts into action his deranged plan to save the factory; doing so involves not
only a way for Tony Durán to launder some money but also a series of conceptual experiments that link his project to those of Carlos Argentino Daneri
and José Arcadio Buendía.

As the reader comes to discover, after the factory's bankruptcy, Luca—
having read Carl Jung's *Man and His Symbols*—had taken it upon himself to
construct a series of conceptual artifacts through which he could finally grasp

the universal nature of dreams; these are no longer mere ideas, but rather objects, in a fascinating inversion that would allow him to sketch out the logic of dreams proposed by Jung's universal archetype theory. Taking Borges's and García Márquez's intuitions to their limits, Piglia therefore proposes an imaginary project in which the possibility itself of a concretization of universality is explored and critiqued with regard to the history of technology. If, as Idelber Avelar has explored in *The Untimely Present*, the authors of the Latin American Boom, most notably Carlos Fuentes and Mario Vargas Llosa, attempted to reassert the universality of their literary projects as a way of compensating for the continent's economic and social backwardness, Piglia critically exposes this project of modernization as fiction. Like Carlos Argentino Daneri's basement and Jose Arcadio Buendía's house, Luca Belladona's factory becomes a laboratory for experimenting with universality and its languages, in this case the language of dreams, which is inevitably linked here to the language of capital.

Luca's dreams become realized in "el mirador," an ironclad dream machine in the form of a pyramidal cone that allows the user to grasp distant objects:

> Hemos pensado llamarla *Nautilus*, a esta máquina, que es la réplica de una nave especial, no es un submarino, es una máquina aérea que sólo produce movimientos en la perspectiva y en la visión de lo que se ve venir. Éste es el anuncio de la nueva época: vehículos quietos que traerán el mundo hacia nosotros en lugar de tener que viajar nosotros hacia el mundo.[29]

> (We are considering calling this machine *The Nautilus*, but it's actually the replica of a spaceship, not a submarine. It's an aerial machine; it produces changes in the perspective and viewpoint of what one comes to see. It's a sign of the times: a stationary vehicle that brings the world to us instead of us having to go to the world.)[30]

Like Macedonio's storytelling machine in Piglia's *La ciudad ausente* (*The Absent City*, 1992) or the perfect map of Buenos Aires sketched in the prologue to *El último lector* (2005, the last reader), "el mirador" remains a rewriting of the aleph: a technology capable of bridging distances and portraying the logic of dreams. Like Borges's characters Carlos Argentino Daneri and José Arcadio Buendía before him, Luca Belladona has paradoxically become an incomprehensible pariah in his project to capture the logic of universality. His universalist project has ironically led him to the realm of illegibility and illegality; he has become a madman and a criminal. However, in a memorable reversal, it is precisely such a contradiction that grants him his status as a symbol for fiction itself, a man who, in his attempt to salvage his family's business, has been made to explore the ways in which value is constructed as a dreamlike simulacrum. In his character one finds a final rewriting of Humboldt's plainsman lost on the Argentine pampa; Piglia's Belladona sketches fiction as the impossible dialectical movement toward a universality that simultaneously casts us off as criminals while allowing us to understand that only from such a peripheral position can our critique of universality persist in its infinite task.

Notes

1. Humboldt, *Personal Narrative*, 168.
2. Humboldt, *Personal Narrative*, 168.
3. Delbourgo, "Electric Machine in the American Garden," 260.
4. Latour, *We Have Never Been Modern*, 119.
5. Borges, "El Aleph," in *Obras completas,* 332.
6. Borges, *The Aleph*, 121.
7. Borges, "El Aleph," in *Obras completas*, 332.
8. Borges, *The Aleph*, 120.
9. Borges, "El escritor argentino y la tradición," in *Obras completas*, 274.
10. Borges, "The Argentine Writer and Tradition," in *Labyrinths*, 216.
11. Borges, "El escritor argentino y la tradición," in *Obras completas*, 273
12. Borges, "The Argentine Writer and Tradition," in *Labyrinths*, 218.
13. Borges, "El Aleph," in *Obras completas*, 343.
14. Borges, *The Aleph*, 124.
15. Balibar, "On Universalism," 2.
16. Balibar, *Saeculum*, 24.
17. Franco, *Decline and Fall of the Lettered City*, 21.
18. García Márquez, *Cien años de soledad*, 81.
19. García Marquez, *One Hundred Years of Solitude*, 2.
20. García Márquez, *Cien años de soledad*, 95.
21. García Marquez, *One Hundred Years of Solitude*, 14.
22. Latour, "On Actor-Network Theory," 372–373.
23. Chakrabarty, *Provincializing Europe*, 83.
24. Balibar, "On Universalism," 3.
25. Piglia, "Roberto Arlt," 25.
26. Piglia, "Roberto Arlt," 27.
27. Piglia, *Blanco nocturno*, 192.
28. Piglia, *Target in the Night*, 170.
29. Piglia, *Blanco nocturno*, 257.
30. Piglia, *Target in the Night*, 227.

References

Adorno, Theodor W., and Max Horkheimer. *Dialectic of Enlightenment*. London: Verso, 2016.

Avelar, Idelber. *The Untimely Present: Postdictatorial Latin American Fiction and the Task of Mourning*. Durham, NC: Duke University Press. 1999.

Balibar, Étienne. "On Universalism: In Debate with Alain Badiou." Transversal Texts, 2007. At European Institute for Progressive Cultural Policies, http://eipcp.net/transversal/0607/balibar/en.html.

———. *Saeculum: Culture, religion, idéologie*. Paris: Galilée, 2012.

Borges, Jorge Luis. *The Aleph*. Translation by Andrew Hurley. New York: Penguin, 2000.

———. *Labyrinths: Selected Stories and Other Writings*. Translation by James East Irby. New York: New Directions, 2007.

———. *Obras completas*. Buenos Aires: Emecé, 1989.

Butler, Judith. "Competing Universals." In *Contingency, Hegemony, Universality: Contemporary Dialogues on the Left*, by Judith Butler, Ernesto Laclau, and Slavoj Žižek, 136–181. London: Verso, 2011.

Chakrabarty, Dipesh. *Provincializing Europe: Postcolonial Thought and Historical Difference*. Princeton, NJ: Princeton University Press, 2008.

Delbourgo, James. "The Electric Machine in the American Garden." In *Science and Empire in the Atlantic World*, edited by James Delbourgo and Nicholas Dew. New York, NY: Routledge, 2008.

Franco, Jean. *The Decline and Fall of the Lettered City: Latin America in the Cold War*. Cambridge, MA: Harvard University Press, 2002.

García Márquez, Gabriel. *One Hundred Years of Solitude*. Translation by Gregory Rabassa. New York: HarperCollins, 2006.

Humboldt, Alexander Von. *Personal Narrative*. Edited by Jason Wilson. New York: Penguin Books, 1995.

Latour, Bruno. "On Actor-Network Theory: A Few Clarifications." *Soziale Welt* 47, no. 4 (1996): 369–381.

———. *We Have Never Been Modern*. Translation by Catherine Porter. Cambridge, MA: Harvard University Press, 1993.

Piglia, Ricardo. *The Absent City*. Translation by Sergio Waisman. Durham, NC: Duke University Press, 2000.

———. *Blanco nocturno*. Barcelona: Anagrama, 2010.

———. *Plata quemada*. Mexico City: Debolsillo, 2017.

———. "Roberto Arlt: La ficción del dinero." *Hispamérica* 3, no. 7 (1974): 25–28.

———. *Target in the Night*. Translation by Sergio Waisman of *Blanco nocturno*. Dallas: Deep Vellum, 2015.

———. *El último lector*. Barcelona: Anagrama, 2005.

14

The Politics of Relativity

Radical Epistemologies and the Revolutionary Potential
of the Scientific Imaginary in José Carlos Mariátegui

BRAIS D. OUTES-LEÓN

In his now classic essay *Ariel* (1900), José Enrique Rodó launched an emphatic critique of the scientific positivism that had guided the ideologies of progress articulated by the Latin American ruling elites during the last decades of the nineteenth century. In contrast with the technological materialism and scientific positivism he associates with the United States, Rodó advocates in his essay for a return to philosophical and aesthetic idealism as a means to regenerate Latin America spiritually and politically. Despite his militant antipositivist agenda, Rodó acknowledges the crucial role played by science in contemporary culture and solemnly proclaims that democracy and science are "los dos insustituibles soportes sobre los que nuestra civilización descansa" (the two undisputable foundations upon which our civilization is built).[1] This grand statement highlights how the Arielista spirit, although antipositivist, was far from being opposed to science. In fact, scientific imaginaries played a major role in the political projects developed by key antipositivist intellectuals in the interwar period such as Alberto Zum Felde and José Vasconcelos. The case of José Carlos Mariátegui is no exception. Although the Peruvian Marxist intellectual was a vocal critic of positivism and his thought was strongly influenced by Henri Bergson's vitalism, Unamuno's religious thought, and Ortega y Gasset's phenomenology, Mariátegui was also deeply interested in science and acknowledged it as the most prestigious discourse of contemporary modernity.

Among the scientific trends of the interwar period, Mariátegui was particularly fascinated by the cultural and political implications of Albert Einstein's theory of relativity. Shortly after his return to Lima in January 1923 after more than three years of exile in Europe, Mariátegui was interviewed for the weekly magazine *Variedades*. To the first question of the interviewer—"What is your philosophy of life?"—Mariátegui replied assertively, "Esta es una pregunta metafísica. Y la metafísica no está de moda. El físico Einstein interesa al mundo mucho más que el metafísico Bergson" (That is a metaphysical question. And metaphysics is not in fashion. The world is far more interested in Einstein the physicist than in Bergson the metaphysician).[2] When asked to identify the three most relevant contemporary personalities, Mariátegui named Albert Einstein, along with the Russian revolutionary Vladimir I. Lenin and the German industrialist Hugo Stinnes.[3] Although the Peruvian intellectual never devoted an entire essay to Einstein's impact on contemporary culture and his grasp of physics theory was flimsy at best, his fascination with relativity manifested in countless references and comments in his newspaper articles and lecture series. For Mariátegui, there was no doubt that the theory of relativity marked a philosophical and cultural turning point in the evolution of contemporary culture. "Einstein," he concludes in an essay from 1929, "ha suministrado a la especulación filosófica con sus descubrimientos de física y matemática, un material tan rico y vasto como imprevisto" (Einstein's contributions to philosophical speculation with his discoveries on physics and mathematics were as rich and vast as they were unforeseen).[4]

Departing from a conception of science as the indisputable pillar of modern culture, Mariátegui conceived of Einstein's theory as the most important manifestation of the cultural and philosophical relativism that defined modern European culture. As the ultimate proof of the relative value of truth, scientific or otherwise, as a social construct, the theory of relativity constituted for Mariátegui the core of a relativist epistemology that fueled his ideological critique of capitalism. Transforming Einstein into a champion of ideological relativism, Mariátegui mobilized in his writings the scientific prestige of relativity to articulate a critique of capitalism and the colonial world order. In contrast with the use of scientific positivism in the late nineteenth century by the Latin American elites to imbue their political agendas and narratives of progress with the intellectual prestige of scientific rationalism, Mariátegui rediscovered in Einstein's theory of relativity the radical potential of scientific epistemologies to further his Marxist project of political emancipation.

An Antipositivist Scientific Imaginary

In his programmatic essay *Defensa del marxismo* (Defense of Marxism, 1928), Mariátegui engages extensively with the crisis of positivism in Latin America, emphasizing how the value of science in contemporary culture has not been invalidated or lessened. Despite "la bancarrota del positivismo y del cientifi-cismo" (the bankruptcy of positivism and scientificism)—understood as the submission of all fields of knowledge to the dictates of science—Mariátegui defends the need of Marxist theorists to be attuned to the latest philosophical and scientific developments, exemplified by Hugo De Vries's gene theory, Sigmund Freud's psychoanalysis, and Albert Einstein's physics. In the eyes of Mariátegui, all political projects of modernity, regardless of their ideology—including Marxism—"se cimentan invariablemente en la ciencia, no en el cientificismo" (invariably have as their foundation not scientificism, but science).[5]

Mariátegui further elaborates on the crucial role of science in his political ideology in *Siete ensayos de interpretación de la realidad peruana* (Seven Interpretive Essays on Peruvian Reality, 1928). In his discussion of the university reform, Mariátegui emphatically defends an educational reform that prioritized the natural sciences and practical knowledge over the humanities with the argument that "el valor de la ciencia como estímulo de la especulación filosófica no puede ser desconocido ni subestimado" (the value of the sciences as stimulus for philosophical speculation cannot be ignored or underestimated).[6] Scientific education was not simply the foundation for a thorough and analytical dissection of reality. Mariátegui goes on to proclaim that "la atmósfera de ideas de esta civilización debe a la ciencia mucho más seguramente que a las humanidades" (the Zeitgeist of this civilization surely owes far more to the sciences than to the humanities).[7]

Most importantly, Mariátegui points to the need for Latin American intellectuals to incorporate into their political projects the scientific and analytical tools developed by European science. "He hecho en Europa mi mejor aprendizaje," Mariátegui confesses, "[y] creo que no hay salvación para Indo-América sin la ciencia y el pensamiento europeos u occidentales" (I received my best training in Europe. And I believe there is no salvation for Indo-America without European or Western science and thought).[8] The religious undertones in this passage are corroborated in *Defensa del marxismo* when Mariátegui asserts categorically that "la religión del porvenir . . . descansará en la ciencia" (the religion of the future . . . will be founded on science).[9] If Marxism was to become the new political religion of tomorrow, it needed to incorporate the scientific imaginary into its political discourse, and there was no prophet of science who commanded as much respect and prestige in the 1920s as Albert Einstein.

As a consequence of the verification of his theories during the solar eclipse of 1919, Einstein became a global icon of the sciences in the 1920s. The far-reaching implications of his theories for the understanding of time and space in particular were constantly discussed in popular media, and despite its esoteric complexity, the theory of relativity captured the imagination of the general public. The popularity of Einstein—only comparable to that of Charles Darwin in the nineteenth century—increased with the award of the Nobel Prize in Physics in 1922. Thanks to his visits to Spain (1923) and Argentina, Uruguay, and Brazil (1925), his status in the Spanish-speaking world grew exponentially in the first half of that decade. Einstein's tour of South America was greeted with unparalleled enthusiasm, and his theories were perceived almost unanimously as a symbol of modernization—the vanguard of a new conception of the universe and its fundamental laws.[10]

Mariátegui interpreted the global popularity of Einstein as the ultimate proof of the profoundly interconnected nature of capitalist modernity. In his view, Einstein's celebrity was a paradigmatic example of the accelerated circulation of ideas in an increasingly globalized world:

> En estos tiempos, la teoría de la relatividad, no obstante su complicación y su tecnicismo, ha dado la vuelta al mundo en poquísimos años. Todos estos hechos son otros tantos signos del internacionalismo y de la solidaridad de la vida contemporánea.[11]

> (In our time, the theory of relativity, despite its complexity and technicality, has reached every corner of the globe in very few years. This event is a sign of the internationalism and solidarity of contemporary life.)

Despite the enthusiastic engagement with Einstein's theories throughout the region in the wake of his visit to South America to which Mariátegui refers in this passage, the highly theoretical nature of his theories made their comprehension by laypersons nearly impossible. In fact, Thomas Glick has observed, the infrastructure of physics departments at Latin American universities was so weak that few local scientists—outside the mathematics departments and schools of engineering—possessed the mathematical level necessary to fully grasp the implications of the theory of relativity.[12] Mariátegui's own scientific grasp of Einsteinian physics was superficial at best. Although he owned Gaston

Moch's *La relativité des phénomènes* (The Relativity of Phenomena, 1922) and Lucien Fabre's *Les théories d'Einstein: Une nouvelle figure du monde* (The Theories of Einstein: A New World Figure, 1921), two popular science monographs on the topic,[13] Mariátegui—a Marxist intellectual with a keen interest in philosophy—was not interested in the scientific foundations of Einstein's work *per se*. What really captured his imagination was how the theory of relativity could help us better understand contemporary culture and politics.

Cultural Relativism

Inspired by José Ortega y Gasset's discussion of relativity in "El sentido histórico de la teoría de Einstein" (The Historical Meaning of Einstein's Theory, 1923), Mariátegui conceived of Einstein's theory of relativity, not as yet another scientific theory, but as the theory that defined modern European culture and gave conceptual unity to its literature, philosophy, and politics.[14] In particular, Mariátegui saw a strong connection between Einsteinian relativity and philosophical relativism, a strand of post-Kantian philosophy that denied the existence of truth as an absolute value and proposed it instead as a relative category. An avid reader of philosophy since his Italian years, Mariátegui explores the nature of epistemological relativism in his essay "Pesimismo de la realidad y optimismo del ideal" (Pessimism of Reality, Optimism of the Ideal, 1925) by engaging with the concept of truth developed by the German pragmatist philosopher Hans Vaihinger in *Die Philosophie des Als Ob* (1911, The philosophy of as if). Taking the Kantian concept of "heuristic fiction" as his starting point, Vaihinger developed the concept of "nützliche Fiktionen" (useful fictions) to refer to concepts like "God" and "Soul," the validity of which, in contrast to hypothesis, cannot be proven and that are used "as if" they were true for pragmatic reasons.[15]

Simplifying and generalizing Vaihinger's philosophical pragmatism to an extreme degree, Mariátegui interprets the German philosopher's concept of "useful truths" as a form of relativism that transforms all truths into fictions devoid of any universal validity. Scientific and philosophical truths are, in Mariátegui's interpretation of Vaihinger's thought, figments of the imagination, comparable to literature or religion due to their fictional nature. Truth becomes, therefore, a discursive construct that is contingent on its historical and ideological points of enunciation. In Vaihinger's pragmatic concept of truth as fiction, Mariátegui finds the great advantage that it allows us to "negar el Absoluto" (deny the Absolute)—that is, the notion of an ahistorical absolute truth—while at the same time bestowing "a la verdad relativa, al mito temporal de cada época, el mismo valor y la misma eficacia que a una verdad absoluta y eterna" (the same value and efficacy to relative truth and the temporal myth of each epoch as to absolute and eternal truth).[16] Truth is defined as a profoundly unstable category that is constantly challenged by the contingency of history since each period seems to give rise to its own fictions. Mariátegui proclaims in "El hombre y el mito" (Man and Myth, 1925), "La verdad de hoy no será la verdad de mañana. Una verdad es válida sólo para una época. Contentémonos con una verdad relativa" (The truth of today will not be that of tomorrow. Truth is valid only for one specific epoch. Let us content ourselves with a relative truth).[17]

The theory of relativity becomes, in this context, the greatest example of the relative and historically contingent nature of truth as fiction. The emergence of the special and general theories of relativity between 1905 and 1917, their verification in 1919, and the subsequent collapse of Newtonian paradigm validated, in Mariátegui's eyes, his radical critique of truth. The Einsteinian "paradigm shift," to use Thomas Kuhn's terminology in *The Structure of Scientific Revolutions*, was a conceptual revolution of cataclysmic proportions that had invalidated a scientific paradigm unquestioned for two centuries and altered the most fundamental laws of physics that governed time, space, mass, and speed. Einstein's special theory of relativity (codified in his famous equation $E = mc^2$) describes a natural world in which space and time not only merge into a continuum, but mass and energy are equivalent as well. The very laws governing time and space were relative to speed. Due to his ability to prove in the most ruthless way that the scientific universal principles that had governed the perception of the universe for centuries were mere fictions, Einstein became in Mariátegui's eyes the apostle of contemporary cultural relativism.[18]

When discussing the work of Luigi Pirandello, for instance, Mariátegui claims that the Italian playwright's aesthetic relativism links him to Vaihinger and Einstein.[19] Similarly, in his essay on Bernard Shaw, Mariátegui emphasizes Shaw's relativism and goes on to tell an anecdote that shows his spiritual connection to Einstein: "La actitud relativista es tan cabal en Bernard Shaw que cuando se divulgó la teoría de Einstein lo único que le asombró fue que se le considerase como un descubrimiento" (the relativist attitude is so thorough in Bernard Shaw that when Einstein's theory was publicized, the only thing that surprised him was the fact that it was considered a discovery).[20] In the eyes of the Peruvian critic, modern culture is relativist, that is to say, Einsteinian.

Relativizing the Bourgeois Absolute

The significance of relativism was not restricted to the realm of philosophy and literature. The scientific revolution initiated by Einstein, the vanguard of scientific modernity at the time, became a powerful inspiration for the Marxist political revolution Mariátegui aspired to unleash. The Peruvian critic often compared Marxism and Einsteinian physics as the conceptual spearheads of a new historical reality. Decrying in a lecture at the Popular University in 1923 the conservatism of Peruvian intellectuals, Mariátegui compares Marxism and the theory of relativity to such modern technologies as the airplane and the wireless telegraphy, before affirming,

> La misma razón para ignorar el movimiento socialista habría para ignorar, por ejemplo, la teoría de la relatividad de Einstein. Y estoy seguro de que al más reaccionario de nuestros intelectuales—casi todos son impermeables reaccionarios—no se le ocurrirá que debe ser proscrita del estudio y de la vulgarización la nueva física, de la cual Einstein es el más eminente y máximo representante.[21]

> (If one ignores the socialist movement, one should ignore on the same grounds Einstein's theory of relativity. And I am sure that not even the most reactionary of our intellectuals—most of them are hopelessly reactionary—would dare ban the study and popularization of the new physics, whose greatest and most eminent representative is Einstein.)

Located at the forefront of intellectual modernity, both Marxism and relativity play the role of revolutionary theories in political economy and physics, respectively. Just as Einstein's physics brought about the collapse of the Newtonian worldview, showing it to be yet another fiction, Marxism would ultimately bring about the demise of capitalism.

This discursive articulation of Einstein's relativity within Mariátegui's cultural and political narratives constitutes a paradigmatic example of his profoundly eclectic and creative form of Marxist criticism that amalgamated seemingly incompatible concepts from the most diverse provenance. By incorporating the scientific imaginary of relativity into his texts, this process of cultural amalgamation—defined by critics as Eclectomarxism (Chang-Rodríguez), Romantic Marxism (Löwy), and heterodox Marxism (Beigel)—subverts the very discursive and political role of science in the contemporary geopolitics of knowledge.[22] In his writings, the theory of relativity—the ultimate proof of the radical relativism that characterizes contemporary culture—ceases to be a theory in the physical sciences to become instead a conceptual tool in the applied sciences of the revolution. Einstein's theory is uprooted from its original discursive point of enunciation in physics theory and is transplanted onto a new discursive ecosystem: a Marxist intellectual project that pursued the revolutionary emancipation of the Peruvian rural and urban proletariat.

In the relativist epistemologies symbolized by Einstein's theories Mariátegui found the discursive tool to carry out a radical critique of capitalist ideology. Contemporary culture becomes, in his eyes, a discursive battlefield in which myriad competing fictions show the contingency of truth as an unstable cultural, ideological, and scientific category. If truths are fictions, if even categories such as space and time (deemed absolute in Newtonian physics) are proven to be devoid of universal validity, why should the capitalist world order and its ideological apparatus be given a free pass?

Following a Spenglerian conception of cyclical history, Mariátegui believed that European civilization and capitalism were coming to an end and that the most remarkable sign of their imminent doom was, in fact, relativism. According to Mariátegui, every civilization in history inhabited a powerful ideological fairytale: "la ilusión de su eternidad" (the illusion of its own eternity).[23] That illusion had been irreparably shattered in European civilization by relativism. Einstein's relativity and Spengler's cyclical history provide Mariátegui with the foundations for an all-out assault on the ideologies of the capitalist world order, or what he calls "el Absoluto Burgués" (the Bourgeois Absolute).[24]

Reminiscent of Antonio Gramsci's concept of hegemony, the Bourgeois Absolute designates in Mariátegui's writing a set of discourses that reproduce and perpetuate the European bourgeois cultural ideology. Because of their hegemonic position within capitalist societies, the discourses that make up the Bourgeois Absolute—from positivism and rationalism in philosophy to naturalist and realist aesthetics—may seem imbued with universal validity. However, Mariátegui suggests that the Bourgeois Absolute, just like the Newtonian paradigm in physics, has entered a period of crisis that will result eventually in its demise. As a consequence of the combined action of historical events such as World War I and the Russian Revolution, the discursive unity among the organic intellectuals that had granted ideological coherence to the capitalist order was now fractured. "La burguesía," he diagnoses, "ha perdido el poder moral que antes le consentía mantener en sus rangos, sin conflicto interno, a la mayoría de los intelectuales" (the bourgeoisie has lost the moral power that allowed it before to keep the majority of intellectuals in its ranks without inner conflict).[25] The result is "una gran crisis de conciencia" (a great crisis of conscience) in contemporary culture that manifested in the emergence of a number of destabilizing cultural discourses such as Marxism, philosophical vitalism and relativism, the historical avant-gardes, psychoanalysis, and last but not least, Einstein's relativity.[26] The resulting loss of discursive hegemony by capitalist ideology signaled, for Mariátegui, the beginning of the end of European capitalism and colonialism's global hegemony and the emergence of a new socialist world order.

For a Decolonization of the Scientific Imaginary

Against the collapse of the Bourgeois Absolute, Mariátegui pits the emergence of what he calls the myth of the revolution in classic essays like "El hombre y el mito." Inspired by Georges Sorel's political thought and Andean syncretic Catholicism, Mariátegui conceives of the revolutionary myth as a political ideology that mobilizes the affects and religious faith of the masses as a means to mobilize them and bring them to political action.[27] By empowering the mystical irrationality of politics as faith, the myth of the revolution in Mariátegui's Indo-American Marxism becomes the engine of history, the element that mobilizes the masses. The revolution becomes an act of faith, as "la fuerza de los revolucionarios no está en su ciencia; está en su fe, en su pasión, en su voluntad. Es una fuerza religiosa, mística, espiritual" (the force of revolutionaries does not reside in their science; it resides in their faith, in their passion, in their will. It is a religious, mystical, and spiritual force).[28]

Although recognizing the preeminence of irrational affects in politics, Mariátegui appeals to the rationality of the scientific discourse symbolized by relativity and to the irrational element of religious faith as two crucial and complementary components of his ideological project. By combining the scientific imaginary with religious faith, Mariátegui attempts to bring about a new form of Latin American Marxism that, by embracing spiritual irrationality and breaking away from the Enlightenment tradition, could engage the masses of Peruvian workers more effectively.

His uneasy combination of scientific imaginaries with religious spirituality can be interpreted as a discursive subversion of the Latin American positivist tradition. By incorporating the scientific imaginary of relativity into his religiously charged discourse of the revolutionary myth, Mariátegui subverts the role played by science within the coloniality of power that defines the contemporary condition of Latin America.[29] Historically, scientific knowledge and its institutions and discursive practices had been a crucial component of the colonial projects of European nations, providing a foundation for the Eurocentric narratives of progress and civilization. Mariátegui's transformation of Einsteinian relativism into a radical epistemology of political emancipation, within his mystical strand of Indo-American Marxism, is a paradigmatic instance of what Walter Mignolo calls "border thinking," an epistemology from a subaltern perspective that articulates new forms of knowledge and knowledge production derived from the critical crossing of several local contexts. What emerges out of Mariátegui's discursive hybridity is, therefore, "an other thinking" that transforms physics theory into a constitutive principle of his critical assault on capitalist ideology and the coloniality of power.[30]

His appropriation of the scientific imaginary thus needs to be understood as a conscious attempt to decolonize science. Ceasing to be one of the defining discourses of Eurocentric modernity, the scientific imaginary functions in Mariátegui as the spearhead of a mystical ideology of revolution that upsets the tight control of European culture over scientific knowledge and the narratives of modernity. By amalgamating the scientific imaginary of relativity into his anti-imperialist intellectual project of revolutionary mysticism, Mariátegui performatively denounces the teleological narratives of progress and the alleged superiority of European civilization as yet other fictions that needed to be deconstructed. Ultimately, the collapse of the Bourgeois Absolute that Mariátegui diagnoses and performatively promotes in his writings is inseparable from what one could call—paraphrasing his own terminology—the collapse of a Colonial Absolute.[31]

The implicit assumption that fuels Mariátegui's critical gesture is that a subversive critique of the Eurocentric capitalist world order can only be truly effective if that critique is carried out using the discursive tools that helped Europe achieve its technological, scientific, and economic hegemony. Only through the active appropriation and subversion of European thought would revolutionary intellectuals be able to articulate an ideological critique that short-circuits the inner logic of the coloniality of power from within. Similarly to Enrique Dussel's concept of transmodernity—defined as an alternative form of modernity that transcends colonial exploitation and allows "both modernity and its negated alterity (the victims) [to] co-realize themselves in a process of mutual creative fertilization"[32]—Mariátegui understands his political and cultural project of liberation as a process of hybridization that turns a critical engagement with European knowledge into the foundation for the political emancipation of Latin America.

In his amalgamation of religiosity and irrationality with the vindication of the theory of relativity as a central component of his critique of contemporary capitalist culture, one can observe the kind of border thinking that makes Mariátegui's transformative appropriation of European scientific imaginaries into an effective strategy of discursive decolonization. In doing so, Mariátegui rediscovers the radical potential of scientific epistemologies as a form of political dissent. By fully unlocking the emancipatory potential of scientific inquiry, Mariátegui transforms Einstein's theory of relativity into a powerful tool of ideological critique capable of subverting the contemporary logic of coloniality. Mariátegui's mobilization of the symbolic power of relativity within his vitalist theory of the revolutionary myth articulates a compelling radical scientific imaginary that subverts the strong links between scientific positivism and the interests and projects of modernity of the Latin American ruling elites at the turn of the twentieth century.

Notes

1. Rodó, *Ariel*, 30.
2. Mariátegui, "Instantáneas," 138.
3. Mariátegui, "Instantáneas," 141.
4. Mariátegui, *Historia de la crisis mundial*, 201.
5. Mariátegui, *Defensa del marxismo*, 38, 41.
6. Mariátegui, *Siete ensayos*, 130.
7. Mariátegui, *Siete ensayos*, 130.
8. Mariátegui, *Siete ensayos*, 6.
9. Mariátegui, *Defensa del marxismo*, 42.
10. Glick, "Science and Society," 494. On the reception of Einstein's theories in Argentina, see Hurtado de Mendoza, "Las teorías de la relatividad." On his visit to Uruguay, see Ortiz and Otero, "Removiendo el ambiente," 1–35. On his visit to Rio de Janeiro, see De Castro Moreira and Passos Videira, *Einstein e o Brasil*. On Einstein's reception in Spain, see Glick, *Einstein in Spain*.
11. Mariátegui, *Historia de la crisis mundial*, 165.
12. Glick, "Science and Society," 494. In the case of Peru, the impact of relativity on the scientific circles of Lima only resulted in regular publications on the topic from the early 1920s onward, among them De Losada y Puga, "Discusión de una fórmula" and "Sobre la adición"; Antúnez de Mayolo, "Significado de la velocidad"; García, "Caída de los cuerpos," "Movimiento relativo," and "La teoría especial"; Corral, "Nueva mecánica

relativista." For the first monograph on Einstein published by a Peruvian, see Miró Quesada, *La relatividad y los quanta.*

13. Vanden, *Mariátegui*, 130, 144.

14. Ortega y Gasset, "El sentido histórico," 213–214.

15. Vaihinger, *Die Philosophie des Als Ob.*

16. Mariátegui, "Pesimismo de la realidad," 30.

17. Mariátegui, "El hombre y el mito," 21. For further comments on the nature of relative truths, see Mariátegui, "La lucha final," 26.

18. In addition, Mariátegui often portrays Einstein as an intellectual leader outside the realm of physics. On Einstein's anti-imperialist stance, see Mariátegui, "Prensa de doctrina," 175–176; on Einstein and Jewish culture in Europe, see Mariátegui, "El semitismo," 213–214; on his pacifism, see Mariátegui, *Historia de la crisis mundial*, 28.

19. Mariátegui, "El caso Pirandello," 99. For further connections between Pirandello and Einstein, see Mariátegui, "Autores y escenarios," 184.

20. Mariátegui, "Bernard Shaw," 140; on the connection between Shaw's relativism and liberalism, see 142–144. In an article published in *Amauta* while Mariátegui was editor of the magazine, Hugo Pesce, one of his closest collaborators, also established a link between Einsteinian physics and Edgar Allan Poe (Pesce, "Poe").

21. Mariátegui, *Historia de la crisis mundial*, 17.

22. Chang-Rodríguez, *Poética e ideología*; Löwy, "Marxism and Romanticism," 88; Beigel, *Itinerario y la brújula.*

23. Mariátegui, "El crepúsculo de la civilización," 79.

24. Mariátegui, "¿Existe una inquietud?," 31.

25. Mariátegui, "¿Existe una inquietud?," 31.

26. This crisis also manifested in the emergence of the historical avant-gardes, since the true revolutionary character of the historical avant-gardes was codified "en el repudio, en el desahucio, en la befa del absoluto burgués" (in the rejection, in the dismissal, in the scorn of the Bourgeois Absolute) (Mariátegui, "Arte, revolución y decadencia," 19).

27. Mariátegui, "El hombre y el mito," 18–19.

28. Mariátegui, "El hombre y el mito," 22.

29. For a definition of the coloniality of power and its connection to science, see Quijano, "Colonialidad del poder," 121–124. On Mariátegui as a precursor of the current debates about postcolonial thought in Latin America, see Moraña, "Mariátegui en los nuevos debates," 43–51.

30. Mignolo, *Local Histories/Global Designs*, 66–68.

31. Influenced by Mariátegui's incorporation of the Einsteinian scientific imaginary into his critique of contemporary imperialism, Víctor Raúl Haya de la Torre, the founder of APRA (Alianza Popular Revolucionaria Americana), developed in the 1930s a philosophy of cultural relativism based on his concept of "espacio-tiempo histórico" (historic space-time), an extravagant combination of Hegel's philosophy of history with Einsteinian physics claiming that each culture inhabits a particular historic space-time distinctly independent and irreducible to other cultures' historic space-times (Haya de la Torre, *Espacio-tiempo histórico*, 383–395). For a brief discussion of Haya de la Torre's concept, see Glick, "Science and Society," 494–495.

32. Dussel, "Eurocentrism and Modernity," 76.

References

Antúnez de Mayolo, Santiago. "Significado de la velocidad de la luz." *Archivos de la Aso-ciación Peruana para el Progreso de la Ciencia* 2 (1922): 102–112.

Beigel, Alejandra. *Itinerario y la brújula: El vanguardismo estético-político de José Carlos Mariátegui*. Buenos Aires: Biblos, 2003.

Chang-Rodríguez, Eugenio. *Poética e ideología en José Carlos Mariátegui*. Madrid: Porrúa, 1983.

Corral, José Isaac. "Nueva mecánica relativista restringida o mecánica no-newtoniana de tipo elíptico." *Revista de Ciencias* 361–63 (1928): 13–20.

De Castro Moreira, Ildeu, and Antonio Augusto Passos Videira. *Einstein e o Brasil*. Rio de Janeiro: UFRJ (Universidade Federal do Rio de Janeiro), 1995.

Dussel, Enrique. "Eurocentrism and Modernity (Introduction to the Frankfurt Lectures)." *Boundary 2* 20, no. 3 (1993): 65–76.

Fabre, Lucien. *Les théories d'Einstein: Une nouvelle figure du monde*. Paris: Payot, 1921.

García, Godofredo. "Caída de los cuerpos de gran altura en la mecánica relativo." *Revista de Ciencias* 330–334 (1925): 5–9.

———. "Movimiento relativo." *Revista de Ciencias* 333–338 (1925): 6–8.

———. "La teoría especial de la relatividad y nuestros puntos de vista." *Revista de Ciencias* 343–345 (1926): 31–42.

Glick, Thomas. *Einstein in Spain: Relativity and the Recovery of Science*. Princeton, NJ: Princeton University Press, 1988.

———. "Science and Society in Twentieth-Century Latin America." In *The Cambridge History of Latin America*, edited by Leslie Bethell, 6:463–535. Cambridge, England: Cambridge University Press, 1994.

Haya de la Torre, Víctor Raúl. *Espacio-tiempo histórico. Obras completas,* vol. 4. Lima: Mejía Baca, 1976.

Hurtado de Mendoza, Diego. "Las teorías de la relatividad y la filosofía en la Argentina (1915–1925)." In *La ciencia en la Argentina entre siglos: Textos, contextos e instituciones*, edited by Marcela Montserrat and Jens Andermann, 36–51. Buenos Aires: Manantial, 2000.

Kuhn, Thomas. *The Structure of Scientific Revolutions*. Chicago: University of Chicago Press, 2012.

De Losada y Puga, Cristóbal. "Discusión de una fórmula de Einstein." *Archivos de la Aso-ciación Peruana para el Progreso de la Ciencia* I (1921): 65–79.

———. "Sobre la adición de velocidades en la cinemática relativista y sobre una definición de la simultaneidad." *Archivos de la Asociación Peruana para el Progreso de la Ciencia* 2 (1922): 17–18.

Löwy, Michael. "Marxism and Romanticism in the Work of José Carlos Mariátegui." *Latin American Perspectives* 25, no. 4 (1998): 76–88.

Mariátegui, José Carlos. "Arte, revolución y decadencia." In *El artista y la época*, 18–22. Lima: Amauta, 1959.

———. "Autores y escenarios del teatro moderno." In *El artista y la época*, 183–188. Lima: Amauta, 1959.

———. "Bernard Shaw." In *El alma matinal*, 139–147. Lima: Amauta, 1964.

———. "El caso Pirandello." In *El alma matinal*, 98–102. Lima: Amauta, 1964.

———. "El crepúsculo de la civilización." In *Signos y obras*, 78–83. Lima: Amauta, 1959.

———. *Defensa del marxismo: Polémica revolucionaria.* Lima: Amauta, 1967.

———. "¿Existe una inquietud propia de nuestra época?" In *El artista y la época*, 29–31. Lima: Amauta, 1959.

———. *Historia de la crisis mundial: Conferencias (1923–1924).* Lima: Amauta, 1971.

———. "El hombre y el mito." In *El alma matinal*, 18–23. Lima: Amauta, 1964.

———. "Instantáneas." In *La novela y la vida*, 138–142. Lima: Amauta, 1959.

———. "La lucha final." In *El alma matinal*, 23–27. Lima: Amauta, 1964.

———. "Pesimismo de la realidad y optimismo del ideal." In *El alma matinal*, 27–31. Lima: Amauta, 1964.

———. "Prensa de doctrina y prensa de información." In *Ideología y política*, 175–178. Lima: Amauta, 1969.

———. "El semitismo y el antisemitismo." In *La escena contemporánea*, 208–218. Lima: Amauta, 1970.

———. *Siete ensayos de interpretación de la realidad peruana.* Caracas: Biblioteca Ayacucho, 2007.

Mignolo, Walter. *Local Histories/Global Designs: Coloniality, Subaltern Knowledges, and Border Thinking.* Princeton, NJ: Princeton University Press, 2000.

Moch, Gaston. *La relativité des phénomènes.* Paris: Flammarion, 1922.

Miró Quesada, Óscar. *La relatividad y los quanta.* Santiago, Chile: Zig-Zag, 1939.

Moraña, Mabel. "Mariátegui en los nuevos debates: Emancipación, (in)dependencia y 'colonialismo supérstite' en América Latina." In *José Carlos Mariátegui y lo estudios latinoamericanos,* edited by Mabel Moraña and Guido Podestá, 41–96. Pittsburgh, PA: Instituto Internacional de Literatura Iberoamericana, 2009.

Ortega y Gasset, José. "El sentido histórico de la teoría de Einstein." In *El tema de nuestro tiempo*, 213–244. Madrid: Calpe, 1923.

Ortiz, Eduardo L., and Mario H. Otero. "Removiendo el ambiente: La visita de Einstein al Uruguay en 1925." *Mathesis* 1, no. 1 (2001): 1–35.

Pesce, Hugo. "Poe, precursor de Einstein." *Amauta* 13 (1928): 24–25.

Quijano, Aníbal. "Colonialidad del poder, cultura y conocimiento en América Latina." In *Capitalismo y geopolítica del conocimiento: El eurocentrismo y la filosofía de la liberación en el debate intelectual contemporáneo*, edited by Walter Mignolo, 117–131. Buenos Aires: Signo, 2001.

Rodó, José Enrique. *Ariel.* Mexico City: Porrúa, 2005.

Sorel, Georges. *Réflexions sur la violence.* Paris: Marcel Rivière, 1946.

Vaihinger, Hans. *Die Philosophie des Als Ob.* Berlin: Reuther and Reichard, 1911.

Vanden, Harry E. *Mariátegui: Influencias en su formación ideológica.* Lima: Biblioteca Amauta, 1975.

15

Beyond Empiricism

Rolando García's Theory of Complex Systems and the
Epistemological Consequences of a Nonlinear Universe

MARA POLGOVSKY EZCURRA

In 1981 the Argentine physicist and epistemologist Rolando García published
a pioneering study discussing the reasons behind the deaths of tens of thousands of people during the 1972 famine affecting the Sahel region of Africa.[1]
Entitled *Nature Pleads Not Guilty: Drought and Man; The 1972 Case History*, the
text became a true epistemological treatise, above and beyond a mere report.
It also proved immensely controversial. According to an informant speaking
under conditions of anonymity, not only did policymakers pay little attention
to the conclusions of the text, but it was also quickly removed from circulation.
The reasons for this controversy were both political and epistemological, García
having posed difficult questions to international politics and science. In a nut-
shell, *Nature Pleads Not Guilty* uproots the received wisdom in the discussion
of famine, which at that time came down to blaming any combination of what
the author calls the four horsemen of the Apocalypse: climate, population explosion, environmental pollution, and the desertification of lands. "These four
'horsemen of the Apocalypse,'" García writes, have been seen "as spectres haunting the path of Man, threatening to drag him into an abyss from which there
is no return. [These] uncontrollable and violent 'natural forces' together with
the excessive human urge to reproduce, serve to hasten the day when an ines-
capable destiny of poverty and death is reached."[2] García takes issue with this
view not only for its determinism and lack of correspondence with empirical

data but also for its simplicity, which fails to account for the various interacting and complex phenomena that ultimately define the relation between "drought and man." García's epistemological reconsideration of this relation cuts through modernity's distancing of culture from nature and what he saw as a prevalent, religiously imbued understanding of natural catastrophe.

These reflections, in turn, set the foundations for García's theory of complex systems (TCS). Along with Edgar Morin's theory of complexity,[3] TCS is one of the few distinctive approaches to the study of complex or nonlinear systems that encompasses both natural and social worlds.[4] García's TCS draws upon Jean Piaget's genetic epistemology—an epistemology focused on the origins or genesis of knowledge—in order to propose a constructivist and interdisciplinary method for understanding self-adaptation, stratification, and emergence in social systems. As opposed to similar and better-known theories, such as the so-called theory of chaos, García's emphasizes the importance of ethics and politics in the thinking of complexity. Moreover, while arguing for the value of adopting complex models to approach sociocultural and socioeconomic questions, García casts a critical eye on what he saw as a misguided belief in the possibility of automating the search of solutions to major human problems, particularly those involving environmental catastrophe and major risk, through computer modeling.[5]

On the basis of these premises, García developed a highly innovative theory and method that Mexican environmental sociologist Enrique Leff describes as one of Latin America's key contributions to ecological thinking, alongside Paulo Freire's eco-pedagogy, José Carlos Mariátegui's ideas around collectivism, and more recently Francisco Varela and Humberto Maturana's work on autopoiesis, among others.[6] The importance of García's TCS is visible in its influence on a variety of spheres that combine social analysis with policy making, including agriculture, education, development, and culture.[7] Yet beyond its strictly utilitarian value, its epistemological significance lies in the way it disarticulates essentialist dichotomies defining the role of nature in modern imaginaries and the disciplinary edifice sustaining those dichotomies. Such disarticulation unfolds from García's interest in nonlinear (socioeconomic and physical) phenomena and the challenges to empiricism and epistemological reductionism resulting from what Lucien Sève describes as "une non-linéarité fondamentale du réel" (a fundamental nonlinearity of the real).[8]

In this essay I discuss García's TCS, outlining its central premises, its relation to ecological thinking, and the ways in which it departs from a mechanistic understanding of reality. I highlight the philosophical dimensions of García's theory, together with his pioneering role in addressing the epistemological demands resulting from the large-scale devastation humans are witnessing in the Anthropocene. My analysis of García's epistemology ultimately addresses how he bridges science and politics. This bridge is counterintuitive, as it is sustained upon the often-paradoxical propositional logic of nonlinearity: both affirming and negating the importance of human action, simultaneously positing determinism and unpredictability. The result is an empirical reading of the workings of biophysical and socioeconomic phenomena rooted in a complexification of modern ways of knowing. This reading is deeply critical of Latin America's condition of dependency and that of the global South more broadly, understood not merely as the consequence of political and economic interests but as rooted in and sustained by the historical imposition and appropriation of reductionist, Eurocentric epistemologies.

Climate Change and Epistemology

One of the premises and driving forces of García's ambitious *Nature Pleads Not Guilty* was the realization that answers to the questions arising from events such as the 1972 famine were strictly unimaginable with the methods and languages of the positivist and discipline-specific science still predominant in the late 1970s. Moreover, he believed that writing about the social effects of climate change under the conditions of a carbon-fueled late capitalism presented not only a sociopolitical challenge but also an epistemological one, given the need to reconsider deep-seated truth claims on the benefits of economic growth in response to the accelerated depletion of planetary resources essential for the survival of humankind.[9] His study therefore counterintuitively concludes that the problem of famine lies above and beyond the meteorological and demographic causes of drought, as had been commonly held. Indeed, solving this problem entailed thinking about drought and its relation to "man" in an entirely new way. García began this task with the publication of his report's three volumes, and then devoted the rest of his life to refining the tenets of this new epistemology and method. From the start it became clear to him that the necessary shift required abandoning the prevalent logical traditions (empiricism and idealism) and moving toward a constructivist epistemology capable of accounting for the dynamism, unpredictability, and recursivity of biosocial systems. By doing so, García redefined famine as a complex phenomenon whose causes outweighed the commonly held blame attributed to climate, population explosion, environmental pollution, and desertification. For the Argentine thinker, understanding and preventing famine involved learning to analyze highly variable human and nonhuman systems in dynamic and often unpredictable interaction. In other words, he resituated famine in the terrains of complexity and nonlinearity, thus provoking a breakthrough, albeit unwelcome, in development thought.

Today it is amply recognized that accelerated resource depletion in capitalist economies and the corresponding identification of a new geological era marked by the influence of human activity on the composition of the earth's atmosphere—known as the Anthropocene—poses epistemological difficulties and controversies. In the past decades, a number of new theoretical models in the sciences and the humanities have sprouted as an attempt to respond to this ever more prescient preoccupation. These include deep ecology (whose main proponent is Arne Naess), ecocriticism (William Rueckert), dark ecology (Timothy Morton), Gaia theory (James Lovelock), eco-Marxism (James O'Connor), eco-anarchism (Murray Bookchin), and the theory of the Anthropocene (Paul J. Crutzen). Despite their markedly distinct orientations, all of these theoretical and disciplinary formulations assert that discussing humans' relations to the environment, as ecological catastrophe becomes increasingly foreseeable, is not only a question of defining a specific object of study but also involves reconsidering deeply and long held beliefs and traditions of thought. Morton writes, "Old ways of thinking, we tell ourselves, are not to be trusted. They helped to get us into this mess in the first place."[10]

As ecological epistemologies proliferate, Latin American sources of ecologism remain little known, and the consequences of what appears to be a process of selective forgetting of thinkers from the South are not insignificant. Claudia A. Baxendale suggests it is common for theoretical or epistemological frameworks to be incorrectly presented as if they were completely new, and that the incorrect attribution of ideas is even more frequent when knowledge originates in the peripheries.[11] This ends up obscuring the moments of true theoretical emergence, the specific conceptual hiatuses they address, and the thinkers who directly or indirectly influenced them. At times these thinkers are scholars based in Latin America, Asia, or Africa whose visibility is limited by the circulation of knowledge. A return to their work by means of a revisionist history of ideas may therefore allow us to grant due recognition to unfairly overlooked thinkers as well as to refine our processes of thought by recognizing previous critical interventions in the dialectical generation of ideas. García's work is illustrative of this situation. The study of his writings illuminates how early he began addressing the relation between human activity and environmental catastrophe from systemic, nonlinear, and complex perspectives that are today rapidly gaining popularity among American and European thinkers. It also unveils the intersections he established between the study of climate change, the development of a theory of complexity, and the need to account for the politics of scientific methods. In García's work, these processes are coeval and organically tied to one another. Later ecocritical thinkers have often attempted to capture this articulation with less rigor, less cultural specificity, and little evidence—as seen in Donna Haraway's cursory *Staying with the Trouble: Making Kin in the Chthulucene* (2016)—while gaining much greater visibility.

García's Theory of Complex Systems

García's TCS is written from a relational, nonanthropocentric, and dynamic perspective. In a complex system, "the whole . . . has properties which are not simply the addition of the properties of the elements."[12] The system is therefore not decomposable while retaining its characteristics; changes in scale lead to qualitative transformations, a phenomenon falling outside of the possibilities of a Newtonian understanding of reality. Moreover, the system's constitutive objects, which García calls "relations and relations of relations,"[13] can equally be human or nonhuman, individual or collective. Indeed, what matters is not their purported nature but the relationships in which they partake, the ways they communicate with one another, and the ways they affect one another. Significantly, this view differs from a structuralist understanding of relations of power and difference, which generally accord with static, synchronic models. In the TCS, systems and hierarchies among subsystems are subject to change over time; these transformations are potentially nonlinear.

This definition of a complex system relates closely to Ludwig von Bertalanffy's general systems theory, which the historian of science Fritjof Capra describes in terms larger than a theory, as "the new vision of reality."[14] This vision focuses not on parts but on systems, not on stasis but on process, not on entities but on relations, not on aggregates but on wholes. According to this vision, the distinction between nature and culture also becomes insignificant, as the humanist concern for the essence or nature of entities is displaced by the question around the purpose and function of open systems. Therefore, systems theory is truly the point of departure from a mechanistic understanding of the world toward the adoption of "epistemological attitudes developed in the natural sciences, particularly in biology and the theory of evolution," according to Dietrich Schwanitz.[15] As a biologist himself, Bertalanffy was one of the first figures to emphasize, already in the 1920s, that prevalent mechanistic approaches in the sciences fell short of being able to explain fundamental aspects of life. As open systems, living organisms are both goal-seeking and in continuous interaction with other systems, with which they exchange energy and information. These are views García shares with Bertalanffy, yet the former moves from the study of live processes to phenomena involving complex interactions between biophysical and socioeconomic systems.

The philosopher and physicist Cliff Hooker describes the impact of complex systems on science as a "recent, ongoing, and profound revolution."[16] This new science is highly ambitious; among other objectives, it seeks to predict future planetary-scale crises and understand human social behavior in large groups, as the research of FutureICT Knowledge Accelerator and Crisis-Relief System shows.[17] Yet Hooker adds that "with a few honourable exceptions," the importance of the rise of complexity in science has been largely ignored as "an object of reflective study" by philosophers and scientists alike.[18] As a physicist specialized in meteorology and fluid dynamics and as a philosopher with an interest in epistemology, García is one of these honorable exceptions; the importance of his TCS lies precisely in the way it intertwines a theory, an epistemological position, and a method.[19]

García defines a complex system as

una *representación* de un recorte de la realidad compleja, conceptualizado como una *totalidad organizada* (de ahí la denominación de sistema) en la cual los elementos no son "separables" y por lo tanto no pueden ser estudiados aisladamente.[20]

(a *representation* of a cut of complex reality, conceptualized as an *organized totality*, hence its denomination as a system, whose elements are not "separable" and therefore cannot be studied in isolation.)

This definition is firmly rooted in a constructivist epistemology, an unusual occurrence in science until the rise of complexity. García's radical constructivism was the result of years of collaboration with Jean Piaget, another dissident scientist whose epistemological writings have been largely disregarded despite being arguably more important than his contributions in the field of education and his pioneering work in biopsychology and complexity.[21] The prolific dialogue between Piaget and García began when the latter was exiled to Geneva in 1967 after a military coup in Argentina the previous year. Together, Piaget and García elaborated a genetic epistemology on the basis of the isomorphic structures of, on the one hand, cognitive development in children and, on the other, the history of science. This led to the publication of their *Psychogenèse et Histoire des Sciences* in 1983 (published in translation as *Psychogenesis and the History of Science* in 1989) and *Towards a Logic of Meanings* in 1991. García, in turn, continued to develop these ideas in *El conocimiento en construcción: De las formulaciones de Jean Piaget a la teoría de sistemas complejos* (Knowledge under Construction: From Jean Piaget's Formulations to the Theory of Complex Systems, 2000) and *Sistemas complejos* (Complex Systems, 2006).

These thinkers' "epistemological framework," in their words,[22] considers both aprioristic idealism and empiricist realism to be insufficient logical models to understand reality, and in particular the reality of biosocial life. The basis for their critical take on the two main epistemological traditions in Western philosophy comes from what García describes as the major scientific revolutions in the fields of physics and logic during the twentieth century: Einstein's theory of general relativity, quantum mechanics, and the recognition of the "limits of empiricism" by such key empiricist logicians as Rudolf Carnap and Bertrand Russell. Discussing the last of these moments of epistemological change in particular, García recalls Willard Van O. Quine's declaration at the 1959 International Congress of Philosophy in Vienna, "hemos dejado de soñar en construir la ciencia a partir de los datos de los sentidos" (we have stopped dreaming of building science from sensory data).[23] However, Piaget and García had been educated in an empiricist tradition and were not ready to see it vanish completely. Rather, they saw their genetic epistemology as a means to confront the challenges resulting from the increasingly acknowledged failure of the senses as all-encompassing sources of knowledge while remaining committed to the empirical tradition. To do so, as Piaget's long-term collaborator Bärbel Inhelder puts it, they "replaced Emmanuel Kant's question 'how is knowledge (e.g., pure mathematics) possible?' with this other question: 'How is knowledge constructed and transformed in ontogenesis?'"[24] They then, in turn, opened this question up to empirical analysis.

The basis of a constructivist epistemology is the idea that "there is no pure reading of reality"; that is, every fact observed always already carries a theory.[25] This proposition may seem unsurprising for a humanist accustomed to self-reflexive analysis. When one situates it within the history of science, it acquires a different dimension, as it points toward the ontogenetic construction of both scientific observation and the mathematical description of physical phenomena. It is precisely within these spheres that Piaget and García posit that neither biophysical systems nor their attributes are given, nor are they subject to neutral observation through direct experience, as logical positivism would suggest. For these thinkers, each observation always presupposes a series of relationships whose consequences for the very act of observation must be recognized.[26] Yet—and this is the key turn of the argument that distances Piaget's and García's thought from the critique of ideology in Marxism and related poststructuralist traditions—they argue that the very relationships that frame observation do not impede but rather facilitate the act of *empirical thinking*. That is, the question is not how to overcome ideology for the study of reality but how to construct the appropriate relational and entangled perspectives in order to approach reality in all its complexity and without any presumption of objective neutrality. Crucially, García and Piaget's constructivism is primarily experimental, rather than hermeneutic, and draws heavily although not exclusively on Piaget's study of cognitive development in children.

To give an example of how this constructivist epistemology is tied up with the empiricist tradition, in *Sistemas complejos*, García writes that

> más de medio siglo de minuciosos trabajos experimentales muestran que la "percepción" de los objetos como tales, distribuidos en el espacio y con una cierta continuidad en el tiempo, si bien es temprana en los niños, no es innata. El espacio y el tiempo "dentro" del cual se ubican los objetos que constituyen nuestra experiencia cotidiana, requieren un largo proceso de elaboración.[27]

> (more than a half century of meticulous experimental work shows that the "perception" of objects as such, namely, as being distributed in space and having certain temporal continuity, develops early in children but is not innate. The space and time "within" which the objects that constitute our daily experience are located require a long process of development.)

In line with this epistemological framework, García describes a system's complexity as in itself a construction rather than an ontological attribute. This does not make complexity any less significant or "real," for a system's complexity has consequences that can be measured, forecast, and even prevented.

Referring to the theoretical biologist Robert Rosen, García argues that "a system is complex if we can interact with it significantly in different ways and if each one of these different forms of interaction requires a different type of description of the system which includes them."[28] These varied forms of interaction depend not on "the thing itself" but on the observer: "Complexity can vary according to the level of description (or the scale of phenomena). A crystallographer will 'typically' interact with a stone in ways which require more than one level of description. The stone of the crystallographer is, therefore, a much more complex system than the stone with which a child breaks a window pane."[29] Aside from highlighting its underlying constructivism, the importance of the observer in García's TCS reveals its proximity to Heinz von Foerster's second-order cybernetics, also a positional and, in this sense, self-reflexive theory.[30] For both García and von Foerster, positionality is crucial to properly define and develop a researcher's *observational method*; and it is this method that ultimately may lead to understanding a system as indeed belonging to a complex order. Epistemologist Gastón Becerra expresses this by suggesting that the simplest definition of complexity in the TCS is as a cognitive relationship between a subject and an object of knowledge.[31] The interesting and perhaps unexpected turn of this radically constructivist but also somewhat circular view is that the consequences of a system's complexity exceed its epistemological description; the value of discussing complexity lies precisely in its palpable consequences, which in many cases, like that of famine, are truly calamitous.

High Complexity: Nonlinear Systems beyond Chaos

In *Nature Pleads Not Guilty*, García describes the impact of climate on the food system of a given society as a complex nonlinear problem in which "the structure of the society receiving the climatic impact is as important—and sometimes more important—than the nature and intensity of the climatic anomaly itself."[32] The causes of famine thus fall into the terrain of what García names "high complexity," a terrain exceeding the already complex dynamics of biophysical processes, given its interest in the social. Highly complex scenarios are solely discernible through meticulously planned, interdisciplinary analysis that involves specialists in the so-called hard sciences, the social sciences, and the humanities—motivated, through their collaboration, to ask each other questions they would not consider in isolation. Before looking at the social and interdisciplinary components of García's theory, though, let us take a closer look at the meaning and consequences of nonlinearity.

For more than two centuries since the publication of Newton's *Principia mathematica* in 1687, the study of linear systems—namely, those systems ruled by fundamental physical laws, whose dynamics are reversible, and that change proportionally in relation to the interacting forces—was the rule in science. This was the case even for biology, in which basic processes such as the irreversibility of life were far from discernible under this epistemological framework. The consequences for biology proved to be notoriously restrictive. As Hooker explains, with a linear approach either life had to be "radically reduced to simple chemical mechanisms and then to applied traditional physics, or it [had] to be taken outside the paradigm altogether and asserted as metaphysically sui generis."[33] This view slowly started to change in the late nineteenth century with the rise of quantum mechanics, general relativity, and the study of nonlinear differential equations by, among others, the French mathematician Henri Poincaré. However, it was not until the second half of the twentieth century that scientists began to realize the paramount importance of complexity and develop the means to model it. Such realization was coupled with the popularization of the theory of chaos by works such as the Nobel laureate Ilya Prigogine and Isabelle Stengers's 1979 book *Order Out of Chaos: Man's New Dialogue with Nature*. From that point on it became clear across a wide disciplinary spectrum that linearity is an exception in nature. "Using a term like nonlinear science," notes the mathematician Stanislaw Ulam, "is like referring to the bulk of zoology as the study of non-elephant animals."[34]

Nonlinearity presents a series of logical puzzles capable of reconstituting the primary forms of knowing in modernity: determinism does not equal predictability, a whole is not equal to the sum of its parts, and chaotic instability coincides with structural stability. All of these propositions, as the philosopher Lucien Sève points out, have the same structure, which is that of a formal logical contradiction: "ils posent p et non-p *ensemble*, ce qui constitue, au regard des règles primordiales de la pensée vraie reconnues depuis Aristotle, l'inadmissible par excellence"[35] (they posit p and not-p together, which constitutes, in view of the primordial rules of true thought recognized since Aristotle, the inadmissible par excellence). The question that follows, writes Sève, is "Pour sauver les faits, va-t-on nier la raison? Pour sauver la raison, va-t-on nier les faits?" (To save the facts, are we going to deny reason? To save reason, are we going to deny the facts?).[36]

Within the formal logic of mathematics there does exist, however, a space for the seemingly convoluted rationality of nonlinearity. In simple terms, nonlinear functions are those in which an interaction in some variable does not vary proportionately to it, leading to extreme variation between cause and effect; a small cause may lead to an enormous effect or vice versa. Nonlinear systems are also irreversible, meaning that in them the arrow of time moves in a single direction. They cannot occur backwards while still satisfying the same laws; their course is path-dependent. These characteristics make nonlinear differential equations mathematically challenging and difficult to approximate without the help of a powerful computer.[37] Moreover, nonlinear systems present situations of emergence (referring both to the appearance of novel structures and to each dimension of the whole exceeding the sum of its parts) and points of bifurcation, in which "a single fluctuation [in the dynamic system] or a combination of them may become so powerful, as a result of positive feedback, that it shatters the preexisting organization."[38] The consequences of these attributes for understanding biophysical and social systems are immense. Interested in the latter, Niklas Luhmann brought a systems perspective, combined with cybernetics and information theory, to the field of sociology.[39] By doing so he attempted to develop a universal sociological theory capable of explaining unpredictable communication processes emerging from complex systemic interactions.[40]

Luhmann's theory proved influential in certain parts of South America, particularly in Chile, where a dialogue was established between Luhmann and Humberto Maturana and Francisco Varela's work on autopoiesis.[41] Yet Luhmann's is not the only systems model that has seen significant developments in Latin American social theory. In fact, García's distinct approach is more attentive not only to individual and collective action but also to human values and systems of (political) representation. For García, such problems as the relation between drought and famine or environmental depletion necessitated, on the one hand, an awareness of cultural patterns, processes of cultural change, and power differentials and, on the other, an understanding of the dynamic interaction of at least three complex subsystems—a physical subsystem (soil, climate, hydrology), an agroproductive subsystem (crops and their associated technology), and a socioeconomic subsystem (the agrarian community, the production relations, the acting economic forces). In such highly complex systems, which necessarily have to be studied as wholes, human values and political (in)action could have effects just as devastating as seemingly erratic climatic fluctuations.

In *Nature Pleads Not Guilty*, García includes a number of biophysical, economic, and social variables in order to break with the humanist presumption that human beings are the motor of history. However, he still concludes that human action was the main cause leading to millions of deaths during the 1972 famine. The explanation of this—again, counterintuitive—conclusion and its underlying complexity and nonlinearity deserve to be quoted at length:

> When we reject the idea that natural disasters or other unplanned catastrophic events are the fundamental reasons for extended malnutrition in the world, this automatically implies that the causes are to be found in the actions of human beings. Nevertheless, it does not imply that whatever happened was planned by groups of people, or by institutions, or by governments, to be as it was. The number of variables at play, nationally and internationally, is so high, and their interactions are so strong, that carefully planned situations easily get out of control. . . . What we maintain is that certain measures implemented by strong economic or political powers (governments, private corporations or whatever) do start socio-economic processes which have a dynamics of their own and which interact with natural processes in a very strong way. Sometimes the whole system will evolve in a manner which makes the processes irreversible or only reversible at a very high cost to the countries or the societies involved.[42]

There are important similarities between this view and the theory of chaos, in the sense that attention is paid not so much to individual actors but to what N. Katherine Hayles describes as "recursive symmetries between different levels of the system," "scaling factors," and "correspondences across scales."[43] García's interest in the political does not easily find a place in other theories, however. Its roots lie deep in the experience of dependency in Latin America. It also reflects García's resolve to articulate an informed critique of the potentially catastrophic effects of poorly grounded good intentions.

Conclusion: The Politics of Complexity

García is widely remembered for his role in arguably the most tragic moment in the history of the Argentine university during the twentieth century, namely, the Noche de los Bastones Largos (Night of the Long Batons). Between 1957 and 1966, in a short democratic interlude, the Universidad de Buenos Aires, the country's most renowned academic institution, went through a process of profound transformation and rapid advancement today known as its "golden age."[44] During those years, García served as dean of the Faculty of Exact and Natural Sciences. In that capacity, he supported a number of innovative initiatives such as the acquisition of the first supercomputer in Latin America (named Clementina), the construction of the university campus (*ciudad universitaria*), and the creation of the National Scientific and Technical Research Council (CONICET); these initiatives placed Argentina at the foreground of scientific development in South America and beyond. In the various institutional roles García occupied during those years, his aim was never scientific innovation per se, but making the university, as he famously said, "the critical and political conscience of society."[45] García was indeed part of a generation of committed intellectuals for whom science could not be understood as politically neutral, since its principal mission in Latin America was, as mathematician Oscar Varsavsky has claimed, "revertir una historia de dependencia externa e interna tanto económica como cultural" (to reverse a history of external and internal economic and cultural dependence).[46]

General Juan Carlos Onganía's military coup in 1966 brought to a sudden and violent halt the possibility of pursuing this goal any further. Immediately after his rise to power, the general abolished the autonomy of the university and then ordered its forceful military occupation as a large-scale yet peaceful, student movement began to be organized to resist the measure. On July 29, 1966, the police entered university buildings, spraying students and dissident professors with tear gas and famously lining them up to hit them with batons.[47] A number of students and faculty members were imprisoned, although this did not prevent them from resisting Onganía's interventionism by quitting the university en masse, with more than a thousand faculty members resigning. Having opposed the military's attempt to abolish the autonomy of the university from the start of the conflict, García then pursued a legal battle against the government, which he lost in a context of diminishing judicial independence. For the scientist, then, this became not only a moment of political *prise de conscience* but also the beginning of a nomadic life and career shaped by unstable political scenarios. After living in Geneva, where he met Piaget, he returned briefly to Buenos Aires in the early 1970s, only to have to leave again, this time to Mexico, after receiving death threats in the newly strained political context leading to the 1976 coup.

These extreme political experiences during the 1960s and 1970s certainly permeated García's politics of method. Yet while some have equated the politicization of science with the adoption of simplistic and biased methodologies that leave the production of reflexive or innovative knowledge aside, for García politicization took a very different route. In his TCS, a reflection on the politics of science and the complex consequences of the different epistemologies framing the production of knowledge is not antithetic to scientific impartiality but one of its preconditions. This theory renders science, philosophy, and politics inseparable while arguably transcending the approaches of Marxism and structuralism.[48] For García, true thinking and true political engagement are complex and unyielding. He did not refrain from endorsing seemingly paradoxical answers to problems that involve mass-scale death and devastation in late modernity. He was more suspicious of the simple answers and an all-too-ingrained tendency to oversee the nonlinear nature of the real.

Acknowledgments

My gratitude goes to José Antonio Amozurrutia for sharing his knowledge on complexity and sociocybernetics during the writing of this text, as well as personal stories of his relationship with García as colleagues at the Centro de Investigaciones Interdisciplinarias en Ciencias y Humanidades (CEIICH) at Universidad Nacional Autónoma de México (UNAM).

Notes

1. García was born in Azul, Argentina, in 1919 and died in Mexico City in 2012.

2. García, *Nature Pleads Not Guilty*, 3.

3. Morin, *La Méthode*.

4. In *Sistemas complejos*, García expresses his distance from Morin's views on complexity. He argues that Morin adopts an obscurantist position that keeps little relation with the historical development of science and does not adhere to a precise methodology for the study of concrete (complex) situations (García, *Sistemas complejos*, 20–21).

5. García, *Sistemas complejos*, 136.

6. Leff, "Latin American Environmental Thought," 4, 10.

7. See, among others, García, *Modernización en el agro*; Amozurrutia, *Complejidad y sistemas sociales*; González, *Entre cultura(s) y cibercultur@(s)*.

8. Sève, *Émergence, complexité et dialectique*, 55. See also Guespin-Michel and Ripoll, "Systèmes dynamiques non linéaires."

9. García, *Nature Pleads Not Guilty*, xi.

10. Morton, *Ecology without Nature*, 26.

11. Baxendale, "El estudio de problemáticas ambientales en América Latina," 2.

12. García, *Food Systems and Society*, 25.

13. Piaget and García, *Psychogenesis*, 123.

14. Capra, *Turning Point*, 263.

15. Schwanitz, "Systems Theory and the Environment of Theory," 266.

16. Hooker, "Introduction to Philosophy of Complex Systems," 3.

17. FuturICT Knowledge Accelerator aims to "understand and manage complex, global, socially interactive systems, with a focus on sustainability and resilience" at a planetary scale (Cordis, https://cordis.europa.eu/project/rcn/99187/factsheet/de). See also Shum et al., "Towards a Global Participatory Platform."

18. Hooker, "Introduction to Philosophy of Complex Systems," 3.

19. Rodríguez Zoya and Rodríguez Zoya, "El espacio controversial de los sistemas complejos," 116.

20. García, *Sistemas complejos*, 21.

21. Hooker, "Introduction to Philosophy of Complex Systems," 19.

22. This term was coined by Piaget and García to distinguish their theory from Thomas Kuhn's notion of "paradigm," which they found to be incapable of accounting for the relation between scientific change and cultural/cognitive change more broadly. They write, "In our view, at each moment in history, as in each society, there exists a dominant framework, a product of social paradigms, which in turn becomes the source of new epistemic paradigms. Once a given epistemic framework is constituted, it becomes impossible to dissociate the contribution of the social component from the one that is intrinsic to the cognitive system" (Piaget and García, *Psychogenesis and the History of Science*, 255).

23. Quoted in García, *Sistemas complejos*, 72.

24. Inhelder, foreword, xi.

25. García, *Sistemas complejos*, 77.

26. García, *Sistemas complejos*, 41.

27. García, *Sistemas complejos*, 42.

28. García, *Food Systems and Society*, 24–25.

29. García, *Food Systems and Society*, 25.

30. Von Foerster, *Understanding Understanding*. See also Krippendorff, "Cybernetics's Reflexive Turns."

31. Gastón Becerra, "Sociocibernética," 88.

32. García, "Annex C: Climate Impacts," 43.

33. Hooker, "Introduction to Philosophy of Complex Systems," 12.

34. Quoted in Campbell et al., "Experimental Mathematics," 374.

35. Sève, *Émergence, complexité et dialectique*, 69–70.

36. Sève, *Émergence, complexité et dialectique*, 71.

37. Campbell et al., "Experimental Mathematics," 374.

38. Toffler, "Foreword: Science and Change," xv.

39. See Luhmann, *Social Systems*.

40. Becerra, "Sociocibernética," 89–90; Farías and Ossandón, "Recontextualizando a Luhmann," 20.

41. Torres Nafarrete and Rodríguez Mansilla, "La recepción del pensamiento de Niklas Luhmann," 60; Maturana and Varela, *Autopoiesis and Cognition*.

42. García, *Nature Pleads Not Guilty*, 52.

43. Hayles, *Chaos Bound*, 170.

44. See Bietenholz and Prado, "Revolutionary Physics," 39; Sigal, *Intelectuales y poder*.

45. Quoted in Aliaga, "El decano de la época dorada."

46. Quoted in Becerra and Castorina, "Una mirada social y política," 332. See Varsavsky, *Ciencia, política y cientificismo*.

47. In a letter published in the *New York Times*, MIT professor Warren Ambrose, then a visiting scholar at the Universidad de Buenos Aires, narrates his experience of being hit by the police and witnessing the fierce repression against the entire academic body. His letter was published on August 3, 1966, under the title "Short Minds, Long Sticks." See Bietenholz and Prado, "Revolutionary Physics," 39.

48. Becerra and Castorina, "Una mirada social y política," 332.

References

Aliaga, Jorge. "El decano de la época dorada." *Página/12*, November 2012.

Amozurrutia, José A. *Complejidad y sistemas sociales: Un modelo adaptativo para la investigación interdisciplinaria*. Mexico City: Universidad Nacional Autónoma de México, 2012.

Baxendale, Claudia A. "El estudio de problemáticas ambientales en América Latina. Revisión de aportes teóricos-epistemológicos: Gallopin, García y Leff." *Fronteras* 11, no. 11 (2012): 1–15.

Becerra, Gastón. "Sociocibernética: Tensiones entre sistemas complejos, sistemas sociales y ciencias de la complejidad." *Athenea Digital* 16, no. 3 (2016): 81–104.

Becerra, Gastón, and José Antonio Castorina. "Una mirada social y política de la ciencia en la epistemología constructivista de Rolando García." *Ciencia, Docencia y Tecnología* 27, no. 52 (2016): 330–350.

Bertalanffy, Ludwig von. *General System Theory: Foundations, Development, Applications*. Harmondsworth, England: Penguin, 1973.

Bietenholz, Wolfgang, and Lilian Prado. "Revolutionary Physics in Reactionary Argentina." *Physics Today* 67, no. 2 (2014): 38–43.

Campbell, David, Jim Crutchfield, Doyne Farmer, and Erica Jen. "Experimental Mathematics: The Role of Computation in Nonlinear Science." *Communications of the Association for Computing Machinery* 28, no. 4 (1985): 374–384.

Capra, Fritjof. *The Turning Point: Science, Society, and the Rising Culture*. London: Bantam, 1983.

Farías, Ignacio, and José Ossandón. "Recontextualizando a Luhmann: Lineamientos para una lectura contemporánea." In *Observando sistemas: Nuevas apropiaciones y usos de la teoría de Niklas Luhmann*, edited by Ignacio Farías and José Ossandón, 16–54. Santiago, Chile: RiL, 2006.

García, Rolando. "Annex C: Climate Impacts and Socioeconomic Conditions; Edited Transcript of Remarks Delivered at Workshop." In *International Perspectives on the Study of Climate and Society: Report of the International Workshop on Climate Issues*, 43–47. Washington, DC: National Academy of Sciences, 1978.

———. *El conocimiento en construcción: De las formulaciones de Jean Piaget a la teoría de sistemas complejos*. Barcelona: Gedisa, 2000.

———, ed. *La epistemología genética y la ciencia contemporánea: Homenaje a Jean Piaget en su centenario*. Barcelona: Gedisa, 1997.

———. *Food Systems and Society: A Conceptual and Methodological Challenge*. Geneva: UN Research Institute for Social Development, 1984.

———. *Modernización en el agro: Ventajas comparativas, ¿para quién? El caso de los cultivos comerciales en El Bajío*. Mexico City: Federación Internacional de Institutos de Estudios Avanzados, 1988.

———. *Nature Pleads Not Guilty: Drought and Man. The 1972 Case History*. Oxford, England: Pergamon, 1981.

———. *Sistemas complejos: Conceptos, método y fundamentación epistemológica de la investigación interdisciFinaria*. Barcelona: Gedisa, 2006.

———. "The Structure of Knowledge and the Knowledge of Structure." In *Piaget's Theory: Prospects and Possibilities*, edited by Harry Beilin and Peter B. Pufall, 21–39. Hillsdale, NJ: Lawrence Erlbaum Associates, 1992.

González, Jorge A. *Entre cultura(s) y cibercultur@(s): Incursiones y otros derroteros no lineales*. Mexico City: Universidad Nacional Autónoma de México, 2015.

Guespin-Michel, Janine, and Camille Ripoll. "Systèmes dynamiques non linéaires, une approche de la complexité et de l'émergence." In Lucien Sève, *Émergence, complexité et dialectique*, coordination by Janine Guespin-Michel, 13–47. Paris: Odide Jacob, 2005.

Haraway, Donna J. *Staying with the Trouble: Making Kin in the Chthulucene*. Durham, NC: Duke University Press, 2016.

Hayles, N. Katherine. *Chaos Bound: Orderly Disorder in Contemporary Literature and Science*. Ithaca, NY: Cornell University Press, 1990.

Hooker, Cliff. "Introduction to Philosophy of Complex Systems." In *Philosophy of Complex Systems*, edited by Hooker, 3–90. Amsterdam: Elsevier/North Holland, 2011.

Inhelder, Bärbel. Foreword to *Piaget's Theory: Prospects and Possibilities*, edited by Harry Beilin and Peter B. Pufall, xi–xiv. Hillsdale, NJ: Lawrence Erlbaum Associates, 1992.

Krippendorff, Klaus. "Cybernetics's Reflexive Turns." *Cybernetics and Human Knowing* 15, no. 3–4 (2008): 173–184.

Leff, Enrique. "Latin American Environmental Thought: A Heritage of Knowledge for Sustainability." *ISEE—International Society for Environmental Ethics* 9 (2010): 1–16.

Luhmann, Niklas. *Complejidad y modernidad: De la unidad a la diferencia*. Translation by Josetxo Beriain and José María García Blanco. Madrid: Trotta, 1998.

———. *Social Systems*. Translation by John Bednarz Jr. and Dirk Baecker. Stanford, CA: Stanford University Press, 1995.

Maturana, Humberto R., and Francisco J. Varela. *Autopoiesis and Cognition: The Realization of the Living*. Dordrecht, Netherlands: Reidel, 1980.

Morin, Edgar. *La Méthode: La Nature de la nature*. Paris: Le Seuil, 2013.

Morton, Timothy. *Ecology without Nature: Rethinking Environmental Aesthetics*. Cambridge, MA: Harvard University Press, 2007.

Piaget, Jean, and Rolando García. *Psychogenesis and the History of Science*. Translation by Helga Feider. New York: Columbia University Press, 1989.

———. *Toward a Logic of Meanings*. Edited by Philip M. Davidson and Jack Easley. Hillsdale, NJ: Lawrence Erlbaum Associates, 1991.

Prigogine, Ilya, and Isabelle Stengers. *Order out of Chaos: Man's New Dialogue with Nature*. London: Flamingo, 1985.

Rodríguez Zoya, Leonardo, and Paula G. Rodríguez Zoya. "El espacio controversial de los sistemas complejos." *Estudios filosóficos* 50 (2014): 103–129.

Schwanitz, Dietrich. "Systems Theory and the Environment of Theory." In *The Current in*

Criticism: Essays on the Present and Future of Literary Theory, edited by Clayton Koelb and Virgil Lokke, 265–294. West Lafayette, IN: Purdue University Press, 1986.

Sève, Lucien. *Émergence, complexité et dialectique*. Coordination by Janine Guespin-Michel. Paris: Odile Jacob, 2005.

Shum, S. Buckingham, K. Aberer, A. Schmidt et al. "Towards a Global Participatory Platform: Democratising Open Data, Complexity Science, and Collective Intelligence." *European Physical Journal Special Topics* 214, no. 1 (2012): 109–152.

Sigal, Silvia. *Intelectuales y poder en la década del sesenta*. Buenos Aires: Puntosur, 1991.

Toffler, Alvin. "Foreword: Science and Change." In *Order out of Chaos: Man's New Dialogue with Nature*, by Ilya Prigogine and Isabelle Stengers. London: Bantam, 1984.

Torres Nafarrete, Javier, and Darío Rodríguez Mansilla. "La recepción del pensamiento de Niklas Luhmann en América Latina." In *Observando sistemas: Nuevas apropiaciones y usos de la teoría de Niklas Luhmann*, edited by Ignacio Fardías and José Ossandon, 55–71. Santiago, Chile: RiL, 2006.

Varsavsky, Oscar. *Ciencia, política y cientificismo*. Buenos Aires: Centro Editor de América Latina, 1986.

von Foerster, Heinz. *Understanding Understanding: Essays on Cybernetics and Cognition*. New York: Springer, 2002.

Contributors

Jens Andermann is professor of Spanish and Portuguese at New York University and editor of the *Journal of Latin American Cultural Studies*. He writes about modern Latin American arts, film, literature, architecture, and material culture and their intersections with extractivism and the legacies of coloniality. His books include *Tierras en trance: Arte y naturaleza después del paisaje*, *New Argentine Cinema*, *The Optic of the State: Visuality and Power in Argentina and Brazil*, and *Mapas de poder: Una arqueología literaria del espacio argentino*.

María del Pilar Blanco is associate professor in Spanish American literature and fellow in Spanish at Trinity College, University of Oxford. She is the author of *Ghost-Watching American Modernity: Haunting, Landscape, and the Hemispheric Imagination* and coeditor, with Esther Peeren, of *The Spectralities Reader: Ghosts and Haunting in Contemporary Critical Theory* and *Popular Ghosts: The Haunted Landscapes of Everyday Culture*.

Edward Chauca is assistant professor of Hispanic studies at the College of Charleston. He has published articles on representations of mental illness and fanaticism in Peruvian literature, and human rights and neoliberalism in Latin American cinema and literature.

Hernán Comastri is a postdoctoral scholar in Argentina's Consejo Nacional de Investigaciones Científicas y Técnicas and professor at the Universidad de Buenos Aires. His main research interests center on social imaginaries regarding science and technology in the second half of the twentieth century.

Miguel de Asúa is professor of the history of science and medicine at the Universidad Nacional de San Martín, a senior member of Argentina's Consejo Nacional de Investigaciones Científicas y Técnicas, and a lifetime member of Clare Hall (Cambridge, England). His books include *Ciencia y literatura: Un relato histórico*, *A New World of Animals: Early Modern Europeans and the Creatures of Iberian America*, in collaboration with Roger French, and *Science in the Vanished Arcadia: Knowledge on Nature in the Jesuit Missions of Paraguay and Río de la Plata*.

Lina del Castillo is associate professor of history and Latin American studies at the University of Texas at Austin. She is the author of *Crafting Republic for the World: Scientific, Geographic, and Historiographic Inventions of Colombia* and of articles in the *Hispanic American Historical Review* and the *Journal of Latin American Studies*.

Carlos Fonseca Suárez is a writer and academic. He is the author of the novels *Coronel Lágrimas* and *Museo animal* and the book of essays *La lucidez del miope*, winner of the National Prize of Literature of Costa Rica. He teaches at the University of Cambridge.

Gabriela Nouzeilles is the Emory L. Ford Professor of Spanish and director of the Program of Latin American Studies at Princeton University. She is the editor of *La naturaleza en disputa: Retóricas del cuerpo y el paisaje*, coeditor of *The Argentina Reader* and the art catalog *The Itinerant Languages of Photography*, and the author of *Ficciones somáticas*.

Brais D. Outes-León is assistant professor of Latin American literature and cultural history at Queens College, City University of New York. His work has appeared in journals such as *Revista Canadiense de Estudios Hispánicos*, *Bulletin of Spanish Studies*, and *Modern Language Notes*.

Joanna Page is professor of Latin American Studies at the University of Cambridge. She is the author of several books, including *Decolonizing Science in Latin American Art*.

Yarí Pérez Marín is assistant professor and deputy director of postgraduate studies in the School of Modern Languages and Cultures at Durham University. Her two main areas of research are sixteenth- and seventeenth-century Hispanic cultures, with an emphasis on the history of medicine, and contemporary Latin American cinema.

Mara Polgovsky Ezcurra is lecturer in contemporary art in the Department of History of Art at Birkbeck, University of London. She is the author of *Touched Bodies: The Performative Turn in Latin American Art* and *Marcos Kurtycz: Corporeality Unbound* and coeditor of *Sabotage Art: Politics and Iconoclasm in Contemporary Latin America*. In 2019 Polgovsky received the *Art Journal* award for the most outstanding contribution to the journal during the previous year.

Julio Prieto is professor of Latin American literature at the University of Potsdam. He is the coeditor, with Ottmar Ette, of *Poetics of the Present: Critical Perspectives on Contemporary Spanish Poetry* and the author of the book of poetry *Marruecos* and of *La escritura errante: Ilegibilidad y políticas del estilo en Latinoamérica*, which won the 2017 Ibero-American Prize of the Latin American Studies Association.

Soledad Quereilhac is a researcher at Argentina's Consejo Nacional de Investigaciones Científicas y Técnicas. She teaches Argentine literature at the Universidad de Buenos Aires and is a member of the Instituto de Historia Argentina y Americana Dr. Emilio Ravignani. She is the author of *Cuando la ciencia despertaba fantasías: Prensa, literatura y ocultismo en la Argentina de entresiglos*.

Heidi V. Scott is associate professor of history at the University of Massachusetts, Amherst. She is the author of *Contested Territory: Mapping Peru in the Sixteenth and Seventeenth Centuries* and has works published in numerous journals including *Hispanic American Historical Review, Journal of Latin American Geography*, and *Environment and Planning B: Society and Space*.

Index

Page numbers in *italics* refer to figures.

Dias de Abreu, Manuel, 5
Díaz, Porfirio, 62, 65–66
Diegues, Carlos, 210
Di Lullo, Orestes, 82–87, 91–92
"Dissertation on the Iron from Tucumán"
  (Luca), 179
Dissipative structures, 259–60
Donde yo no estaba (Cohen), 260, 262
Duchamp, Marcel, 242
Durkheim, Emile, 162–63
Dussel, Enrique, 6, 300

Eberhard, Hermann, 41
Eco-anarchism, 308
Ecocriticism, 308
Ecological epistemologies, 308–9
Eco-Marxism, 308
Eielson, Jorge Eduardo: *Alphabet,* 244, *245;*
  avant-garde and, 235; cultural history and,
  240; *Homage to Leonardo,* 236–37, *237;* hybrid
  assemblages and, 209, 211, 236–37, 246;
  integration in, 210; *khipu* and, 7, 210, 235,
  237, 244; *Khipu series,* 237, *239;* knot poetics
  of, 209, 211, 234, 238–46; *Knots as Stars,*
  Stars as Knots, 241; knowledge sharing and,
  234; *Leonardo's Codex on the Flight of Birds*
  and on Knots, 235–36, *236,* 237, *237, 238,* 240,
  243–44, 247; as poetic-scientific mediator,
  235; scientific diction of, 235; topology of
  reading and, 240–46
Einstein, Albert, 11, 208–9, 292–301, 311
En busca de Klingsor (Volpi), 257–58
Engels, Friedrich, 166
Enríquez, Martín, 104
Ensayo sobre la expresión popular artística en
  Santiago (Canal Feijóo), 80, 82
Ensayo sobre las revoluciones políticas y la
  condición social de las Repúblicas colombi-
  anas (hispano-americanas) (Samper), 163
Environmental writing, 12, 308
Epel, David, 265
Escomel, Edmundo, 136
Escuela Nacional Preparatoria, 63
Esposito, Roberto, 92
Esquivel, Laura, 210
Estrella, Eduardo, 7
Estudios icnológicos (Casamiquela), 53
Europe: modernity discourses and, 6–7;
  popular science publications in, 68; scientific
  exchange with Latin America, 7, 26, 110, 140,
  144; scientific knowledge and, 99; traditional

medicine and, 144; views on Colombian
  republics, 155, 163; views on indigenous
  peoples, 133–36; views on Spanish knowledge
  production, 134–35
Evidential paradigms, 32, 57n6
Evolution: arts and, 257; cooperation and, 257,
  259; mirror neurons and, 256; natural drift
  theory and, 255–56; paradigms of, 264–65;
  political metaphors in, 265; post-Darwinian,
  254–56; symbiosis and, 259, 261–63, 265

Fabre, Lucien, 295
"La fantasia y la ciencia" (Moreno), 222
Farfán, Agustín, 8, 103–5, 107–12
Felde, Alberto Zum, 291
Fernández, Cristina Beatriz, 15
Fiction. *See* Literature
El fin de lo mismo (Cohen), 265
Finlay, Carlos Juan, 5
Fishburn, Evelyn, 14
Flora española (Quer), 178
Flores Galindo, Alberto, 136
Fond, Sigaud de la, 278
Foucault, Michel, 234
Franco, Jean, 283
Franklin, Benjamin, 9, 278–79
Freire, Paulo, 306
Freud, Sigmund, 293
Fuente, Juan de la, 104
Fuentes, Carlos, 208, 288
FutureICT Knowledge Accelerator and Crisis-
  Relief System, 310, 319n17

Gaia theory, 308
Galilei, Galileo, 230
Gallese, Vittorio, 256
Gamio, Manuel, 8
García, Rolando: academic career of, 316–17;
  on climate change, 307–8; *El conocimiento*
  en construcción, 311; constructivist epistemology of, 311–12; epistemological framework of, 13, 311, 319n22; on famine, 305–8,
  313, 316; genetic epistemology and, 311;
  human impact on environment and, 309;
  Nature Pleads Not Guilty, 305, 307, 313, 315;
  Noche de los Bastones Largos and, 316–17;
  nonlinear phenomena and, 306, 313–14, 316;
  philosophy of science and, 273; Piaget and,
  306, 310–12, 317; on science and politics,
  273, 307, 316, 318; *Sistemas complejos,* 311–12,

of, 179; patriotic marches of, 173–74; scientific knowledge and, 173–75

Lorena, Antonio, 136

Lost worlds, 42, 58n25

Lotus Bleu, 226

Lovelock, James, 308

Lozoya, Xavier, 98

Luca, Esteban de, 173–74, 176, 179–80, 182

Lugones, Leopoldo, 208–9, 222, 225–26, 254

Luhmann, Niklas, 315

Luis de Velasco, Don, 121

Lyell, Charles, 32

Lyotard, François, 3

Madison, James, 179

Magazines, Argentine: *modernismo* and, 216–17, 220–21, 226–27; scientific discourse in, 215–26; scientific imaginaries in, 207–8, 216, 227; social art and, 222; spiritualism in, 215, 217, 222–26; theosophy and, 225–26. *See also Constancia; Philadelphia; La Quincena: Revista de Letras*

Malabou, Catherine, 265

Maldonado, Ángel, 133, 136, 143

Man and His Symbols (Jung), 287

Manifesto antropófago (Andrade), 242

Los maravillosos secretos de los naipes (Moreno), 222

Marcet, Jane, 68

"Marcha patriótica" (Luca), 174

"Margot: Boceto naturalista" (Piquet), 221

Margulis, Lynn, 262

Mariátegui, José Carlos: antipositivism of, 291, 293; border thinking and, 300–301; Bourgeois Absolute and, 298–300; collectivism and, 306; critique of capitalism and, 292, 294, 298, 300–301; decolonizing science and, 273, 300; *Defensa del marxismo,* 293; discursive hegemony and, 298–99; on Einstein, 292, 294–98, 302n18; "El hombre y el mito," 299; Marxism and, 291–93, 295, 297, 299; on modernity, 293; on myth of the revolution, 299, 301; philosophical relativism and, 295–96; on Pirandello, 296; relativism and, 298; revolutionary mysticism and, 10–11; science and, 291–92; on science and politics, 273; scientific imaginaries and, 299–301; on Shaw, 296; *Siete ensayos de interpretación de la realidad peruana,* 293; theory of relativity and, 292, 295–301

Mariño, Cosme, 222

Marques, João Benedicto de Azevedo, 225

Martel, Julián. *See* Miro, José María

Martínez, Guillermo, 258

Martínez, Juan Ángel, 218

Marx, Karl, 166, 286

Marxism: critique of ideology in, 312; demise of capitalism and, 297; Mariátegui and, 291–93, 295, 297, 299; political emancipation and, 273, 292, 297–99; politicization of science and, 318

Mathematics: Anglo-American/European, 99; knowledge and, 230, 235; land surveying and, 162; literary innovation and, 208, 210, 232; native American cultures and, 7; nonlinear functions in, 314–15; physical phenomena and, 312; topology and, 238, 241; universality and, 272

Maturana, Humberto, 15, 255–56, 259, 262, 306, 315

Maximilian I, 62–63, 67

Medical anthropology, 136, 144–45

Medical literature: in early modern Latin America, 103–5; empathy in, 111–12; illustration in, 105, *106,* 113n11; indigenous knowledge and, 8, 108–10, 133–34; knowledge production and, 135; mental health and, 133. *See also Tractado breue de Anothomia y Chirvgia* (Farfán)

Medical narratives, 15

La medicina popular en Santiago del Estero (Di Lullo), 82

La medicina popular peruana (Valdizán and Maldonado), 133, 136, 140, 143

Medina, Eden, 12

Megatherium, 36, 39, 43–45, 54

Mémoires (Franklin), 278

Memorial del engaño (Volpi), 265

Memoria sobre la dilatación progresiva del aire atmosférico (Redhead), 173

Mente abierta/ Open Mind (Capote), 16

Mercante, Víctor, 224

Mexico: archaeological wealth in, 64–66; autochthonous contributions to science in, 64, 67; development of archaeology in, 65–66; indigenous rights and, 8; literacy rates in, 69; popular science publications in, 68–72, 75; scientific knowledge and, 63, 66; scientific writing in, 15, 64–67, 75; tianguis of, 109–10. *See also* República Restaurada

Mignolo, Walter, 300

Milstein, César, 5

www.ingramcontent.com/pod-product-compliance
Lightning Source LLC
Chambersburg PA
CBHW020822270326
41928CB00006B/409